physica pss solidi c

www.physica-status-solidi.com

conferences and critical reviews

Editor-in-Chief

Martin Stutzmann, Garching

Regional Editors

Martin S. Brandt, Garching
Peter Deák, Budapest
José Roberto Leite, Saõ Paulo
John I. B. Wilson, Edinburgh

Managing Editor

Stefan Hildebrandt, Berlin

Proceedings

5th International Conference on
Optics of Surfaces and Interfaces (OSI-V)

Léon, México
26–30 May 2003

Guest Editor

Bernardo S. Mendoza

0 · 8 · 2003

physica status solidi (c) – conferences and critical reviews

Editor-in-Chief:	Martin Stutzmann
Managing Editor:	Stefan Hildebrandt
Production Editors:	André Danelius, Heike Höpcke, Irina Juschak
Editorial Assistance:	Katharina Fröhlich, Margit Schütz
Editorial Office:	physica status solidi Bühringstr. 10, 13086 Berlin, Germany Telephone: +49 (0) 30/47 03 13 31, Fax +49 (0) 30/47 03 13 34 e-mail: pss@wiley-vch.de
Publishers:	WILEY-VCH Verlag GmbH & Co. KGaA
Postal Address:	Bühringstr. 10, 13086 Berlin, Germany
Publishing Director:	Alexander Grossmann
Ordering:	Subscription Service, WILEY-VCH Verlag GmbH & Co. KGaA Postfach 10 11 61, 69451 Weinheim, Germany Telephone +49 (0) 62 01/60 64 00, Fax +49 (0) 62 01/60 61 84 e-mail: subservice@wiley-vch.de or through a bookseller
Printing House:	Druckhaus Thomas Müntzer GmbH, Bad Langensalza, Germany Printed on chlorine- and acid free paper.

physica status solidi (c) – conferences and critical reviews is published several times per year by WILEY-VCH Verlag GmbH & Co. KGaA.

Subscription:

Volume **0** (2002/2003) of physica status solidi (c) – conferences and critical reviews is available online at Wiley InterScience (www.interscience.wiley.com). Please register for a free trial access.

Single print issues may be ordered by ISBN at www.wiley-vch.de or through your local bookseller.

Regular print and online subscriptions will be offered from 2004 in connection with subscriptions to physica status solidi (a) and (b).

ISSN 1610-1634

ISBN 3-527-40494-5

Contents

Visit our homepage on: http://www.physica-status-solidi.com
Full text on: http://www.interscience.wiley.com

This Table of Contents is organized according to the topics presented at the conference. Articles with page numbers marked (b) are reprinted from phys. stat. sol. (b) **240**, No. 3, 469–536 (2003). You may find papers with phys. stat. sol. (b) and phys. stat. sol. (c) citations in the two sections of this volume, separated by coloured sheets for easy orientation.
Note from the Publisher: This issue of physica status solidi (c) has been produced from publication-ready manuscript files, written by the authors using the provided Word or LaTeX templates.

Infrared spectroscopy

Reflectance anisotropy spectroscopy (RAS)

Reflectance difference spectroscopy (RDS)

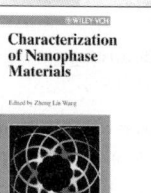

Preface

This issue of physica status solidi (c) contains most of the papers presented at the *Optics of Surfaces and Interfaces: From Basic Research to Applications* (OSI-V) conference, held in the Centro de Investigaciones en Optica (CIO), in León, Guanajuato, México, from 26 to 30 May 2003.

OSI-V is the merging of the previous Epioptics workshops, which have been arranged in Berlin, Germany (1992), Dublin, Ireland (1993), Rome, Italy (1995), Halle, Germany (1994), Åalesund, Norway (1997), Saint-Maxims, France (1999), and Bad Honnef, Germany (2001), and the summer schools of Erice, Italy in 1996, 1998, 2000, and 2002. Starting with OSI-V, the new series of OSI will be organized every two years and will alternate with the Erice school devoted to the same topics.

Surface optical spectroscopy comprises a variety of quite different linear and non-linear techniques which have the potential to study, among others, electronic, magnetic, vibronic, and morphologic properties at surfaces and interfaces of several systems. Some of these techniques have been developed in the last decade to a stage of surface sensitivity comparable to the classical surface science methods based mainly on electrons. Optical surface science besides having different interactions than electrons with the interface, offers in addition the advantage of being in general not restricted to solid–vacuum interfaces. Thus solid–gas phase, solid–liquid and solid–solid interfaces can be analyzed. Consequently, surface optical spectroscopies can be used in a wide range of applications, from basic research at ideal surfaces, well defined on an atomic scale, to real surfaces in gas phase or liquid environments (epitaxy, catalysis, etc.), where process monitoring becomes possible.

Therefore, the conference main topic was the study of surfaces and interfaces through the use of optics. This international meeting brought together 60 researchers and students from universities and institutions working at the front edge of optical spectroscopies at surfaces and interfaces. The goal was to evaluate the status of surface optical spectroscopy, in order to obtain a closer connection between theory and experiment, and their interconnection with applied research. The meeting focused on several properties of surfaces and interfaces, ranging from magnetic, electronic, vibronic, to morphologic characterization of semiconductors, metals and organic materials. The ubiquitous presence of the bulk signal was suppressed by the use of symmetry reduction, by well established optical techniques, like reflectance anisotropy spectroscopy (RAS), second harmonic generation (SHG) or sum frequency generation (SFG).

The academic spirit and scientific excellence was no exception in OSI-V, through the presentation of 10 invited talks, 35 contributed talks, and 14 posters, with the usual atmosphere that stimulated many discussions during and after the sessions. As a result of this, the present issue of pss (c) contains 7 invited papers and 32 contributed papers that cover the main topics of OSI-V.

I would like to express my thanks to the CIO who partially supported the conference and, in particular, to Dr. Fernando Mendoza, General Director and to Dr. Ramón Rodríguez, Director of Research. I would like to mention some people that made the meeting ran smoothly so that every participant had the peace of mind to enjoy the scientific and social program, they are: Brenda Martínez, Annette Torres, Lety González, Patty Ramírez, Tere Avalos, Pancho Huerta, and Aracely and Gabriela Mendoza. To all of them and those not mentioned goes not only my deep appreciation, but that of all the participants. Finally, I thank Drs. Oracio Barbosa, Eugenio Méndez and Alfonso Lastras-Martínez, for their support and help in the local organization.

As an outlook of what is to come in future OSI conferences, one can envision the following: site and species selective materials alteration, bond-breaking and bond-making processes (growth), involving multi-beam time correlated laser systems; dynamics studies involving quantum control of spin-carrier injection, transport, and relaxation processes in nanostructured systems; wavelength selective, laser-assisted nanofabrication using quantum control techniques; and wavelength–phase and site selective bond breaking in biological systems, on cellular and sub-cellular levels. The use of lasers in electronic-vibrational spectroscopy, carrier and spin dynamics, and materials alteration, will be employed in the study of surface or interface state characterization, of above examples, using combined linear and nonlinear optical techniques like RAS, SHG, SFG, and non-linear optical microscopy.

This volume will certainly convey the fact that optics is entering to a stage of maturity in the fascinating study of surfaces and interfaces, which will impact on the multidisciplinary field of surface science.

León, October 2003
The Editor
Bernardo S. Mendoza
bms@cio.mx

Committees

Organizing Committee

Dr. Bernardo Mendoza – Chairman
and
Dr. Oracio Barbosa
Centro de Investigaciones en Optica – CIO
León, México

Dr. Alfonso Lastras
Instituto de Investigaciones en Comunicaciones Opticas – IICO
San Luis Potosí, México

Dr. Eugenio Méndez
Centro de Investigaciones Científica y Estudios Superiores de Ensenada – CICESE
Ensenada, México

International Committee

Y. Borensztein, France
O. Hunderi, Norway
A. Cricenti, Italy
R. Del Sole, Italy
J. McGilp, Ireland
Th. Rasing, Holland
P. Weightman, UK
L. Mochán, México
W. Richter, Germany
D. Aspnes, USA

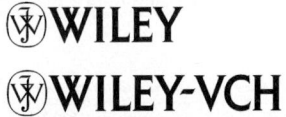

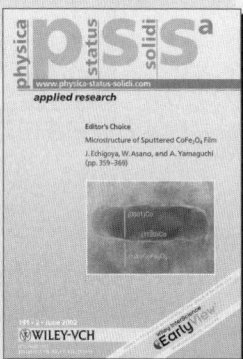

physica **p** status **s** solidi **s** b

www.physica-status-solidi.com

basic research

WILEY-VCH

phys. stat. sol. (b) **240**, No. 3, 469–479 (2003) / **DOI** 10.1002/pssb.200303827

Many-body and overlayer effects on surface optical properties

F. Bechstedt[*, 1], **R. Del Sole**[2], **S. Glutsch**[1], **P. H. Hahn**[1], **O. Pulci**[2], and **W. G. Schmidt**[1]

[1] Institut für Festkörpertheorie und Theoretische Optik, Friedrich-Schiller-Universität, Max-Wien-Platz 1, 07743 Jena, Germany

[2] Istituto Nazionale per la Fisica della Materia, Dipartimento di Fisica dell' Universita' di Roma Tor Vergata, Via della Ricerca Scientifica 1, 00133 Roma, Italy

Received 30 May 2003, revised 4 August 2003, accepted 11 August 2003
Published online 25 November 2003

PACS 68.35.Bs, 71.15.Qe, 73.20.At, 78.68.+m

We demonstrate the potential of recently developed total-energy and electronic-structure methods for the calculation of the optical properties of real surfaces. The many-body effects are fully taken into account by a solution of the combined Dyson and Bethe-Salpeter equations. We show that an initial-value formulation of the polarization function allows an efficient numerical calculation of the optical susceptibility for large slabs consisting of many atoms. As examples we investigate GaP(001) and Si(001) surfaces covered by hydrogen. In the case of P-rich GaP(001)2 × 2–H surfaces the low-energy region of the reflectance anisotropy (RA) is dominated by electron–hole pair excitations in surface states. Surface-induced modifications of bulk excitons near the E_1 and E_2 transitions are responsible for the RA of the monohydride Si(001)2 × 1–H surface.

1 Introduction

Recent years have seen impressive methodological progress in the accurate numerical modeling of optical properties from first principles using the many-body perturbation theory (MBPT) [1]. It has become possible to compute single-particle electronic excitations in an accurate manner using Hedin's GW approximation (GWA) [2]. In addition, the Bethe-Salpeter equation (BSE) for electron–hole pair excitations can be solved in the framework of the same approximation, in order to account for excitonic and local-field (LF) contributions to the polarization function [3–5]. However, the large numerical effort required to solve the BSE has restricted such calculations to the interaction of relatively few electron–hole pairs. Therefore, the calculations of optical properties are usually limited to bulk semiconductors [6–10] or to semiconductor surfaces with strongly localized and energetically well separated states [11, 12]. There are only first trials to include more pair states in surface calculations, for instance to describe surface-modified bulk excitons [13].

At the same time, surface reflectance spectroscopies in the visible to near-UV spectral range have been successfully developed for monitoring surfaces during film growth by molecular beam epitaxy (MBE) and metalorganic vapor phase epitaxy (MOVPE) in real time. Special techniques such as reflectance anisotropy spectroscopy (RAS) are now frequently used not only for in situ diagnostic probes but also to obtain important information about the atomic structure of surfaces in various environments and with adsorbed species [14–16]. However, since this method gives only indirect information via the spectral variation of the RAS signal, a careful theoretical modeling is required and also possible to do [17, 18].

---

[*] Corresponding author: e-mail: bech@ifto.physik.uni-jena.de, Phone: +49 3641 947150, Fax: +49 3641 947152

Coverage by hydrogen gives stable overlayers of semiconductor surfaces, which may represent important intermediate steps of growth using MOVPE, chemical beam epitaxy (CBE) or gas-phase MBE. This holds for (001) substrates of both III–V semiconductors and silicon. Examples are the phosphorous-rich 2×1 or $2 \times 1/2 \times 2$ reconstructions of InP(001) and GaP(001) surfaces [19–25] as well as the single-domain monohydride-terminated Si(001)2×1 surface [26]. In discussing spectroscopic results for such surfaces, the role of hydrogen is sometimes overlooked. The zig-zag chains which have been clearly resolved by scanning tunneling microscopy (STM) on P-rich InP(001)2×1 surfaces [19, 21] have been interpreted as a violation of the electron counting principle [18, 27]. Strong electron correlation effects have been suggested to explain the insulating character of the surface by opening of a Mott-Hubbard gap [19]. An extensive computational search for H-free 2×1 geometries found symmetric, rather than asymmetric P dimers to be energetically favored for both InP(001)2×1 and GaP(001)2×1 [28–30]. The presence of hydrogen gives a natural explanation of the experimental findings concerning the insulating behavior [22]. On the other hand, there are also indications that Si-terminated SiC(001) surfaces show the opposite behavior and become metallic in the presence of hydrogen [31, 32].

In this paper the progress in the calculation of surface optical properties is discussed with a focus on the RAS. This not only concerns the computation of the geometries, thermodynamic phases, and electronic structures but also the inclusion of many-body effects in the framework of the MBPT. We show that the reflectance anisotropy (RA) of complicated surfaces can now be calculated from first principles. Hydrogen-covered (001) surfaces of semiconductors are used as examples. In Section 2 the computational methods are described. We present a novel numerically efficient approach to solve the BSE for the polarization function on the time domain using an initial-value formulation. The surface examples, P-rich GaP(001)$2 \times 1/2 \times 2$–H and Si(001)2×1–H, are discussed in detail in Section 3. A brief summary concludes the paper in Section 4.

2 Computational methods

2.1 Modeling of surfaces

In order to calculate the surface optical spectra we proceed in three steps. First, the energetically favored surface phase is identified and the equilibrium geometry is determined. The required total-energy and electronic-structure calculations are based on the density functional theory (DFT) [33] and the local density approximation (LDA) [34]. The electron-ion interaction is described by nonlocal norm-conserving pseudopotentials [35]. Semicore states are taken into account by nonlinear core corrections to the exchange and correlation energy. A massively parallel, real-space finite-difference method [36] is used to deal with the large unit cells needed to describe the surfaces. A multigrid technique accelerates the convergence. The spacing of the finest grid used to represent the electronic wave functions and charge density is about 10% of a bulk bond length.

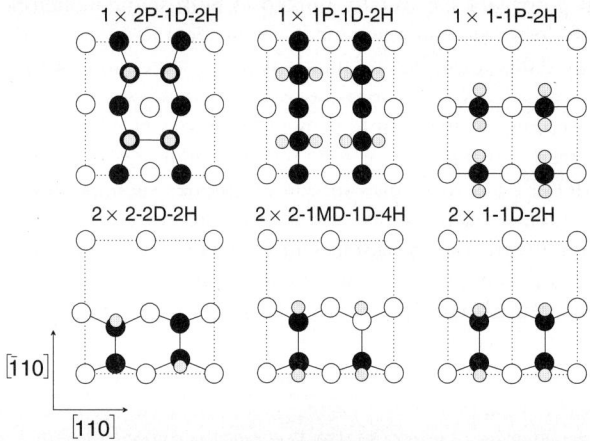

Fig. 1 The energetically most favorable GaP(001)–H surface structures (top view). Empty (filled, grey) circles represent Ga (P, H) atoms. A 2×2 area is shown.

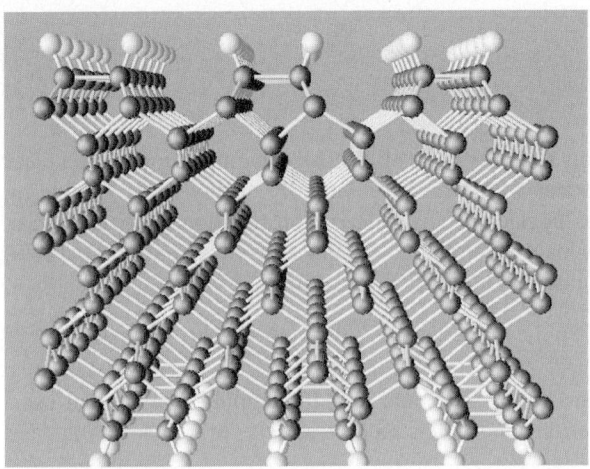

Fig. 2 Slab representing the monohydride phase of the Si(001)2 × 1–H surface.

The surfaces are modeled by periodic arrangements of supercells consisting of 16 (12) atomic layers in the GaP(001) [Si(001)] case. They are separated by vacuum regions large enough to decouple the surfaces. The computational details are like those in Ref. [37]. We use symmetric and asymmetric slabs to model the Si and GaP surface structures, respectively. In the latter case atoms in the lowest bilayer are kept frozen during the structural optimization, and the surface dangling bonds at the bottom layer are saturated by pseudohydrogen (hydrogen). We use a linear cutoff function to suppress the optical signal from the bottom layers in order to avoid spurious effects on the calculated spectra.

The possibility of hydrogen adsorption on gas-phase grown GaP(001) surfaces leads to a large number of conceivable models. We investigated more than 40 plausible structures [38], which differ with respect to their geometry, their Ga/P ratio, and the number of adsorbed hydrogen atoms. The energetically favored hydrogen-induced surface reconstructions are shown in Fig. 1. The notation is such that a leading P indicates adsorption on top of a P-terminated substrate and a hyphen followed by P, D or MD denotes the adsorption of P atoms, P dimers or mixed Ga–P dimers, respectively. The number of hydrogen atoms per surface unit cell concludes the notation. In the case of Si(001)2 × 1 we only study a monohydride coverage, i.e., pairs of surface Si atoms form σ bonds and the second dangling bond of each Si atom is saturated by a H atom (cf. Fig. 2).

2.2 Reflectance anisotropy

The surface optical property of interest is the RA for normal incidence. For (001) surfaces the optical anisotropy between the light polarization directions $x \equiv [1\bar{1}0]$ and $y \equiv [110]$ is studied. The frequency-dependent RA can be obtained from slab calculations as [17, 18, 39]

$$\frac{\Delta R(\omega)}{R(\omega)} = \frac{8\pi\omega}{c} \text{Im} \frac{\alpha_{xx}(\omega) - \alpha_{yy}(\omega)}{\varepsilon_b(\omega) - 1}, \tag{1}$$

where $\alpha_{jj}(\omega)$ ($j = x$; y) are the diagonal components of the half-slab polarizability tensor, and $\varepsilon_b(\omega) = 1 + 4\pi\alpha_b(\omega)$ is the corresponding bulk dielectric function.

We take advantage of the repeated-slab approximation. The Bloch picture is valid with wave vectors k from the surface Brillouin zone (BZ) and band indices v as good quantum numbers. Within the LDA to exchange and correlation the single-particle Kohn–Sham (KS) equation yields eigenfunctions $|vk\rangle$ (or $\psi_{vk}(x) = \langle x | vk \rangle$) and eigenvalues $\varepsilon_v(k)$. The polarizability is related to the polarization function P. Within the Bloch picture one has [5, 18, 40]

$$\alpha_{jj}(\omega) = -\frac{2e^2\hbar^2}{V} \sum_{c,v,k} \sum_{c',v'k'} \{M^j_{cv}(k) M^{j*}_{c'v'}(k) P(cvk, c'v'k'; \omega) + \text{c.c. and } \omega \leftrightarrow -\omega\} \tag{2}$$

with matrix elements of the velocity operator v

$$M_{cv}^{j}(\boldsymbol{k}) = \frac{\langle ck| v_j |vk\rangle}{\varepsilon_c(\boldsymbol{k}) - \varepsilon_v(\boldsymbol{k})} \tag{3}$$

and V as the normalization volume. In (2) the sums run over pairs of electrons in empty conduction band states $|ck\rangle$ and holes in occupied valence band states $|vk\rangle$, which are virtually or physically excited by photons. The effect of the photon wave vector is neglected. The polarization function P obeys a BSE. Neglecting the coupling of resonant and antiresonant electron–hole pairs as well as the non-particle-conserving contributions to the electron–hole interaction, the BSE is of the form

$$\sum_{c'',v'',k''} \left\{ H(cvk, c''v''k'') - \hbar(\omega + i\gamma)\,\delta_{cc''}\delta_{vv''}\delta_{kk''} \right\} P(c''v''k'', c'v'k'; \omega) = -\delta_{cc'}\delta_{vv'}\delta_{kk'} \tag{4}$$

with the effective electron–hole pair Hamiltonian $H(cvk; c'v'k')$ and a small damping γ of the pair excitations. The Hamiltonian of pairs of excited electrons and holes, more precisely, of quasielectrons and quasiholes, is given by [3, 5, 18, 40]

$$H(cvk, c'v'k') = [\varepsilon_c^{QP}(\boldsymbol{k}) - \varepsilon_v^{QP}(\boldsymbol{k})]\,\delta_{cc'}\delta_{vv'}\delta_{kk'} + W(cvk, c'v', \boldsymbol{k}') + \bar{v}(cvk, c', v', \boldsymbol{k}') \tag{5}$$

with the matrix elements

$$W(cvk, c' v' k') = -\int d^3x \int d^3x' \psi_{ck}^*(\boldsymbol{x})\ \psi_{c'k'}(\boldsymbol{x})\, W(\boldsymbol{x}, \boldsymbol{x}')\, \psi_{vk}(\boldsymbol{x}')\, \psi_{v'k'}^*(\boldsymbol{x}') \tag{6}$$

and

$$\bar{v}(cvk, c'v'k') = 2\int d^3x \int d^3x' \psi_{ck}^*(\boldsymbol{x})\, \psi_{vk}(\boldsymbol{x})\, \bar{v}(\boldsymbol{x} - \boldsymbol{x}')\, \psi_{c'k'}(\boldsymbol{x}')\, \psi_{v'k'}^*(\boldsymbol{x}') \tag{7}$$

of the (statically) screened Coulomb interaction $W(\boldsymbol{x}; \boldsymbol{x}')$ [41] and a bare Coulomb interaction $\bar{v}(\boldsymbol{x} - \boldsymbol{x}')$. Only the short-range part of the latter is taken into account in agreement with the physical character of expression (7) as electron–hole exchange [4].

The screened contribution W (6) to the total electron–hole interaction includes the classical attraction of electron and hole as represented by the diagonal elements $c = c'$ and $v = v'$. It is responsible for the electron–hole binding in the Wannier-Mott exciton [42]. The other contributions represent the mixing of electron–hole pairs which is responsible for the redistribution of oscillator strength in optical spectra [6, 7, 40]. The electron–hole exchange term $\propto \bar{v}$ (7) [3, 43, 44] corresponds to the inclusion of LF effects [45, 46]. Indeed, in the bulk case it has been shown [44, 47] that the inclusion of $\bar{v}$ in the BSE (4) gives a polarization corresponding to the macroscopic dielectric susceptibility.

The excited electrons and holes also interact with the inhomogeneous electron gas of the system. This leads to an exchange-correlation self-energy Σ of the system beyond the exchange-correlation potential V_{XC} that is already used in the Kohn-Sham equation of the DFT-LDA. As a consequence, this renormalization gives rise to quasielectrons and quasiholes [1, 2, 18]. Assuming the same energetical ordering of the positions of the quasiparticle peaks and of the KS energies, the quasiparticle effects are included within first-order perturbation theory [48, 49]. The KS wave functions are not updated [50] and the DFT-LDA eigenvalues are corrected according to

$$\varepsilon_v^{QP}(\boldsymbol{k}) = \varepsilon_v(\boldsymbol{k}) + \Delta_v(\boldsymbol{k}), \tag{8}$$

$$\Delta_v(\boldsymbol{k}) = \langle vk| \Sigma(\varepsilon_v^{QP}(\boldsymbol{k})) - V_{XC} |vk\rangle. \tag{9}$$

The exchange-correlation self-energy operator Σ is taken within the GWA [2]. In the explicit calculations we introduce further approximations following the schemes by Hybertsen and Louie [51] and Bechstedt et al. [52] or Cappellini et al. [53]. For several semiconductor surfaces this approximate treatment of self-energy corrections (9) has been shown to result in excitation energies which are within 0.1 eV of the experimental values [13, 54].

2.3 Time-dependent formulation

Since the atomic geometry of a surface is known from the total-energy minimization procedure (cf. Sect. 2.1), the optical properties of a surface (2) can be calculated by solving the BSE (4) starting in the next step with a pair Hamiltonian (5) in DFT-LDA quality, $H(cvk, c'v'k') = [\varepsilon_c(k) - \varepsilon_v(k)]\,\delta_{cc'}\delta_{vv'}\delta_{kk'}$. Then the many-body effects can be added by improvement of the Hamiltonian in the third step according to the inclusion of quasiparticle shifts (9) and electron–hole interactions W (6) and $\bar{v}$ (7). In principle, the last step (including W and $\bar{v}$) can be done by diagonalizing the Hamiltonian matrix (5) [6, 8–10, 55]. However, in contrast to surface optical features to which only a few band pairs contribute [11, 12, 55], the dimension of the matrices for the H-covered (001) surfaces becomes rather large. The number of pair states cvk is dominated by typically more than 100 k points needed to sample the surface BZ. At least two valence and conduction bands per atom have to be taken into account in order to cover the spectral region of several eV. Each slab contains about 24 atoms or more. This results in a dimension of the exciton Hamiltonian of about $N = 10^5 \ldots 10^6$. Even with powerful supercomputers, the diagonalization of matrices of this dimension, which scales as $O(N^3)$, is prohibitively slow.

Therefore, we follow an earlier idea by Glutsch et al. [56] to calculate the polarizability from the time evolution of a vector $|\Psi(t)\rangle$ with N elements, which are defined as

$$\sum_{c',v',k'} M^j_{c'v'}(k')\, P(cvk, c'v'k'; \omega) = \frac{i}{\hbar} \int_0^\infty dt\, e^{i(\omega+i\gamma)t} \Psi_{cvk}(t). \tag{10}$$

The evolution of the elements of $|\Psi(t)\rangle$ are driven by the pair Hamiltonian (5)

$$\sum_{c',v',k'} H(cvk, c'v'k')\, \Psi_{c'v'k'}(t) = i\hbar \frac{\partial}{\partial t} \Psi_{cvk}(t) \tag{11}$$

with their initial values

$$\Psi_{cvk}(0) = M^j_{cv}(k). \tag{12}$$

We solve the initial-value problem by the central difference method [40, 57]. The upper limit of the Fourier integral in (10) can be truncated due to the exponential $\exp(-\gamma t)$. The number of time steps, i.e., the matrix-vector multiplications, is nearly independent of the dimension of the system. The operation count for this method scales thus quadratically with the rank of the pair Hamiltonian, N, and is therefore particularly suitable for complex systems such as surfaces. The time-dependent polarization function $\alpha_b(t) \propto \mathrm{Im}\langle M^j | \Psi(t)\rangle$ resulting for bulk silicon is plotted in Fig. 3. The CPU time savings resulting from using the novel time-dependent method versus the direct diagonalization of the Hamiltonian is demonstrated in Fig. 4.

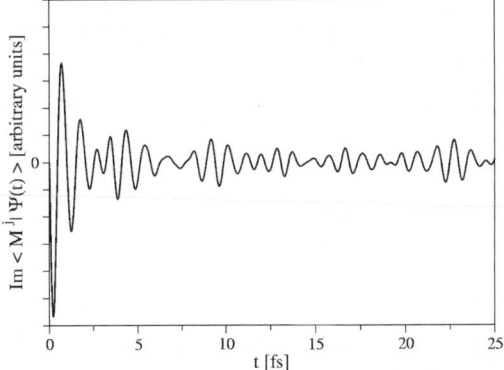

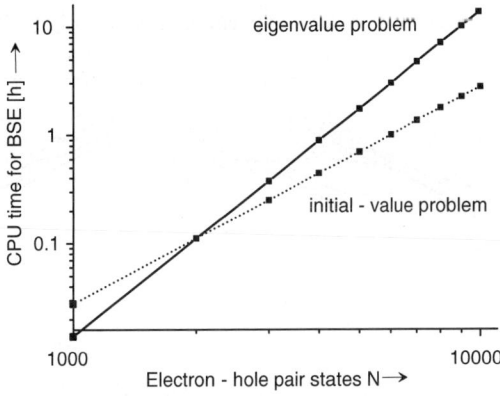

Fig. 3 Time-dependent polarization $\mathrm{Im}\langle M^j | \Psi(t)\rangle$ calculated for bulk Si [57].

Fig. 4 CPU time needed to solve the BSE for bulk Si using the direct diagonalization of H (5) (eigenvalue problem) and the time-dependent formulation (11) (initial-value formulation) on a single Pentium PC.

3 Electronic structure and reflectance anisotropy

3.1 Hydrogen-covered P-rich GaP(001)2 × 1/2 × 2 surface: Surface-state transitions

Due to the varying surface stoichiometry, the thermodynamic grandcanonical potential Ω has to be studied in dependence on the chemical potentials μ of the surface constituents in order to determine the surface ground state for given preparation conditions. We focus our attention to the range of intermediate to slightly more P-rich preparation conditions for the substrate. The hydrogen chemical potential is fixed such as to describe MOVPE growth conditions. Then the 2 × 2–2D–2H structure (Fig. 1) represents the most favorable surface. It represents a periodic arrangement of oppositely buckled P dimers with buckling amplitude of 0.32 Å on top of a cation-terminated substrate. One H atom is bonded to the "down" atom of the P dimer.

The energetical arguments alone are only one indication that the 2 × 2–2D–2H structure corresponds to the surface observed experimentally. Its formation may be kinetically hindered. In order to clarify whether this structure indeed corresponds to the experimentally observed one, spectroscopic signatures have to be discussed. The surface bands calculated long high-symmetry lines of the 2 × 2 surface BZ are shown in Fig. 5. The surface band gap computed within DFT-LDA near Γ is slightly larger than 1 eV. The inclusion of quasiparticle effects results in its opening of about 0.6 eV [54]. Based on the calculated surface electronic structure (Fig. 5) we have also simulated STM images for bias voltages of ±3.5 V. The calculated filled-state image shows excellent agreement with experimental images measured for the P-terminated surface at sample bias 3.35V [25]. Zig-zag chains running along the [110] direction are observed. The bright spots forming the zig-zag pattern are mainly due to the lone-pair state V1 localized at the "up" atom of the P dimer (cf. Fig. 1). This state essentially forms the uppermost occupied surface band near K. This explains why only one P atom is seen in STM experiment. Arguments related to a violation of the electron counting rule or to the occurrence of strong electron correlation effects are not needed to explain the STM findings.

The electronic bands and wave functions in DFT-LDA quality represent the basis for the calculation of the surface optical properties according to expression (1). According to the experience with the RA spectra of many other polar (001) surfaces of III−V semiconductors [22, 23, 54, 58, 59], we assume that excitonic effects are negligible and replace the pair Hamiltonian (5) for a moment by the diagonal one, $H(cv\mathbf{k}, c'v'\mathbf{k}') = [\varepsilon_c^{QP}(\mathbf{k}) - \varepsilon_v^{QP}(\mathbf{k})] \delta_{cc'}\delta_{vv'}\delta_{\mathbf{kk'}}$. In this way we were even able to explain the finestructure of the measured low-temperature RA spectra of InP(001)2 × 4 [59]. In the explicit computations done here we approximate the self-energy corrections (9) by using a rigid shift of 0.8 eV, which is the value we obtained for the GW corrections in GaP bulk [54, 60]. We use an energy broadening of $\hbar\gamma = 0.15$ eV in order to account for the lifetime broadening and the finite number of $\mathbf{k}$ points. A set equivalent to 1024 points in the full 1 × 1 surface BZ is used. The resulting RA spectrum is represented in Fig. 6 and compared with several experimental spectra [23−25].

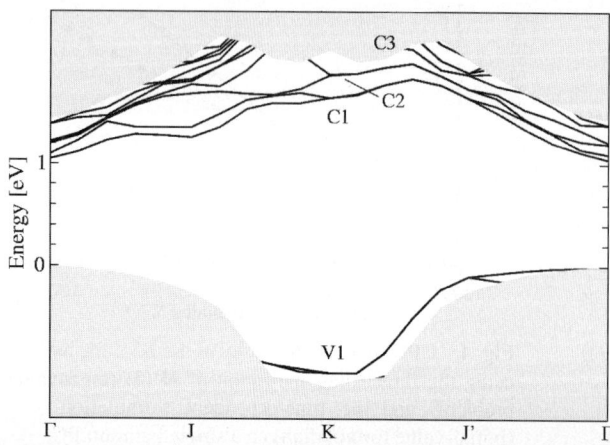

Fig. 5 (online colour at: www.interscience. wiley.com) Surface band structure (solid lines) of the GaP(001)2 × 2–2D–2H surface [38]. Grey regions indicate the projected bulk band structure.

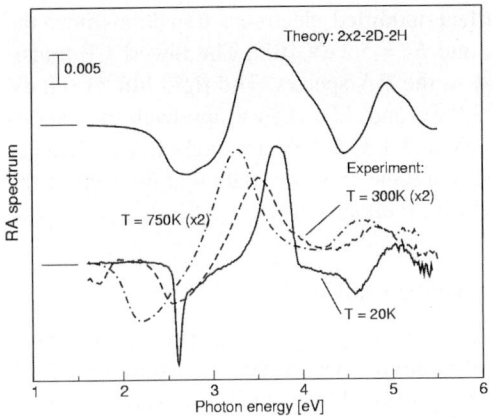

Fig. 6 Reflectance anisotropy spectrum of the GaP(001) 2 × 2–2D–2H structure [38]. The calculated one is compared with spectra for P-terminated GaP(001)2 × 1/2 × 2 surfaces measured at 20 K [25], 300 K [23], and 750 K [24].

The computed spectrum in Fig. 6 exhibits two significant negative and positive features. The broad minimum at around 2.5–2.9 eV is a clear fingerpoint of the H- and P-dimer-covered GaP(001) surface. This is clearly demonstrated in Fig. 7. For the low-energy range this figure indicates strong contributions of optical transitions from the uppermost occupied lone-pair states V1 which are localized at the "up" atom in each of the P dimers. The final states of the optical transitions are the lowest surface conduction band states C1, C2, and C3. They include strong contributions of σ^{*} antibonding states of the P dimers. However, in addition contributions from deformed Ga–P backbond states localized in the first bulklike Ga–P bilayer appear, in particular for C2 and C3. Taking into account the spectral broadening the position of this negative RA feature agrees reasonably with the corresponding negative peaks measured for room temperature [23] or low temperature [25]. This is valid in particular considering an overestimation of the quasiparticle shift of the surface-state transitions of about 0.2 eV. A detailed study of the quasiparticle effects for GaP(001)2 × 4 found the shift to be nonuniform: It amounts to about 0.6 eV for the surface-state-related features and 0.8–1.0 eV for features near the bulk critical points E_1 and E_2 [54], somewhat in contrast to the assumed rigid shift of 0.8 eV. The measured high-temperature phase [24] is probably remarkably disturbed with respect to the ideal 2 × 2–2D–2H structure. The narrowness of the negative peak at 2.6 eV in the measured low-temperature spectrum of Fig. 6 [25] may be interpreted as an indication for strong excitonic effects below the bulk absorption edge E_0.

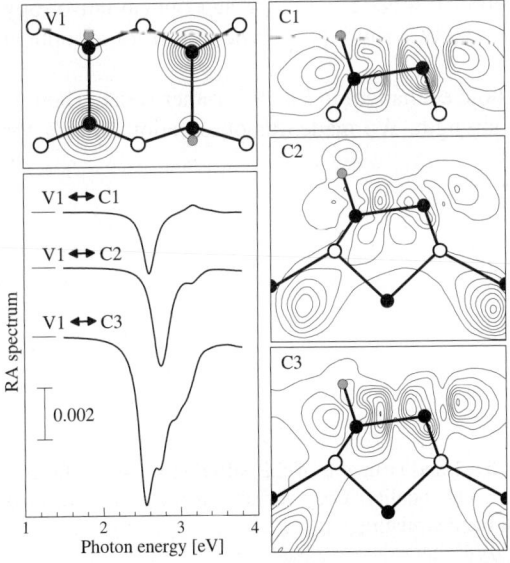

Fig. 7 Calculated RA contributions of optical transitions between certain surface bands as indicated in Fig. 5. The orbital character of the corresponding states at the K point of the surface BZ is shown (V1: top view, C1–C3: side views).

© 2003 WILEY-VCH Verlag GmbH & Co. KGaA, Weinheim

The RA in the high-energy region is dominated by surface-modified electronic transitions near the bulk critical-point (CP) energies $E_1 = 3.8$ eV, $E_0' = 4.8$ eV, and $E_2 = 5.2$ eV [61]. The lowest CP transitions with energies $E_0 = 2.9$ eV cannot really be identified in the RA spectra. The rigid hift of 0.8 eV guarantees that the theoretical (quasiparticle) values $E_0 = 2.9$ eV and $E_0' = 4.8$ eV approach the experimental ones. The two positive RA features around 3.4–3.8 eV and 4.8–5.2 eV are likely to be related to anisotropically deformed bulk Bloch states in the surface region. One may speculate that the sign of the negative RA feature around 4.5 eV indicates an anisotropy due to P dimer states.

3.2   Monohydride Si(001)2 × 1 surface: Surface-modified bulk excitons

The saturation of one dangling bond per surface Si dimer atom by a hydrogen atom (Fig. 2) generates symmetric dimers (more strictly speaking, σ-bonded pairs of Si atoms) with a bond length $d_{\mathrm{dim}} = 2.38$ Å and a Si–H bond length of 1.48 Å. The vanishing tilting is accompanied by a small inward relaxation of the first atomic layer. As a consequence of the H adsorption the π and π^* bands (more strictly, D_{up} and D_{down} bands [18]) are removed from the fundamental gap. This passivation behavior is clearly demonstrated in Fig. 8. No surface bound states appear. Only surface resonance states can be observed. Occupied surface resonance bands are visible near the K point about 2 eV below the valence-band maximum. There are also indications for these surface bands along the KJ' and KJ lines in the 2 × 1 surface BZ.

After fixing the atomic geometry and the calculation of the accompanying band structure in DFT-LDA quality, most interesting is the development of the optical RA spectrum (1) with the inclusion of the various many-particle effects according to four approximations of the two-particle Hamiltonian (5). They are the independent-particle approximation (i.e., DFT-LDA), the independent-quasiparticle approximation (i.e., DFT-LDA with inclusion of self-energy effects in GW approximation), the independent-quasiparticle approach with electron–hole exchange (i.e., local-field effects), and the approach of quasiparticles with the full Coulomb correlation (i.e., screened electron–hole attraction and electron-hole exchange) of the quasielectrons and quasiholes. The corresponding resulting spectra are represented in Fig. 9. They are compared with measured RA data [26]. The RA spectrum in DFT-LDA quality exhibits several structures. Most significant are a positive RA feature close to the E_1 and E_0' transitions near 3 eV (called also A') and a negative peak at about the E_2 transitions near 3.6 eV (called B'). The significance of these spectral structures is underlined after inclusion of quasiparticle effects (i.e., within GWA). Due to the state dependence of the quasiparticle shifts the two peaks are enhanced and shifted to energies of about 3.65 and 4.25 eV. The other spectral features widely disappear or occur as shoulders (e.g., the structure at about 4 eV). Only a negative peak near 3.3 eV remains. In any case the inclusion of self-energy corrections calculated within the GW approximation (9) gives rise to a greater optical anisotropy than the less sophisticated DFT-LDA calculations. The model of a rigid quasiparticle shift for all optical transitions does not work for the Si(001)2 × 1–H surface.

The influence of LF effects (or equivalently electron–hole exchange) in Fig. 9 is rather weak. There is no significant energy shift or redistribution of oscillator strength. We made this observation already for

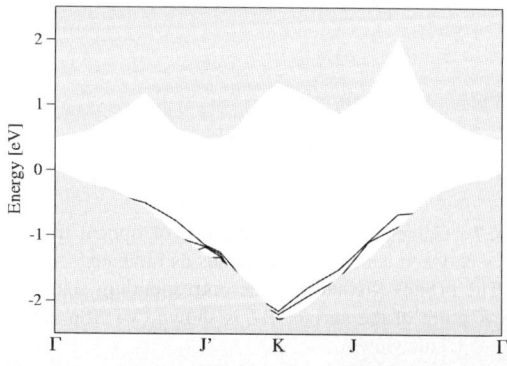

Fig. 8   Band structure of the Si(001)2 × 1–H surface in DFT-LDA quality. Grey regions indicate the projected bulk band structure.

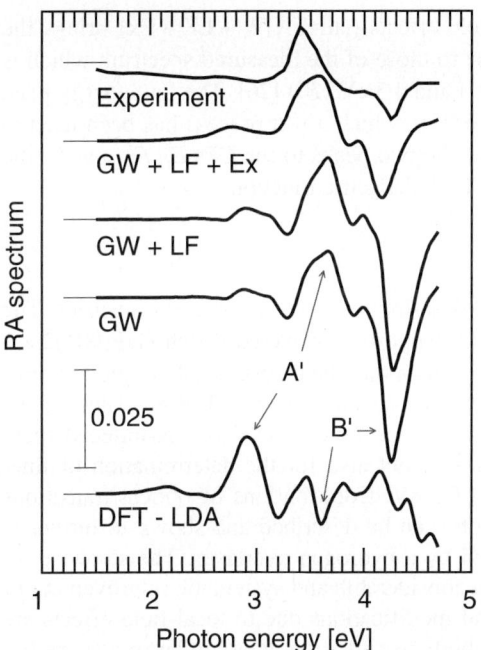

Fig. 9 RA spectrum calculated in different approximations with respect to the many-body effects for the monohydride Si(001)2 × 1–H surface [57]: within DFT-LDA (i.e., using KS eigenvalues $\varepsilon_v(\mathbf{k})$; in the GW approximation (i.e., shifted by quasiparticle shifts (9)); in the GWA with LF effects (i.e., $\bar{v}$ (7) is included in the Hamiltonian (5)), and in the GWA with LF effects and screened electron–hole attraction (i.e., the interactions $\bar{v}$ (7) and W (6) are fully included in the Hamiltonian (5)). The computed spectra are compared with measured data (i.e., experiment) [26].

the RA of the Si(110)1 × 1–H surface [13]. This finding somewhat contradicts the common believe that LF effects due to the presence of the surface should be important for the correct description of the surface RA [62–64]. Much more important are excitonic effects. The negative peak near 3.3 eV almost vanishes and oscillator strength is redistributed from the E_2 to the E_1/E_0' energy region. The later effect has been also observed for bulk Si [6, 57] and the Si(110)1 × 1–H surface [13]. It is mainly due to the coupling of electron–hole pairs (more strictly, quasielectron-quasihole pairs) from different energy regions by the screened Coulomb interaction W (6). In addition this effect also shifts the peaks towards smaller photon energies by about 0.1 eV.

The main contributions to the two resulting peaks A' and B' near photon energies of about 3.5 eV and 4.2 eV follow from the difference of the imaginary part of the two slab polarizabilities (2) for light polarizations $j = x$ and $j = y$. The difference effect is reduced by the factor $\mathrm{Re}\{1/[\varepsilon_b(\omega) - 1]\}$ which is of the order of 10. This is clearly demonstrated in Fig. 10. Interestingly the two polarizabilities for $j = x$ and y only show one pronounced peak with a maximum near E_2. A peak or a structure near the F_1 transition is widely missing for both polarization directions. This is in clear contrast to the findings for the Si(110)1 × 1–H surface [13].

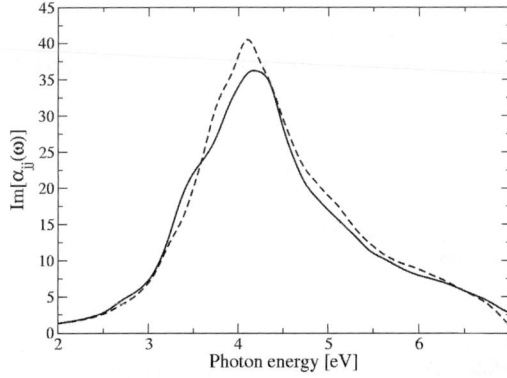

Fig. 10 Imaginary part of the slab polarizability (2) of Si(001)2 × 1–H for two directions of the light polarizations. Solid line: $j \equiv x$, dashed line: $j \equiv y$ The calculations were performed including quasiparticle, LF and electron–hole attraction effects.

The final RA spectra in Fig. 9 with all the many-particle effects (curve GW + LF + Ex) brings the lineshape and the peak positions of the theoretical RA close to those of the measured spectrum which is characterized by positive and negative peaks around 3.4 (A') and 4.3 eV (B') [26]. The low-energy peak around 3.4 eV has also been observed in an earlier study [65], in which a wet process has been used to prepare the surface. As a consequence of the relationship of the two peaks to the CPs E_1, E_0', and E_2 the surface optical anisotropy is explained as modulation of the bulk dielectric function.

4 Summary

We have demonstrated the recent progress in the theoretical description of surface optical properties. The reflectance anisotropy is considered as model property and the hydrogen-covered P-rich GaP(001)2 × 2 and Si(001)2 × 1 surfaces as prototypical systems. We have shown that the theory is able to predict optical properties also for surfaces relevant in gas-phase or chemical-beam epitaxies. The precision of the calculations allows not only for the identification of specific surface structures or surface-induced spectral features for the optimized geometry of a certain overlayer but also for the determination of lineshapes and peak positions with an accuracy of about 0.1–0.2 eV. Contributions of optical transitions between surface states and of surface-modified bulk excitons can be described and shown to influence the reflectance anisotropy spectra of (001) surfaces covered by a hydrogen overlayer. The stepwise inclusion of many-particle effects in the calculation leads to a considerable and systematic improvement of the agreement with the experiment. In contrast, the spectral modifications due to local-field effects are negligible for these surfaces. The applicability of the many-body perturbation theory to large systems has been explicitly demonstrated by calculating the optical anisotropy of the monohydride Si(001)2 × 1 surface. The progress has been made possible by the availability of powerful, massively parallel computers and the development of algorithms which allow for the treatment of many electron–hole-pair states in an effcient, yet accurate manner.

Acknowledgement Grants of computer time from the Leibniz-Rechenzentrum München, the Höchstleistungsrechenzentrum Stuttgart, the John von Neumann-Institut, and the INFM under 'Iniziativa Transversale Calcolo Parallelo' at CINECA are gratefully acknowledged. This work has been supported by INFM-PRA 1MESS, MIUR COFIN 2002, INFM-PAIS- Celex and by the EU through the NANOPHASE Research Training Network (Contract No. HPRN-CT-2000-00167). P. H. H. has been supported by a Marie Curie fellowship of the EU programme MAPS under the contract number HPMT-CT-2001-00242.

References

[1] G. Onida, L. Reining, and A. Rubio, Rev. Mod. Phys. **74**, 601 (2002).
[2] L. Hedin, Phys. Rev. 139, A796 (1965).
    L. Hedin and S. Lundqvist, Solid State Phys. **23**, 1 (1969).
[3] L. J. Sham and T. M. Rice, Phys. Rev. **144**, 708 (1966).
[4] W. Hanke and L. J. Sham, Phys. Rev. B **21**, 4656 (1980).
[5] G. Strinati, Riv. Nuovo Cimento **11**, 1 (1988).
[6] S. Albrecht, L. Reining, R. Del Sole, and G. Onida, Phys. Rev. Lett. **80**, 4510 (1998).
[7] L. X. Benedict, E. L. Shirley, and R. B. Bohn, Phys. Rev. Lett. **80**, 4514 (1998); Phys. Rev. B **57**, R9385 (1998).
[8] M. Rohlfing and S. G. Louie, Phys. Rev. Lett. **81**, 2312 (1998).
[9] B. Arnaud and M. Alouani, Phys. Rev. B **63**, 085208 (2001).
[10] P. Puschnig and C. Ambrosch-Draxl, Phys. Rev. B **66**, 165105 (2002).
[11] M. Rohlfing and S. G. Louie, Phys. Rev. Lett. **83**, 856 (1999).
[12] M. Rohlfing, M. Palummo, G. Onida, and R. Del Sole, Phys. Rev. Lett. **85**, 5440 (2000).
[13] P. H. Hahn, W. G. Schmidt, and F. Bechstedt, Phys. Rev. Lett. **88**, 016402 (2002).
[14] J. F. McGilp, Prog. Surf. Sci. **49**, 1 (1995).
[15] W. Richter and J. F. Zettler, Appl. Surf. Sci. **100/101**, 465 (1996).
[16] D. E. Aspnes, Solid State Commun. **101**, 85 (1997).
[17] R. Del Sole, in: Photonic Probes of Surfaces, edited by P. Halevi (Elsevier Sci. Publ. Co., Amsterdam, 1995).

[18] F. Bechstedt, Principles of Surface Physics (Springer, Berlin, 2003).
[19] L. Li, B.-K. Han, Q. Fu, and R. F. Hicks, Phys. Rev. Lett. **82**, 1879 (1999).
[20] L. Li, B.-K. Han, D. Law, C. H. Li, Q. Fu, and R. F. Hicks, Appl. Phys. Lett. **75**, 683 (1999).
[21] P. Vogt, Th. Hannappel, S. Visbeck, K. Knorr, N. Esser, and W. Richter, Phys. Rev. B **60**, R5117 (1989).
[22] W. G. Schmidt, P. H. Hahn, F. Bechstedt, N. Esser, P. Vogt, A. Wange, and W. Richter, Phys. Rev. Lett. **90**, 126101 (2003).
[23] A. M. Frisch, W. G. Schmidt, J. Bernholc, M. Pristovsek, N. Esser, and W. Richter, Phys. Rev. B **60**, 2488 (1999).
[24] M. Zorn, B. Junno, T. Trepk, S. Bose, L. Samuelson, J. T. Zettler, and W. Richter, Phys. Rev. B **60**, 11557 (1999).
[25] L. Töben, Th. Hannappel, K. Mölker, H.-J. Crawack, C. Pettenkofer, and F. Willig, Surf. Sci. **494**, L755 (2001).
[26] R. Shioda and J. van der Weide, Appl. Surf. Sci. **130–132**, 266 (1998).
[27] M. D. Pashley, Phys. Rev. B **40**, 10481 (1989).
[28] O. Pulci, K. Lüdge, W. G. Schmidt, and F. Bechstedt, Surf. Sci. **464**, 272 (2000).
[29] O. Pulci, W. G. Schmidt, and F. Bechstedt, phys. stat. sol. (b) **184**, 105 (2001).
[30] O. Pulci, K. Lüdge, P. Vogt, N. Esser, W. G. Schmidt, W. Richter, and F. Bechstedt, Comput. Mater. Sci. **22**, 32 (2001).
[31] V. Derycke, P. G. Soukiassian, F. Amy, Y. J. Chabal, M. D. D'Angelo, H. B. Enriquez, and M. G. Silly, Nature Mater. **2**, 253 (2003).
[32] V. M. Bermudez, Nature Mater. **2**, 218 (2003).
[33] P. Hohenberg and W. Kohn, Phys. Rev. B **136**, 864 (1964).
[34] W. Kohn and L. J. Sham, Phys. Rev. A **140**, 1133 (1965).
[35] M. Fuchs and M. Scheffler, Comput. Phys. Commun. **119**, 67 (1999).
[36] E. L. Briggs, D. J. Sullivan, and J. Bernholc, Phys. Rev. B **54**, 14362 (1996).
[37] W. G. Schmidt, Appl. Phys. A **75**, 89 (2002).
[38] P. H. Hahn, W. G. Schmidt, F. Bechstedt, O. Pulci, and R. Del Sole, Phys. Rev. B **68**, 033311 (2003).
[39] F. Manghi, R. Del Sole, A. Selloni, and E. Molinari, Phys. Rev. B **41**, 9935 (1990).
[40] F. Bechstedt, W. G. Schmidt, and P. H. Hahn, phys. stat. sol. (a) **188**, 1383 (2001).
[41] F. Bechstedt, K. Tenelsen, B. Adolph, and R. Del Sole, Phys. Rev. Lett. **78**, 1528 (1997).
[42] G. D. Mahan, Many-Particle Physics (Plenum Press, New York, 1990).
[43] W. Hanke and L. J. Sham, Phys. Rev. B **12**, 4501 (1975).
[44] R. Del Sole and E. Fiorino, Phys. Rev. B **29**, 4631 (1984).
[45] S. Adler, Phys. Rev. **126**, 413 (1962).
[46] N. Wiser, Phys. Rev. **129**, 62 (1963).
[47] P. Hahn, Diploma thesis, Friedrich-Schiller-Universität Jena 2001.
[48] M. S. Hybertsen and S. G. Louie, Phys. Rev. **34**, 5390 (1986).
[49] F. Bechstedt, Festkörperprobleme/Adv. Solid State Phys. **32**, 161 (1992).
[50] O. Pulci, F. Bechstedt, G. Onida, R. Del Sole, and L. Reining, Phys. Rev. B **60**, 16758 (1999).
[51] M. S. Hybertsen and S. G. Louie, Phys. Rev. B **37**, 2733 (1988).
[52] F. Bechstedt, R. Del Sole, G. Cappellini, and L. Reining, Solid State Commun. **84**, 765 (1992).
[53] G. Cappellini, R. Del Sole, L. Reining, and F. Bechstedt, Phys. Rev. B **47**, 9892 (1993).
[54] W. G. Schmidt, J. L. Fattebert, J. Bernholc, and F. Bechstedt, Surf. Rev. Lett. **6**, 1159 (1999).
[55] M. Rohlfing and S. G. Louie, Phys. Rev. B **62**, 4927 (2000).
[56] S. Glutsch, D. S. Chemla, and F. Bechstedt, Phys. Rev. B **54**, 11592 (1996).
[57] W. G. Schmidt, S. Glutsch, P. H. Hahn, and F. Bechstedt, Phys. Rev. B **67**, 085307 (2003).
[58] W. G. Schmidt, E. L. Briggs, J. Bernholc, and F. Bechstedt, Phys. Rev. B **59**, 2234 (1999).
[59] W. G. Schmidt, N. Esser, A. M. Frisch, P. Vogt, J. Bernholc, F. Bechstedt, M. Zorn, Th. Hannappel, S. Visbeck, and W. Richter, Phys. Rev. B **61**, R16335 (2000).
[60] O. Pulci, M. Palummo, V. Olevano, G. Onida, L. Reining, and R. Del Sole, phys. stat. sol. (a) **188**, 1261 (2001).
[61] S. Zollner, M. Garriga, J. Kirchner, J. Humliček, M. Cardona, and G. Neuhold, Phys. Rev. B **48**, 7915 (1993).
[62] B. S. Mendoza and W. L. Mochán, Phys. Rev. B **53**, R10473 (1996).
[63] D. Herrendörfer and C. H. Patterson, Surf. Sci. **375**, 210 (1997).
[64] B. S. Mendoza, R. Del Sole, and A. I. Shkrebtii, Phys. Rev. B **57**, R12709 (1998).
[65] A. B. Müller, F. Reinhard, U. Resch, W. Richter, K. C. Rose, and U. Rossow, Thin Solid Films **233**, 19 (1993).

phys. stat. sol. (b) **240**, No. 3, 480–489 (2003) / **DOI** 10.1002/pssb.200303837

Optical reflectance of a composite medium with a sparse concentration of large spherical inclusions

A. García-Valenzuela[*, 1] and **R. G. Barrera**[**, 2]

[1] Centro de Ciencias Aplicadas y Desarrollo Tecnológico, Universidad Nacional Autónoma de México, Apartado Postal 70-186, México D.F., 04510, México

[2] Instituto de Física, Universidad Nacional Autónoma de México, Apartado Postal 20-364 , México D.F., 01000, México

Received 30 May 2003, revised 4 August 2003, accepted 11 August 2003
Published online 25 November 2003

PACS 42.25.Dd, 42.70.–a, 78.20.Ci, 78.68.+m, 78.90.+t

We analyze the coherent reflection of a half-space of a composite material consisting of an homogeneous matrix with spherical inclusions. We pay attention to the case where the radius of the particles is comparable to the wavelength of the incident radiation. We consider a simple model for the interface of the composite and present numerical calculations to illustrate the contribution of the embedded particles to the optical reflectance and discuss surface effects.

1 Introduction

Composite and colloidal materials are becoming increasingly important in modern technology. These materials are designed and constructed to have physical properties that cannot be found in homogeneous materials. When two or more materials are mixed together at a micro- or nano-scale, the resulting composite has physical properties that are different from those of the constituent materials. In particular, their optical properties might be substantially different.

Here we will consider a simple type of composite which consists of discrete spherical inclusions embedded within an otherwise homogeneous matrix of a different material. When the size of the particles is small compared to the wavelength of the incident radiation, λ, one can use well-established approximations to calculate an effective index of refraction, such as the Maxwell Garnett or Bruggeman effective-medium theories [1, 2]. When the wavelength of light within the inclusions is no longer very large compared to the size of the particle, extended-effective medium theories are available [3]. Extended-medium theories include a dynamical correction to the calculation of the polarization and magnetization of the isolated particles providing effective optical coefficients, that is, an effective electrical permittivity and an effective magnetic permeability. Whenever these extended-effective-medium theories provide an adequate approximation to the effective optical coefficients, one may use them freely in the laws of continuous electrodynamics. In this case the effective-medium theory is called: unrestricted. For instance, one may use the Fresnel reflection coefficients with the resulting effective optical coefficients to calculate the reflection of a plane wave from a flat surface of the composite. However when the size of the particles is comparable to the wavelength of incident radiation, extended-effective-medium theories are

---

[*] Corresponding author: e-mail: garciaa@aleph.cinstrum.unam.mx, Phone: +5255 5622 8602 ext. 1118, Fax: +5255 5622 8651.
[**] Consultant at Centro de Investigación en Polímeros (grupo COMEX).

no longer valid and it might not be possible to use, in the usual way, the effective-optical coefficients within the laws of continuous electrodynamics. In the case of large particles, one must solve the wave multiple-scattering problem and take configurational averages to calculate the macroscopic fields (also called average or coherent fields) and quantities such as their effective propagation wave vector or the reflection coefficient of a half space of the composite.

In previous works [4, 5] we have derived the reflection coefficient of the coherent field from a half-space with a plane interface of a sparse, random, uniformly-distributed ensemble of spherical particles in vacuum by two different procedures. In Ref. [4] we first calculate the average scattered field from an ensemble of particles located at random within a thin slab using the single scattering approximation. Then we took account of multiple scattering by modelling the half space as a semi-infinite pile of thin slabs to determine the propagation wave vector of the average wave travelling between the thin slabs, as well as the reflected average field. We found that in order to describe correctly the reflection of light from a half-space with effective optical properties, the effective medium should have besides an effective electric permittivity an effective magnetic permeability. Furthermore both of these effective optical coefficients turned out to depend on the polarization of the incident beam and on the angle of incidence. We reached the same conclusions and derive the same expressions for the optical effective coefficients in Ref. [5], where we start by setting up the multiple-scattering system of equations and solving for the average propagating and reflected fields using the effective-field approximation. Some examples were investigated numerically for particles in vacuum. One can show that when the radius of particles becomes smaller than the incident wavelength, the reflection coefficients found in Refs. [4, 5] reduce to the Fresnel reflection coefficients for an effective medium in which the effective magnetic permeability reduces to the one of vacuum. This means that in this limit the effective medium corresponding to a composite with nonmagnetic components is nonmagnetic, that is, it does not posses a dynamic effective magnetic response. Although this is true for small inclusions, what we have shown in Refs. [4, 5] is that this assessment is no longer true when the size of the inclusion is comparable to the wavelength of the incident radiation.

More accurate approximations to the effective propagation constant and half-space reflection coefficients from a random distribution of particles, based on multiple scattering theory and the so called quasi-crystalline approximation, have been put forth [6]. The quasi-crystalline approximation can be used to model dense ensembles of particles. However calculations with the quasi-crystalline approximation are quite more complicated. Our formulas are substantially simpler and allow us to investigate the main effects related to the coherent reflectance of a composite with spherical particles in a more transparent way, although we are limited to a low concentration of particles.

In this paper we use the previously derived formulas to analyze and discuss the reflectance from a half-space of a composite material consisting of a homogeneous matrix with a sparse, randomly-distributed ensemble of large spherical-particles, embedded within the matrix. First we review briefly the main results regarding the coherent reflectance of a half-space of a random ensemble of particles. Then we discuss the surface model for a composite material and then we present some numerical calculations of a system of particles with a high refractive index embedded within a polymeric-type matrix. Finally we present our conclusions.

2 Reflectance from a half-space of spherical particles

2.1 In vacuum

In previous works, we have derived the coherent reflection coefficients for a plane wave incident at oblique angles on a half space of a sparse concentration of identical spherical particles with refractive index n_p, embedded in vacuum. The reflection coefficient of the coherent wave can be written as

$$r_{hs} = \frac{-i\gamma S_m(\pi - 2\theta_i)}{\cos^2\theta_i + i\gamma S(0) + \sqrt{\cos^4\theta_i + 2i\gamma S(0)\cos^2\theta_i}}, \qquad (1)$$

where θ_i is the angle of incidence, $S_m(\theta)$ are the elements of the 2×2 amplitude scattering matrix of an isolated particle [7], $m = 1$ for TE polarization, and $m = 2$ for TM polarization. $S(0) = S_1(0) = S_2(0)$ is the forward scattering amplitude, and

$$\gamma \equiv \frac{3f}{2x^3}, \tag{2}$$

where $f = \rho 4\pi a^3/3$ is the volume filling fraction of the spheres, ρ is the number density of spheres, a is the radius of the particles and $x = ka$ is the size parameter. This reflection coefficient is a good approximation for all angles of incidence but for a low density of particles (in the range of a few percent of volume fraction, depending on radius of the particles). The region of validity of this approximation in the space of variables (θ_i, f, n_p) is not well known to date, but a simple test to estimate the confidence of the approximation has been described in Ref. [4]. The effective index of refraction of the coherent wave travelling within the ensemble of particles was found to be, to lowest order in γ,

$$n_{\text{eff}} \approx 1 + i\gamma S(0), \tag{3}$$

which is the same expression as the one derived by van de Hulst, long time ago [8]. As mentioned above, we have also found that if one wants to interpret the ensemble of particles as a half-space of an effective (homogeneous) medium, then one is forced to accept an effective magnetic permeability in addition to an effective electric permittivity in order to reproduce Eq. (1) from the Fresnel reflection coefficients. These effective optical coefficients depend on both, the state of polarization, and the angle of incidence. Nevertheless, it is not necessary to interpret the ensemble of particles as an effective medium. One can always work directly with the expression for the reflection coefficient given by Eq. (1) and with the expression for the effective index of refraction given by Eq. (3).

2.2   In a matrix

The results obtained for the reflection coefficient of a half space of particles in vacuum can be easily extended to a half space of particles embedded in a boundless homogeneous matrix with optical coefficients ε_m and μ_m. We assume that both coefficients ε_m and μ_m are real, so that there is no absorption within the matrix. Then one can evaluate the scattering amplitudes for a particle considering that it is surrounded by a homogeneous medium. To recall this fact we will denote the scattering amplitude elements for particles embedded in a medium other than vacuum by $S'_m(\theta)$. We should also replace k for $n_m k$, with $n_m = \sqrt{\varepsilon_m \mu_m}$ in the expression for γ. Again we will use a prime to remind us about this fact. Then we have,

$$\gamma' \equiv \frac{3f}{2x'^3} = \frac{3f}{2(n_m ka)^3}. \tag{4}$$

Thus, the half space reflection coefficient for a plane wave traveling within the matrix and incident at an angle θ_m to a half space of particles embedded in the same matrix, is written as

$$r'_{hs} = \frac{-i\gamma' S'_m(\pi - 2\theta_i)}{\cos^2 \theta_i + i\gamma' S'(0) + \sqrt{\cos^4 \theta_i + 2i\gamma' S'(0) \cos^2 \theta_i}}. \tag{5}$$

The effective index of refraction to first order in γ is,

$$n_{\text{eff}} \approx n_m [1 + i\gamma' S'(0)]. \tag{6}$$

3   Surface model for a half space of a composite material

If we consider the coherent reflection of a plane wave travelling in a homogeneous medium of refractive index n_0 that is incident at an angle θ_i on a half space of a composite material consisting of a homogeneous material with spherical inclusions, we must take care of modelling the surface carefully. We may

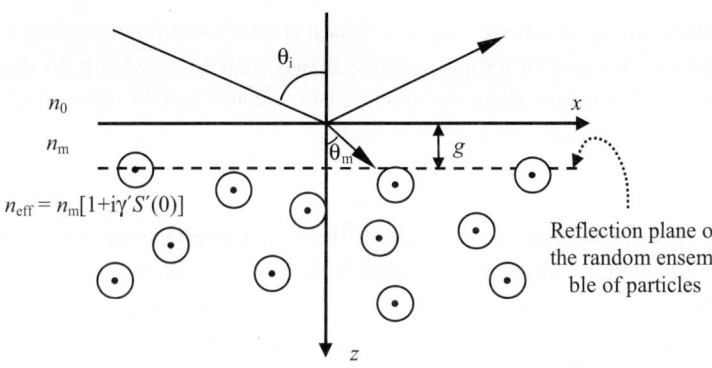

Fig. 1 Geometry of the problem.

have different situations. For instance, the particles may reach the surface of the homogeneous matrix and be partially outside the matrix, or the particles may be repelled from the surface leaving a slab of homogeneous matrix between the vacuum and the half space of the composite. Also, the surface of the matrix may have some roughness which will also affect the coherent reflection coefficient. All these effects may have a strong influence on the coherent reflectance from a composite that might be comparable or might be even larger than the contribution from the embedded particles. Therefore, in general, one must specify the surface conditions in order to have a reliable model for the coherent reflectance from a composite material.

Here we will assume that all the particles are completely embedded within the matrix. Thus, the minimum possible distance from the centre of the particles to the surface of the matrix is one particle radius. We will then model the system as a three-layered system as shown in Fig. 1. That is, we consider the composite material as a slab of homogeneous matrix of width g in contact with a half-space of homogeneous matrix with particles embedded in it. We must recall that the reflection coefficient of a half space of particles in Eq. (5) assumes that the centre of the particles are uniformly distributed on one side of a mathematical plane. This mathematical plane may be regarded as the reflection plane of the half space of particles, that is, as the surface of an effective homogeneous medium.

The relation between the angle of incidence to the composite, θ_i, and the angle of incidence to the half space of particles embedded in the matrix, θ_m, is given by Snell's law at the interface outside-medium/matrix, that is, $n_0 \sin \theta_i = n_m \sin \theta_m$. The coherent reflection coefficient from a half-space of the composite matrix material, r, is obtained by calculating the reflection from the system: outside-medium/homogeneous-matrix/composite-matrix. This corresponds to a thin slab of homogeneous matrix on a composite-matrix substrate. The reflection coefficient is

$$r = \frac{r_m + r'_{hs} \exp\left(2ikn_m \cos\theta_m \, g\right)}{1 + r_m r'_{hs} \exp\left(2ikn_m \cos\theta_m \, g\right)} \tag{7}$$

where r_m is the reflection coefficient of the outside-medium / homogeneous-matrix interface, and g is the width of the homogeneous-matrix slab (i.e., the distance to the reflection plane from the half space of particles). The latter equation may be modified without much difficulty to include, when important, the effect of roughness on the matrix interface by using a suitable model (see for example Ref. [9]), and g may be adjusted to accommodate on the boundary conditions some of the specific features of the density of particles. If the matrix interface is considered flat, then r_m is calculated with the Fresnel reflection coefficients. As already said, we will consider that the matrix has a real index of refraction and use Eq. (5) in Eq. (7). In general, when the particle radius is not small compared to the wavelength, the oscillating phase term: $\exp\left(2ikn_m \sqrt{n_m^2 - n_0^2 \sin^2\theta_i}\, g\right)$, will have a noticeable effect and will be mixed with the behaviour of r'_{hs} which is mostly dependent on the particles properties.

In this paper we will consider only the case of external reflection, that is when $n_0 < n_m$. An example could be $n_0 = 1$ (air) and $n_m = 1.33 - 1.6$ (water, glass, or some kind of polymer). In this case, the refraction angle θ_m never approaches grazing, and r'_{hs} is always very small for a sparse concentration of particles. We will also restrict our analysis to the case of a non-magnetic matrix, that is, $\mu_m/\mu_0 = 1$. In this case, there is a Brewster's angle only for TM polarization. At an angle of incidence equal to Brewster's angle of the outside-medium/matrix interface we have that $r_m^{TM} = 0$ and

$$r^{TM} = r_{hs}'^{TM} \exp\left(2ikn_m \cos\theta_m \, g\right).$$

The reflectance is given by $R = |r|^2$. Thus at the Brewster's angle of the matrix interface we have that $R^{TM} = R_{hs}'^{TM}$ is due to the presence of the particles alone. On the other hand, with the exception of angles near Brewster's angle, corresponding to the n_0/n_m interface in TM polarization, we can expand r as

$$r \simeq \left[r_m + r'_{hs} \exp\left(2ik\sqrt{n_m^2 - n_0^2 \sin^2\theta_i}\, g\right) \right]\left[1 - r_m r'_{hs} \exp\left(2ik\sqrt{n_m^2 - n_0^2 \sin^2\theta_i}\, g\right) \right]$$

$$\simeq r_m + (1 - r_m^2)\, r'_{hs} \exp\left(2ik\sqrt{n_m^2 - n_0^2 \sin^2\theta_i}\, g\right). \tag{8}$$

The reflectance is given by

$$R = R_m + (1 - r_m^2)^2 \, R'_{hs} + 2r_m(1 - r_m^2)\, \mathrm{Re}\left[r'_{hs} \exp\left(2ik\sqrt{n_m^2 - n_0^2 \sin^2\theta_i}\, g\right) \right]. \tag{9}$$

For a composite with a sparse distribution of particles, $|r'_{hs}| \ll 1$, we may neglect the second term on the right hand side, which is proportional to $R'_{hs} = |r'_{hs}|^2$ and keep only the third term on the right hand side which is proportional to r'_{hs}. Thus, having the particles embedded in a matrix may yield a larger contribution to the coherent reflection from the particles than when the particles are in vacuum, because in the reflectance there is now a linear term in r'_{hs}, which for a dilute system is much larger than $|r'_{hs}|^2$. Now, for grazing incidence $r_m \to 1$, thus the factor, $(1 - r_m^2)$, that appears in the second and third terms on the right hand side of Eq. (9) goes to zero. Therefore, the relative importance of the contribution to the reflection from the particles decreases as the angle of incidence increases towards grazing incidence.

4   Numerical results

We will restrict our calculations to a specific system in which the particles are non-magnetic, with a real refractive index $n_p = 2.8$ and embedded in a homogeneous, non-magnetic ($\mu_m = 1$) matrix with a real refractive index $n_m = 1.45$. These values of refractive index are close to those in a typical white paint. We will also consider a volume filling fraction of spheres of $f = 0.1$. It is always interesting to compare our results with an heuristic model obtained by substituting the effective refractive index given by Eq. (6) in the Fresnel reflection coefficient and ignoring any possible magnetic effect. That is, by considering that the half space of particles behaves as an effective medium equivalent to an ordinary, homogeneous, non-magnetic, material where $\varepsilon_{eff} = n_{eff}^2$. We will refer to this heuristic approximation as the isotropic approximation. For a meaningful comparison, the isotropic model should also consider the surface model discussed above. Thus, the isotropic approximation consists of using Eq. (7) with r'_{hs} calculated with the Fresnel coefficients upon assuming $\varepsilon_{eff} = n_{eff}^2$. One can show that as the particle radius decreases, the isotropic approximation coincides with our reflection formula. Thus, the difference between our results and those of the isotropic approximation will give us an idea of the magnetic effects inherent to our approach.

4.1   Shift of Brewster's angle

First, let us consider the effect of the embedded particles on the value of the Brewster's angle of the composite material. In Fig. 2 we plot the reflectance for TM polarization versus the angle of incidence in the vicinity of Brewster's angle of the matrix interface for different particle radii. We assume that the particles can barely touch the matrix interface, that is we used $g = a$. We show the reflectance curves for the matrix alone and for the composite using our formulas and the isotropic approximation. We can ap-

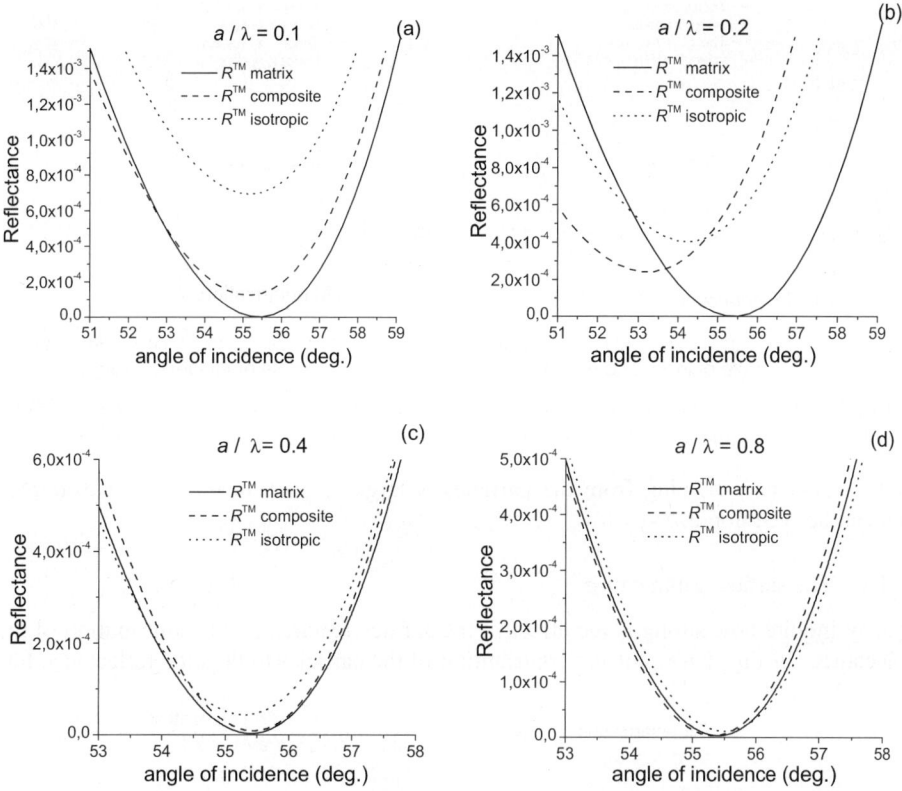

Fig. 2 TM reflectance versus angle of incidence about Brewster's angle of the matrix interface for different particle's radius, a) $a/\lambda = 0.1$, b) $a/\lambda = 0.2$, c) $a/\lambda = 0.4$, and d) $a/\lambda = 0.8$. The full curve is the reflectance for a homogeneous matrix, the dashed curve is the reflectance of the composite system for $g = a$, and the dotted line is for the isotropic model also with $g = a$.

preciate that the contribution of the particles shift the Brewster's angle from it's value for the matrix alone. The isotropic approximation predicts a different shift and overestimates the contribution of the particles to the reflectance. These plots also show that the contribution of the particles to the reflectance decreases as the particle radius increases, while keeping the filling fraction constant.

Surface roughness of the matrix interface would also result in a shift of the Brewster's angle [9]. Thus, if surface roughness is important, the shift of the Brewster's angle will have a contribution from both, the interface roughness and the embedded particles in the composite. Detailed modelling of the surface would be required in order to discern one effect from the other.

4.2 Contribution of the particles to the reflectance

In Fig. 3 we plot the fraction of the reflectance that is due to the presence of the particles as a function of the angle of incidence for two different values of the radius of the particles, $a/\lambda = 0.2$ and $a/\lambda = 0.4$, and for the case $g = a$. Fig. 3a is for TE polarization and Fig. 3b is for TM polarization. We can appreciate that the fraction of the reflectance due to the presence of the particles is larger for smaller angles of incidence. As already said, this fact can be understood from inspection of Eq. (9). In the graphs of Fig. 3 we also plot the curves predicted by the isotropic approximation and we can see that they are different in magnitude and even in sign for some angles of incidence. Thus, the isotropic approximation predicts already qualitatively erroneous results for particles as small as $a/\lambda = 0.2$. The plots in Fig. 3b reach a value of one at Brewster's angle of the matrix/air interface. This does not mean, however, that the contri-

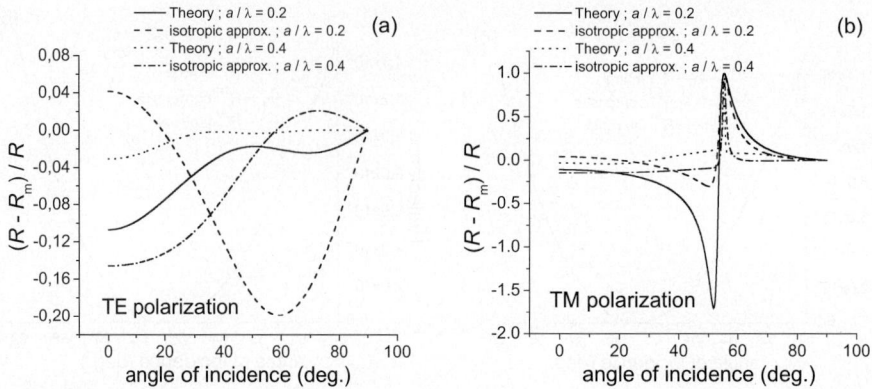

Fig. 3   Fraction of the reflectance due to the random ensemble of particles for two different particle radius, $a/\lambda = 0.2$ and $a/\lambda = 0.4$, a) for TE polarization and b) for TM polarization. All curves are for $g = a$.

bution to the reflectance coming from the particles is large, in fact, in the present example it is very small, as it can be seen in Fig. 2.

4.3   The surface parameter g

Now, we may inquire how strong is the effect of the surface features on the contribution of the particles to the reflectance. In Fig. 4 we plot the contribution of the particles to the total reflectance for three dif-

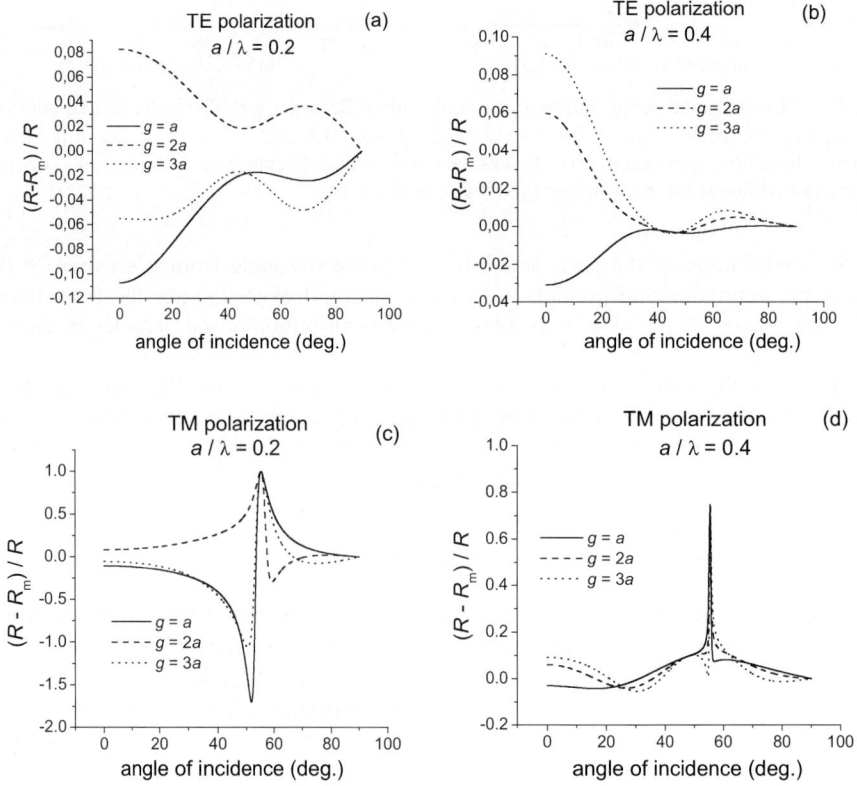

Fig. 4   Fraction of the reflectance due to the particles for different values of the g parameter, $g = a$, $2a$, and $3a$ and for two values of the particle's radius. a) $a/\lambda = 0.2$ and b) $a/\lambda = 0.4$ for TE polarization; c) $a/\lambda = 0.2$ and d) $a/\lambda = 0.4$ for TM polarization.

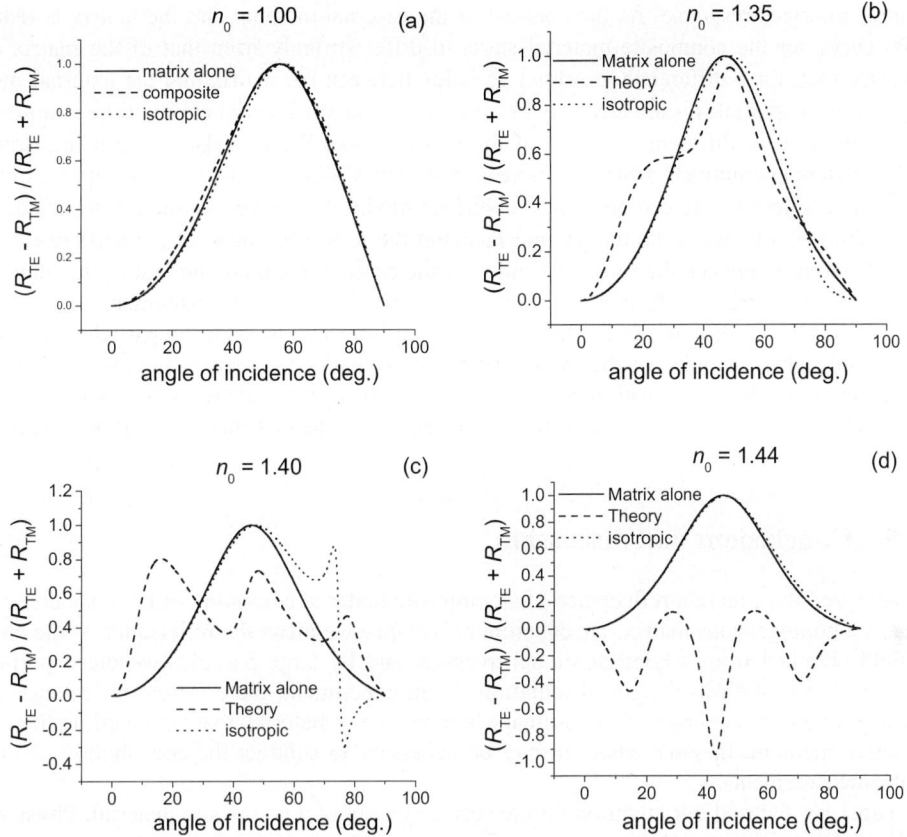

Fig. 5 Differential polarized reflectance [$(R_{TE} - R_{TM})/(R_{TE} + R_{TM})$] for particles of radius $a/\lambda = 0.3$ and for different contrast between the outside medium and the matrix. a) $n_0 = 1.00$, b) $n_0 = 1.35$, c) $n_0 = 1.40$ and d) $n_0 = 1.44$. Curves were generated with $g = a$.

ferent values of g. The minimum value of g is a when one assumes that all particles are completely embedded within the matrix. However, due to surface conditions one may expect a transition region with lower or higher density of particles. Fig. 4 shows that the contribution of the particles to the reflectance changes significantly for most angles of incidence as g increases. This is an example of how the surface conditions affect the coherent reflectance in a composite.

4.4   Differential reflectance

One may be interested in obtaining information about the particles within the matrix from reflectance measurements. We have already shown in Ref. [5], that differential measurements of polarized reflectance, specifically $(R^{TE} - R^{TM})/(R^{TE} + R^{TM})$, of a half space of particles as a function of the angle of incidence can be qualitatively very different from measurements of half space of an ordinary homogeneous material. However, when the particles are embedded in a matrix, the contribution of the matrix interface to the differential measurements may be so large as to mask completely the effect of the particles on the differential measurements. This, of course, will be a function of the contrast between the external medium and the matrix. In Fig. 5 we plot the differential reflectance, $(R^{TE} - R^{TM})/(R^{TE} + R^{TM})$, versus the angle of incidence for different values of the refractive index of the external medium and for particles of radius $a/\lambda = 0.3$, and $g = a$. We also plot the curves for the isotropic approximation and for the matrix/air interface. For air as the external medium, $n_0 = 1$, the contribution of the particles is strongly masked by the matrix interface response. As the contrast of the external medium and the matrix is reduced, the differential curve for the composite material starts to differ strongly from that of the matrix alone. In Fig. 5c we see that, for a difference in refractive index between the matrix and the external medium of 0.1, the effect of the particles is already very noticeable and the differential curve of the composite material is qualitatively very different than that of the matrix alone. We can also see that the isotropic approximation differs qualitatively from the curve of the matrix alone. In the isotropic approximation this difference is due solely to the geometry of the surface model where we assume a slab of homogenous matrix of width $g = a$ between the matrix interface and the reflection plane of the particle ensemble. In Fig. 5d the difference between the refractive index of the external medium and that of the matrix is 0.01, and the effect of the matrix interface is now negligible on the differential measurement.

Figure 5 indicates that it may not be difficult to obtain information from the particles in a composite by matching the refractive index of the external medium to that of the matrix. In practice this should be possible by immersing the composite in a liquid mixture and adjusting its refractive index to be close to that of the matrix. Also, in this case the effect of surface roughness, if there is, will be reduced considerably.

5   Conclusions and discussion

We have analyzed the coherent reflectance of a composite material consisting of large spherical particles embedded in a homogeneous matrix. By coherent reflectance we mean the reflectance of the coherent or average field obtained from a configurational average, and by large particles we mean particles with radius comparable to the wavelength of radiation. In an experiment, the coherent reflectance would be measured as if one were dealing with an ordinary homogeneous material, that is, regarding the composite as an effective medium. In some cases, it may be necessary to subtract the contribution of the diffuse fields to the measurements.

In this paper we focused our attention on the surface model of a composite material. Physical restrictions in the surface region may have a strong influence on the coherent reflectance of a composite material. We considered a surface model in which the particles are completely surrounded by the matrix material. When the particles are located at random, except close to the surface, we can model the surface as slab of homogeneous matrix in-between the external medium and an effective medium consisting of the random ensemble of particles embedded in the matrix. The width of the slab should be specified based on a physical model of the interaction of the particles with the matrix interface.

In the graphs presented above, we assumed a flat matrix interface. In practice, when one deals with solid composites, the effect of the surface roughness may often be comparable or larger than the effect of the embedded particles. In our model of the surface of the composite, the surface roughness could be easily incorporated.

We showed that the presence of the embedded particles shifts the Brewster's angle from its value for the matrix/air interface. From this shift one cannot obtain the effective index of refraction in the usual way, that is, by assuming the isotropic approximation. Instead, one must compare the experimental data with the results of the reflection model presented here and the half-space reflection in Eq. (5). From this comparison one may obtain the values of the scattering matrix elements of the particles and the density of particles.

We showed that the width of the homogeneous layer on the surface model can have a strong influence on the contribution of the particles to the total reflectance of the composite. Although here we were limited to our specific surface model, the latter conclusion tells us that, in general, one must model with great care the surface of a composite in order to be able to relate reflectance measurements to the particles in the composite. Of course, as the contrast between refractive indexes of the external medium and the matrix material is lowered, the relative importance of the contribution of the particles to the reflectance increases and the surface model becomes less determinant. We found in our example, that a contrast in refractive index of about 0.1 is enough to unmask the contribution of the particles from the reflectance of the matrix interface in differential polarized reflectance measurements. This result indicates the possibility of performing experimental measurements of random ensemble of particles embedded in a solid matrix by roughly matching the refractive index of the external medium (e.g., with a liquid) with that of the matrix.

All calculations presented in this paper were for a volume filling fraction of 0.1. According to the confidence test suggested in Ref. [4], the plots presented in Sect. 4 may incur in relatively large errors for some angles of incidence and particle's radius. In the worst case, we expect the error to be of about 20% and only within limited intervals of the angle of incidence. Thus, the qualitative behaviour of the curves presented here is correct and the observations made from them are correct.

Finally, in some cases, one may need to consider the possibility of having particles partly embedded in the matrix and partly on the outside medium. One way of taking this possibility into account may be to consider the protruding particles on the surface as surface roughness. Nevertheless, experimental results on actual physical samples are very much needed.

Acknowledgement   We acknowledge support from Dirección General de Asuntos del Personal Académico of Universidad Nacional Autónoma de México through grants IN-104201 and IN-108402.

References

[1] J. C. Maxwell Garnett, Philos. Trans. R. Soc. Lond. **203**, 385 (1904).
[2] D. A. G. Bruggeman, Ann. Phys. (Leipzig) **24**, 636 (1935).
[3] R. Ruppin, Opt. Commun. **182**, 273 (2000).
[4] R. G. Barrera and A. García-Valenzuela, J. Opt. Soc. Am. A **20**, 296 (2003).
[5] A. García-Valenzuela and R. G. Barrera, J. Quant. Spectrosc. Radiat. Transf. **79–80**, 627 (2003).
[6] L. Tsang and J. A. Kong, Scattering of electromagnetic waves: Advanced topics (J. Wiley & Sons, New York, 2001).
[7] C. F. Bohren and D. R. Huffman, Absorption and Scattering of Light by Small Particles (J. Wiley & Sons, New York, 1983).
[8] H. C. van de Hulst, Light Scattering by Small Particles (J. Wiley & Sons, New York, 1957).
[9] C. Baylard, J.-J. Greffet, and A. A. Maradudin, J. Opt. Soc. Am. A **10**, 2637 (1993).

phys. stat. sol. (b) **240**, No. 3, 490–499 (2003) / **DOI** 10.1002/pssb.200303861

Spin/carrier dynamics at semiconductor interfaces using intense, tunable, ultra-fast lasers

Y. Jiang[1], **R. Pasternak**[*, 1], **Z. Marka**[1], **Y. V. Shirokaya**[1], **J. K. Miller**[1], **S. N. Rashkeev**[1], **Yu. D. Glinka**[1], **I. E. Perakis**[1], **P. K. Roy**[2], **J. Kozub**[3], **B. K. Choi**[4], **D. M. Fleetwood**[4], **R. D. Schrimpf**[4], **X. Liu**[5], **Y. Sasaki**[5], **J. K. Furdyna**[5], and **N. H. Tolk**[1]

[1] Department of Physics and Astronomy, Vanderbilt University, Nashville, TN 37235, USA
[2] Agere Systems, Orlando, FL 32819, USA
[3] Free-Electron Laser Center, Vanderbilt University, Nashville, TN 37235, USA
[4] Department of Electrical Engineering, Vanderbilt University, Nashville TN, USA
[5] Department of Physics, University of Notre Dame, Notre Dame, Indiana 46556, USA

Received 30 May 2003, revised 4 August 2003, accepted 11 August 2003
Published online 25 November 2003

PACS 42.65.Ky, 72.25.Fe, 73.40.Qv, 73.50.Gr, 77.55.+f

We review recent advances in spin/carrier dynamics at semiconductor interfaces using intense, tunable, ultrafast lasers, involving (a) carrier dynamics at Si/SiO$_2$ interfaces with different oxide thicknesses, (b) radiation enhanced electron transport in ultra-thin oxides, and (c) ultrafast spin dynamics in semiconductor heterostructures, probed by second-harmonic generation (SHG).

1 Introduction

Second Harmonic Generation (SHG) has been shown to be an effective technique for characterizing carrier and spin dynamics at semiconductor interfaces. In particular, we have developed a contactless two-color optical technique that allows us to monitor carrier transport (injection, tunneling) by multiphoton internal-photoemission induced second-harmonic generation. We have used this technique to measure the Si/SiO$_2$ band-offset. One- and two-photon internal-photoemission thresholds were measured to be 4.5 eV and 2.25 eV respectively. In addition, we applied this technique to measure X-ray irradiation enhanced electron transport across thin oxides Si/SiO$_2$ samples. Measured electron transport rates across an irradiated oxide were found to be substantially higher in comparison to unirradiated oxides. This effect is attributed to the presence of X-ray irradiation induced defects that act as intermediate trapping sites facilitating enhanced electron tunneling through the oxide. Finally, we report the application of pump–probe second harmonic generation to monitor spin dynamics in nonmagnetic semiconductor heterostructures. Spin-polarized electrons were selectively excited by a pump beam in the GaAs layer of GaAs/GaSb/InAs structures. However, the induced magnetization manifests itself through the SHG probe signal from the GaSb/InAs interface, thus indicating a spin-polarized electron transport. We find that the magnetization dynamics is governed by interplay between the spin density evolution at the interfaces and the spin relaxation.

2 SHG studies of silicon interfaces

Second-harmonic generation has proved itself to be a novel, nondestructive probe of Si/oxide interfaces [1]. We have developed a contactless two-color optical technique that allows us to monitor carrier trans-

---

[*] Corresponding author: e-mail: robert.pasternak@vanderbilt.edu, Phone: +1 615-343-2957

port (injection, tunneling) processes at Si/oxide interfaces [2]. This method involves two steps: (1) optically stimulated electron injection into the oxide by a high intensity pump laser and (2) detection of transport, trapping and recombination rates using time-dependent electric-field induced second-harmonic generation arising from charge separation at the interface by a less intense probe laser. A pump beam from tunable Optical Parametric Generator (1–6 eV) is used to inject carriers at different power and photon energy. A probe beam from Ti:sapphire laser at 800 nm with low intensity is used to monitor the time-dependent electric field via SHG. The radiated SH intensity can be described by

$$I^{2\omega}(t) = |\chi^{(2)} + \chi^{(3)} E(t)|^2 \, (I^\omega)^2 \,, \tag{1}$$

where I^ω and $I^{2\omega}(t)$ are the intensities of the fundamental and the time-dependent SHG beams, $\chi^{(3)}$ is the third-order nonlinear susceptibility, $\chi^{(2)}$ is the effective second-order susceptibility and $E(t)$ is a interface electric field. The quasistatic electric field in Eq. (1) is proportional to the density of charged surface electron traps, which can be given through the solution of the rate equation:

$$\frac{dn_e}{dt} = (n_{0e} - n_e)/\tau^e_{\text{PUMP}} + (n_{0e} - n_e)/\tau^e_{\text{PROBE}} - n_e/\tau^e_{\text{detrap}} \,, \tag{2}$$

where n_{0e} describes the initial number of unfilled electronic traps, $1/\tau^e_{\text{PUMP}}$ and $1/\tau^e_{\text{PROBE}}$ give the rates of filling up the surface electronic trap states due to the pump and probe beams (the second one is negligible). τ^e_{detrap} characterizes lifetime of surface electronic trap states; it contains contributions from the rate describing the detrapping of surface electrons, the tunneling rate through the oxide and their recombination with holes at the interface.

2.1   Band offset studies of Si/SiO$_2$ interfaces

We apply the two-color technique, involving time-dependent EFISH combined with a tunable laser source for carrier injection, to determine band offsets at semiconductor interfaces [2]. One sample studied was a conventional, thermally grown 42 Å thick SiO$_2$ deposited on Si(100) by Lucent Technologies. At this oxide thickness, the injected electrons reach the surface with a high probability, giving rise to a large field and therefore to a large easily detectable time-dependent SHG signal [3]. This oxide is also thick enough that electron tunneling from the surface back to the interface is negligible [4]. We found that τ^e_{detrap} is of the order of several hours for a ~40 Å oxide. This is not significant under the timescale of our experiment. Since $1/\tau^e_{\text{PUMP}}$ is significantly greater then the other two rate constants in Eq. (2), the solution can be easily found:

$$n_e(t) = n_{0e}(1 - e^{-t/\tau^e_{\text{PUMP}}}) \,. \tag{3}$$

Therefore our time-dependent EFISH data can be fitted by

$$I^{(2\omega)} = |y_0 + a(1 - e^{-t/\tau^e_{\text{PUMP}}})|^2 \,, \tag{4}$$

where y_0 and a are phenomenological values related to the initial and saturation SHG levels.

Figure 1a shows typical example of a time-dependent pump–probe type SHG measurement taken on our 42 Å thick Si/SiO$_2$ sample. When the injection beam was turned on, the signal rose rapidly indicating the creation of a time-dependent quasistatic electric field. This field originates from charge separation across the interface due to trapping of injected electrons at the oxide surface by the ambient oxygen, while the holes remain in the Si. After a steady-state signal was reached, we blocked the injection beam and observed that the signal remained constant. This is in contrast with our observations on a 17 Å sample (Fig. 1b). For this ultrathin sample case, when we blocked the pump beam, the SHG signal was observed to slowly decrease in intensity indicating an increasingly lessened electric field across the interface. This observation may be explained by noting for oxide thicknesses below a critical value (~30 Å), electron tunneling rates become significant, leading to faster detrapping from the O$_2$-rich surface and subsequent recombination with holes at the interface.

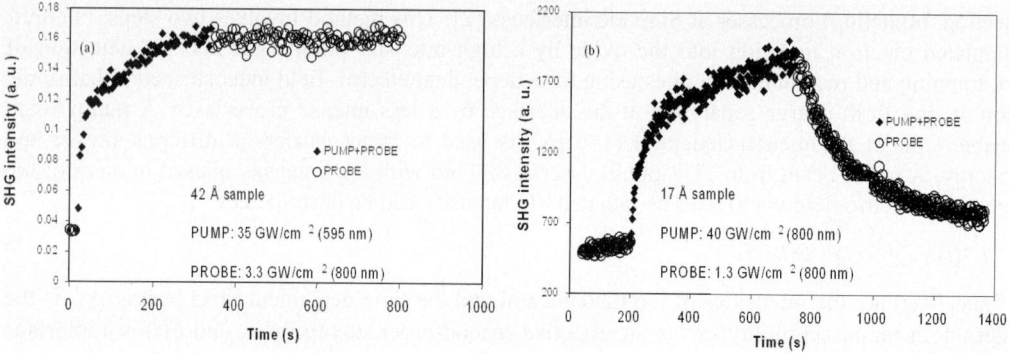

Fig. 1 a) Two-color type time-dependent SHG data taken on a 42 Å thick oxide sample compared to b) data taken on a 17 Å sample. After the pump laser is turned off the SHG signal decreases for the thinner sample, indicating that electrons tunnel back to the interface.

Figure 2a shows the SHG signal measured for different pump beam intensities, at the same wavelength 516 nm (2.4 eV). As the injection beam intensity increases the SH signal reaches the saturation level at an earlier time indicating an enhancement in the injection rate.

We have determined $1/\tau^e_{\text{PUMP}}$ for a large range of incident pump beam intensities at several different wavelengths between 258 nm and 663 nm. Fitting dependence of $1/\tau^e_{\text{PUMP}}$ rate on the pump laser intensity gives a slope, which we associate with the number of photons required in the injection given wavelength. Figure 2b shows the dependence of $1/\tau^e_{\text{PUMP}}$ rate constants on the pump laser intensity at 516 nm. Fitting this intensity dependent data to $1/\tau^e_{\text{PUMP}} \propto (I_{\text{PUMP}})^n$ gives a slope of 2.26 ± 0.10, which we associate with the number of photons required in the injection process at 516 nm (2.4 eV). Here we assumed that the measured trap-filling rate ($1/\tau^e_{\text{PUMP}}$) is proportional to the probability of the n-photon interband transition given by [5],

$$W_n = K_n(I^{(\omega)})^n/(n\hbar\omega) , \qquad (5)$$

where K_n is the n-photon absorption coefficient, $I^{(\omega)}$ is the incident beam intensity at energy $\hbar\omega$.

The number of photons required to inject electrons from silicon to SiO_2 was determined for several different energies in a range of pump photon energies (1.9–4.8 eV) (Fig. 3). We observe the stepwise jumps from one- to two-photon (between 4.56 and 4.50 eV) and then from two- to three-photon (between 2.30 and 2.20 eV) processes, as the incident pump energy decreases. The energies for stepwise jumps are

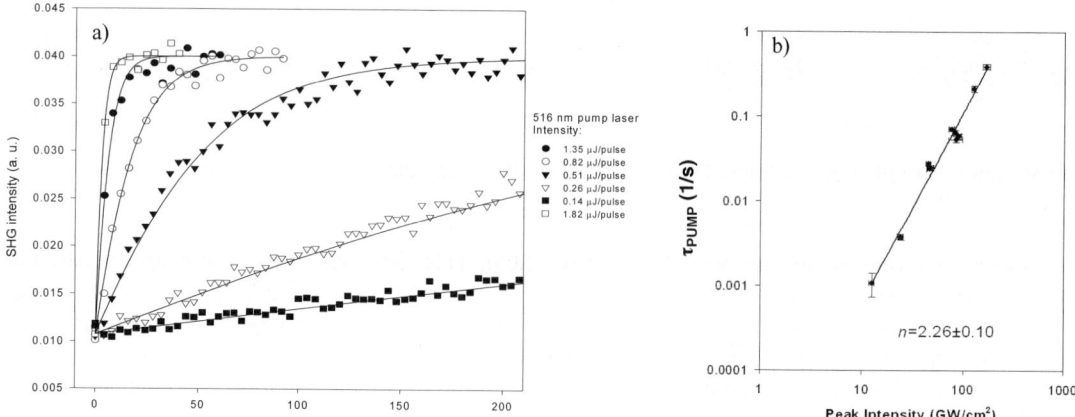

Fig. 2 a) Time-dependent electric field induced SHG signals for varying pump intensities at 516 nm. The black lines show the fits to the data using the rate Eq. (4). b) Electron trap-filling rate constants versus pump laser (516 nm) intensity, determined by fitting time-dependent EFISH data.

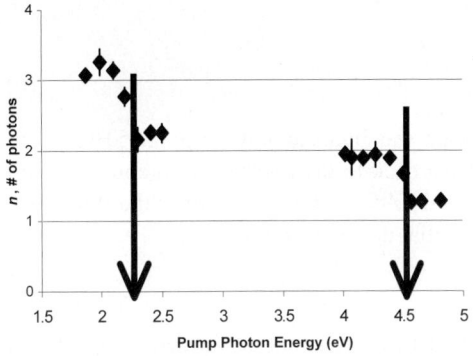

Fig. 3 Order of electron injection process from Si into SiO$_2$ versus pump photon energy.

identified as the thresholds for one-photon (~4.53 eV) and two-photon (~2.25 eV) electron injection from Si valence band into SiO$_2$ conduction band. To our best knowledge this is the first time when multiphoton internal photoemission thresholds were observed.

Published values for the Si(VB)-SiO$_2$(CB) offset are generally in the 4.05–4.6 eV [6, 7] range. The wide variation is mostly attributed to excitations to overlapping excitonic levels with the SiO$_2$ conduction band. Measured thresholds in internal photoemission (IPE) [8] studies are in the range of 4.05–4.35 eV [6], where the lower end of the range is attributed to mobile ion contamination in the oxide. IPE measurements are generally performed on MOS structures, where the oxide thickness is in the order of 1 μm. In a recent study Afanasev et al. found that the Si(100)(VB)–SiO$_2$(CB) barrier of 4.25 eV is unchanged down to oxide thicknesses of ~10 Å (IPE) [9]. A recent experimental study by the Lucovsky's group determined 4.35 eV for the Si(100)(VB)–SiO$_2$(VB) offset via X-ray photoemission method [10]. This gives 4.6 eV for Si(VB)–SiO$_2$(CB) difference if one uses 8.95 eV for SiO$_2$ bandgap.

2.2 Two-color optical technique for characterization of X-ray radiation enhanced electron transport in SiO$_2$

Carrier movement (injection, transport, tunneling, recombination) and charge trapping in gate oxides are essential factors in understanding semiconductor device performance and degradation, especially in a radiation environment. In addition to the population of precursor defects with trapped charge and the possible generation of new traps, X-ray irradiation also may influence the carrier dynamics at interfaces. It has been shown recently that high doses of ionizing radiation in thin oxides (40–60 Å) may cause Radiation Induced Leakage Current (RILC) [11]. The conduction mechanism in RILC (as well as in Stress Induced Leakage Current, SILC) has been attributed to neutral oxide defects, which mediate electron tunneling across the oxide [12].

Presently, characterization of radiation damage in Si/SiO$_2$ systems is usually accomplished with electrical methods such as capacitance–voltage ($C–V$) and current–voltage ($I–V$) measurements. We use a novel two-color technique, based on time-dependent electric field induced SHG, for direct measurements of changes in electron transport characteristics due to X-ray irradiation in thin oxides to characterize the radiation response of a 42 Å SiO$_2$ film on Si(100) [13]. We find that the detrapping rate of surface charge in the X-ray irradiated devices is much higher than that of unirradiated devices.

The samples were cut from a wafer of thermally grown 42 Å SiO$_2$ film on Si(100) produced by Lucent Technologies. The irradiation was carried out with a 10-keV X-ray source, at a dose rate of ~1 krad(SiO$_2$)/s. Since the samples had no gate, they were irradiated without electrical bias.

The following experimental procedure was used in the measurements presented here:

(1) Initially the pump laser was off and the probe laser measured a constant background SH signal from the sample. After turning on the pump beam, a fast rise in the probe signal was observed, indicating that electrons were injected into the oxide, transported to the surface and trapped by the ambient surface oxygen. The holes remained at the interface thus resulting in an electric field arising from charge separation. The solution of Eq. (2) is an exponential rise to a maximum which under the conditions of

$1/\tau^e_{PUMP} \gg 1/\tau^e_{PROBE}$, $1/\tau^e_{detrap}$ becomes

$$n_e(t) = n_{0e}(1 - \exp(-t/\tau^e_{PUMP})) . \tag{6}$$

(2) After the signal reached a saturation level, the pump beam was blocked and the probe SHG signal decreased. In this case, the surface trapped electrons transported back to the interface to recombine with the holes remaining at the interface. Since now $1/\tau^e_{PUMP} = 0$ and $1/\tau^e_{detrap} \gg 1/\tau^e_{PROBE}$ only the third term in Eq. (2) is present, and the solution becomes a simple exponential decay for $n_e(t)$:

$$n_e(t) = n_{0e}\exp(-t/\tau^e_{detrap}) . \tag{7}$$

Figure 4 compares time-dependent pump–probe type EFISH measurements performed on irradiated [15 Mrad(SiO$_2$)] and non-irradiated Si/SiO$_2$ samples with 42 Å oxides. Electron tunneling across a non-irradiated oxide at this thickness has a very low probability; it only becomes significant below thicknesses of ~30 Å [14]. In both irradiated and non-irradiated samples the electric field increases rapidly after the pump laser is turned on. For the non-irradiated sample the EFISH signal stays at a constant saturation level after the pump laser is blocked, indicating that the electrons in the surface traps have a long lifetime. In contrast, the electric field across the X-ray irradiated sample decreases steadily, implying that the trapped electrons tunnel back to the interface and recombine with the holes, thereby decreasing the electric field. This suggests that the X-ray irradiation increases the rate of electron transport through the oxide. Systematic dose dependent measurements were not carried out on these samples, but we have seen similar time-dependent behavior (shown in Fig. 4) for similar samples irradiated up to 2 Mrad(SiO$_2$) and also 20 Mrad(SiO$_2$) doses via a 10 keV X-ray source.

We have also studied the long-term time dependence of the observed leakage in our X-ray irradiated samples at a dose of 20 Mrad(SiO$_2$). Figure 5 shows the $1/\tau_{det rap}$ (Eq. (7)) values (presumably associated with trap-assisted tunneling rates) deduced from our measurements versus time after irradiation. The figure depicts that the rate at which the electrons tunnel back to the interface decreases with time after X-ray irradiation. These rates were several orders of magnitude lower 2–3 days after irradiation. We suggested that the defects mediating the electron transport across the oxide may anneal at room temperature. In our experiments the OPA pump laser should not give rise to any significant heating due to its low repetition rate. In addition, we reduced the duty factor of this laser to 1/8th using a shutter. Also, each data point was taken on a fresh spot on the sample, at least 500 μm away from the previous measurement

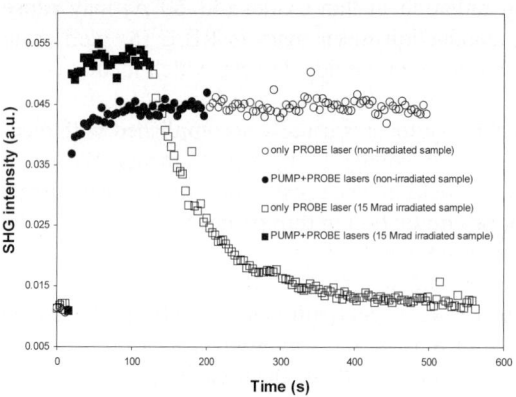
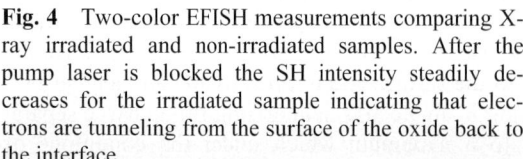
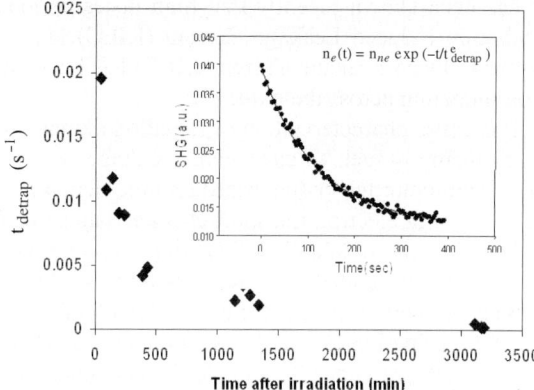

Fig. 4   Two-color EFISH measurements comparing X-ray irradiated and non-irradiated samples. After the pump laser is blocked the SH intensity steadily decreases for the irradiated sample indicating that electrons are tunneling from the surface of the oxide back to the interface.

Fig. 5   Measured detrapping rates on fresh spots on a 42 Å oxide as a function of time past after X-ray irradiation up to 20 Mrad dose. The inset shows an example of an exponential decay fit for obtaining detrapping rates.

point. With these precautions, we believe laser-heating induced annealing should not be a significant factor in these experiments.

One way of finding out whether our laser itself gives rise to the observed decrease in the electron tunneling rate with time is to perform consecutive measurements on the same laser spot. We irradiated the same spot several times on the sample with the pump laser after the probe signal reached a minimum steady-state level, measured the tunneling rates versus time after irradiation, and compares it with measurements taken on fresh sample spots. The continued irradiation by the pump laser on the same spot enhances the decrease in the observed X-ray irradiation effect probably by altering the concentration of irradiation-induced trapping sites.

There are two possible mechanisms to account for our observations of pump-laser-induced reduction of the radiation effect. The first is optically induced local healing of the defect responsible for the leakage in X-ray irradiated samples. Since the laser peak intensities are high and multiphoton processes are possible, the pump laser may break bonds at the defect sites (such as the H-bridge site) and lead to the local relaxation of the structure and consequent removal of the defect.

3 Ultrafast spin dynamics in GaAs/GaSb/InAs heterostructures probed by SHG

Ultrafast spin-sensitive spectroscopy provides unique information about spin relaxation in semiconductor heterostructures as well as spin-polarized electron transport across interfaces [15, 16]. Knowledge of the processes governing spin dynamics is essential for designing novel multifunctional electronic and opto-electronic devices, including base components for quantum computing [17]. Among the wide variety of multilayer semiconductors, GaSb/InAs heterostructures are especially promising [18].

The excitation of an ensemble of spins by a circularly polarized laser light tuned just above the band gap gives rise to a net magnetization. The typical time-resolved techniques, such as polarized photoluminescence spectroscopy [19, 20], pump–probe transmission/reflection [21–24], and Faraday or Kerr rotation [17, 25], all rely on the *linear* response of the spin subsystem to a probing light, and are well suited for monitoring dynamics in the *bulk* of semiconductor structures. On other hand, the *nonlinear* optical effects, such as SHG, are known to be highly sensitive to local magnetic fields occurring at magnetized surfaces at interfaces in magnetic-semiconductor-based multilayers [26]. Therefore, the application of SHG in the pump–probe configuration is a promising method for studying the dynamics of optically excited spins at the semiconductor *interfaces*.

We apply the pump–probe SHG technique to study ultrafast spin dynamics in nonmagnetic heterostructures. Spin-polarized electrons were selectively excited in the GaAs layer of GaAs/GaSb and GaAs/GaSb/InAs heterostructures. Only the GaAs/GaSb/InAs samples showed a significant induced magnetization indicating interlayer spin-polarized electron transport from GaAs to InAs. The dominant contribution to the magnetic-field induced SHG signal results from high local density of spins accumulated at the semi-metallic GaSb/InAs interface. Temperature dependence of induced SHG signals in the range from 4.3 to 300 K revealed two distinct mechanisms governing magnetization dynamics: the evolution of the local spin density at the interfaces and the spin relaxation.

We have investigated two heterostructures grown by molecular beam epitaxy: (1) GaAs(100 nm)/GaSb(400 nm) and (2) GaAs(100 nm)/GaSb(500 nm)/InAs(20 nm). The initial beam of 150 fs pulses from a mode-locked Ti : Al_2O_3 laser (Mira 900) at the wavelength of 800 nm (1.55 eV) and a repetition rate of 76 MHz was split into pump and probe beams. The probe beam of 120 mW average power has passed through an optical delay stage. The pump beam was chopped at a frequency of 400 Hz and, after that, had the same average power. The overlap spot of the pump and probe beams on the sample was ~100 μm in diameter. The pump beam was incident normally on the sample with either left or right-handed circular polarization (σ^+ or σ^-, respectively). The probe beam was linearly polarized (*p* or *s*), and directed to the sample surface at the angle of 75°. The pump-induced SHG signal was monitored as a function of probe-to-pump delay times. Note that only p linearly polarized probe light contributes to the induced signal, similarly as that for SHG measurements of magnetized surfaces is typically observed

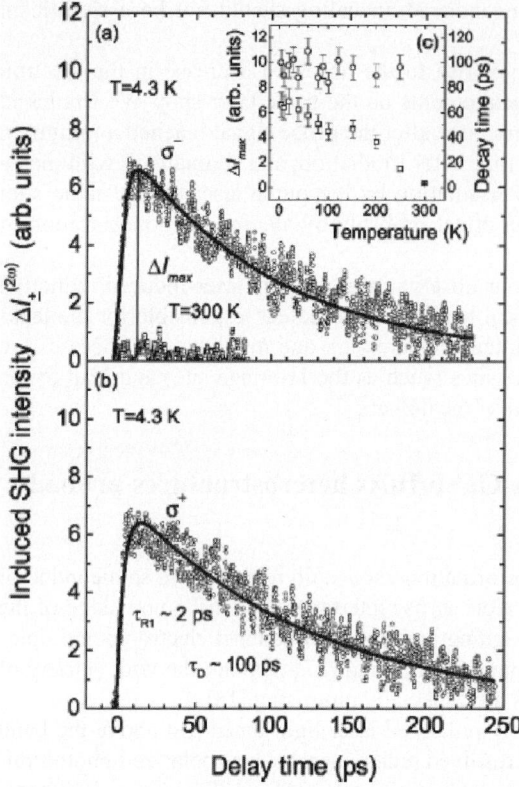

Fig. 6 Pump-induced SHG signal from GaAs/GaSb sample measured at 4.3 and 300 K for σ polarized a) and σ^- polarized b) pump. The fit with exponential rise/decay functions is shown by solid curves. Inset (c): temperature dependence of maximum signal intensity, ΔI_{max} (squares), and of decay-time constant, τ_D (circles).

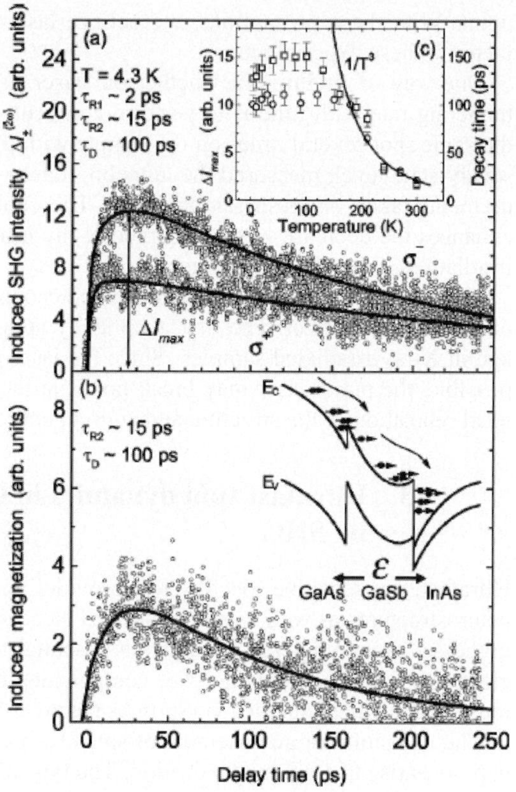

Fig. 7 Pump-induced SHG signal from GaAs/GaSb/InAs sample measured at 4.3 K with σ^- and σ pump polarization a) and extracted magnetization b) obtained as a difference between signals in (a). The fit with exponential rise/decay functions is shown by solid curves. Inset (c): temperature dependence of maximum signal intensity, ΔI_{max} (squares), and of decay-time constant, τ_D (circles). Band realignment due to interfacial electric fields for GaAs/GaSb/InAs heterostructure is schematically shown in (b).

[27]. The SHG signal was optically separated from the reflected fundamental probe beam and measured by a photomultiplier tube through a "lock-in" amplifier triggered by chopped pump pulses.

Figure 6 shows the pump-induced SHG signals taken on GaAs/GaSb heterostructure (sample 1) at a temperature $T = 4.3$ K. No significant difference was observed between signals measured with σ^+ or σ^- polarized pump light [Figs. 6a and 6b, respectively], indicating that the signal is due to the induced electric field at the interface. The interfacial electric fields caused by charge separation between photoexcited carriers are known to strongly enhance the SHG response [28]. The measured signal was fitted by a combined exponential rise/decay function. The signal intensity increases with a time constant of $\tau_{R1} \sim 3$ ps, followed by a decay with $\tau_D \sim 100$ ps. The induced signal completely disappears at room temperature [Fig. 6a].

The induced SHG signal from GaAs/GaSb/InAs samples is shown in Fig. 7a. A new striking feature is a long-lived $\tau_{R2} \sim 15$ ps rise-time component, which results in a shift of the signal peak towards longer times with respect to those for the GaAs/GaSb samples. Moreover, the SHG signal intensities for σ^- and

σ^+ pump polarizations are different, indicating an induced magnetization, which we ascribe to the presence of spin-polarized electrons in the InAs layer. Since the spins were excited in the GaAs layer, this indicates an interlayer spin-polarized electron transport. Correspondingly, the long-lived $\tau_{R2} \sim 15$ ps rise-time component characterizes the rate of spin transfer to the InAs layer.

Retaining only linear terms in the induced electric field, $\varepsilon(t)$, and magnetic field, $M(t)$, the nonlinear pump–probe polarization can be presented as [26, 28],

$$P_{\pm}^{NL}(2\omega, t) = [\chi^{(2)} + \chi_e^{(3)}\varepsilon(t) \pm \chi_m^{(3)}M(t)][E(\omega)]^2, \tag{8}$$

where $E(\omega)$ is the electric field component of the incident probe light, and $\chi^{(2)}$, $\chi_e^{(3)}$, and $\chi_m^{(3)}$, are the corresponding nonlinear susceptibilities. The alternate signs in Eq. (8) indicate two possible directions of the induced magnetic field normal to the interface. The magnetic- and electric-field induced contributions were then extracted from the pump induced SHG signal intensity, $\Delta I_{\pm}^{(2\omega)}(t)$ as

$$\Delta I_{-}^{(2\omega)} - \Delta I_{+}^{(2\omega)} \propto M(t), \qquad \Delta I_{-}^{(2\omega)} + \Delta I_{+}^{(2\omega)} \propto \varepsilon(t). \tag{9}$$

Figure 7b shows the extracted induced magnetization for GaAs/GaSb/InAs heterostructure, whereas the extracted electric-field-induced signal closely follows that for the GaAs/GaSb sample.

Because the laser light was tuned just above the GaAs band gap, spin-polarized electrons excited in the smaller band gap GaSb and InAs layers are much more energetic (0.74 and 1.11 eV, respectively) and lose their spin polarization as they relax to lower energy states [30]. These unpolarized electrons accumulate in the GaAs and InAs regions while the holes are amassed in the GaSb layer. The resulting charge separation at the interfaces [inset in Fig. 6b] gives rise to the interfacial electric fields resulting in the initial growth of the SHG signal (~2 ps at 4.3 K). This rise time decreases to ~300 fs for $T \sim 250$ K, matching the typical room temperature values for carrier thermalization [21–23]. The induced electric fields at the interfaces bend the initial energy profile and lower the barrier at the GaAs/GaSb interface [inset in Fig. 7b]. A subsequent relaxation of the interfacial electric fields manifests itself as the $\tau_D \sim 100$ ps decay of the induced SHG signal [30]. The appearance of longer decay time in Fig. 7a originates from constant background due to a residual electric field at the GaSb/InAs interface. This background is present only in the GaAs/GaSb/InAs sample and reflects the semimetallic nature of the GaSb/InAs interface which does not fully "unbend" as the induced interfacial fields relax. The 100 ps decay-time constant (common for both samples) for interfacial electric fields is obtained after extracting this constant background. Importantly, the interfacial electric fields are known to be much stronger at the semimetallic GaSb/InAs interface as compared those at the GaAs/GaSb interface [18]. This leads to higher local spin density at the GaSb/InAs interface and, hence, to significant induced magnetization in the GaAs/GaSb/InAs sample.

The temperature dependence of the peak intensity for the GaAs/GaSb sample exhibits a sharp decrease in the range of 4.3–100 K, while for the GaAs/GaSb/InAs sample, the signal first grows and then stabilizes [Figs. 6c and 7c]. We attribute this behavior to thermally activated electrons in GaAs overcoming the interfacial barrier. In the GaAs/GaSb/InAs sample, the initial signal increase is attributed to the arrival of additional spins at the GaSb/InAs interface. Note that electrons with uncompensated spin in GaAs are activated first since they occupy states with higher energies. Subsequent signal stabilization in the range from 50 to 100 K is due to the competing process involving a decrease in the interfacial electric field as the unpolarized electrons begin to pass through the barrier. Further intensity decrease in the range from 100 to 300 K indicates a weakening of the interfacial fields as the electron wave functions become more extended, effectively reducing the carrier density at the interface. Note that the induced signal in GaAs/GaSb samples shows a plateau in the range from 100 to 170 K, which we attribute to the thermal activation of electrons previously trapped at the impurity centers in bulk GaAs.

The temperature dependence of decay-time constants is similar for both samples in the range from 4.3 to 180 K [Figs. 6c and 7c] staying at $\tau_D \sim 100$ ps. At higher temperatures, however, τ_D decreases as T^3 for the GaAs/GaSb/InAs sample, while remaining unchanged for the GaAs/GaSb sample. The T^3 dependence is consistent with Dyakonov–Perel (DP) mechanism [29], and the measured room temperature value $\tau_D \sim 20$ ps matches that in InAs measured previously using pump–probe spectroscopy [24]. Note, however, the DP mechanism is expected to dominate for temperatures down to ~50 K. The regime change at $T \sim 180$ K indicates indicates a crossover to interface dominated dynamics. As the electric field at the GaSb/InAs interface relaxes, the time evolution of spin density follows that of charge density, leading to a reduction of the magnetic-field-induced SHG signal.

4   Summary

In this paper, we review recent advances in spin/carrier dynamics at semiconductor interfaces using intense, tunable, ultrafast lasers, including carrier dynamics at Si/SiO$_2$ interfaces, radiation enhanced electron transport in ultra-thin oxides, and ultrafast spin dynamics in semiconductor heterostructures, probed by SHG. We have made first measurements of the Si valence to SiO$_2$ conduction band offset via multiphoton internal-photoemission induced second-harmonic generation. We have showed that multiphoton internal-photoemission induced second-harmonic generation promises to become a valuable experimental tool in determining band offsets in wide variety of semiconductor interfaces including many new alternate and chemically modified oxides under investigation. This novel contactless technique can be easily applied for most systems where the two materials in contact are isotropic or amorphous. We determined that the electron-tunneling rate decreases with time at room temperature after X-ray irradiation, indicating that the electron transport-mediating traps may be annealed at room temperature. This is in contrast with annealing studies performed at substantially higher temperatures on RILC, suggesting that the defects responsible for the Laser Interrogated Leakage Current (LILC) observed in our experiments are either (1) defect precursors (observable at lower dosages due to increased sensitivity), or (2) different than, the defects responsible for RILC. It is likely these defects are O vacancies (E' centers) or hydrogen-related. Pump–probe SHG measurements for nonmagnetic GaAs/GaSb/InAs semiconductor heterostructures revealed interlayer spin-polarized electron transport. We found that the optically induced magnetization dynamics in such structures originates from two distinct sources: one of them related to the evolution of the local spin density at the interfaces, and the other one arising from the spin relaxation. The extreme sensitivity of the SHG to the interfacial fields, which allowed us to distinguish between these two mechanisms, makes it a unique tool for studying the spin and carrier dynamics in multilayer semiconductors.

References

[1] O. A. Aktsipetrov, A. A. Fedyanin, A. V. Melnikov, E. D. Mishina, A. N. Rubtsov, M. H. Anderson, P. T. Wilson, M. ter Beek, X. F. Hu, J. I. Dadap, and M. C. Downer, Phys. Rev. B **60**, 8924 (1999).
[2] Z. Marka, R. Pasternak, S. N. Rashkeev, Y. Jiang, S. T. Pantelides, P. K. Roy, J. Kozub, and N. H. Tolk, Phys. Rev. B **67**, 045302 (2003).
[3] N. Sahmir, J. G. Mihaychuk, H. M. van Driel, and H. J. Kreuzer, Phys. Rev. Lett. **82**, 359 (1999).
[4] M. L. Green, E. P. Gusev, R. Degraeve, and E. L. Garfunkel, J. Appl. Phys. **90**, 2057 (2001).
[5] B. S. Wherret, J. Opt. Soc. Am. B **1**, 67 (1984).
[6] V. K. Adamchuk and V. V. Afanas'ev, Prog. Surf. Sci. **41**, 111 (1992).
[7] J. Robertson, J. Vac. Sci. Technol. B **18**, 1785 (2000).
[8] R. Williams, Phys. Rev. **140**, A569 (1965).
[9] V. V. Afanas'ev, M. Houssa, A. Stesmans, and M. M. Heyns, Appl. Phys. Lett. **78**, 3073 (2001).
[10] J. W. Keister, J. E. Rowe, J. J. Kolodziej, H. Niimi, T. E. Madey, and G. Lucovsky, J. Vac. Sci. Technol. B **17**, 1831 (1999).
[11] A. Scarpa, A. Paccagnella, F. Montera, G. Ghibaudo, G. Pananakakis, G. Ghidini, and P. G. Fuochi, IEEE Trans. Nucl. Sci. **44**, 1818 (1997).
[12] M. Ceschia, A. Paccagnella, A. Cester, A. Scarpa, and G. Ghidini, IEEE Trans. Nucl. Sci. **45**, 2375 (1998).
[13] Z. Marka, R. Pasternak, R. G. Albridge, S. N. Rashkeev, S. T. Pantelides, N. H. Tolk, B. K. Choi, D. M. Fleetwood, and R. D. Schrimpf, J. Appl. Phys. **93**, 1865 (2003).
[14] M. L. Green, E. P. Gusev, R. Degraeve, and E. I. Garfunkel, J. Appl. Phys. **90**, 2057 (2001).
[15] Yu. D. Glinka, T. V. Shahbazyan, I. E. Perakis, N. H. Tolk, X. Liu, Y. Sasaki, and J. K. Furdyna, Appl. Phys. Lett. **81**, 220 (2002).

[16] Yu. D. Glinka, T. V. Shahbazyan, I. E. Perakis, N. H. Tolk, X. Liu, Y. Sasaki, and J. K. Furdyna, Surf. Interface Anal. **35**, 146 (2003).

[17] S. A. Wolf, D. D. Awschalom, R. A. Buhrman, J. M. Daughton, S. Von Molnar, M. L. Roukes, A. Y. Chtchelkanova, and D. M. Treger, Science **294**, 1488 (2001).

[18] I. Vurgaftman, J. R. Meyer, and L. R. Ram-Mohan, J. Appl. Phys. **89**, 5815 (2001).

[19] J. F. Smyth, D. A. Tulchinsky, D. D. Awschalom, N. Samarth, H. Luo, and J. K. Furdyna, Phys. Rev. Lett. **71**, 601 (1993).

[20] A. P. Heberle, W. W. Ruhle, and K. Ploog, Phys. Rev. Lett. **72**, 3887 (1994).

[21] A. Tackeuchi, O. Wada, and Y. Nishikawa, Appl. Phys. Lett. **70**, 1131 (1997).

[22] Y. Ohno, R. Terauchi, T. Adachi, F. Matsukura, and H. Ohno, Phys. Rev. Lett. **83**, 4196 (1999).

[23] K. C. Hall, S. W. Leonard, H. M. van Driel, A. R. Kost, E. Selvig, and D. H. Chow, Appl. Phys. Lett. **75**, 3665 (1999).

[24] T. F. Boggess, J. T. Olesberg, C. Yu, M. E. Flatte, and W. H. Lau, Appl. Phys. Lett. **77**, 1333 (2000).

[25] I. Malajovich, J. J. Berry, N. Samarth, and D. D. Awschalom, Phys. Rev. Lett. **84**, 1015 (2000).

[26] K. H. Bennemann, J. Magn. Magn. Mater. **200**, 679 (1999).

[27] J. Reif, J. C. Zink, C. M. Schneider, and J. Kirschner, Phys. Rev. Lett. **67**, 2878 (1991).

[28] Yu. D. Glinka, W. Wang, S. K. Singh, Z. Marka, S. N. Rashkeev, Y. Rogachyova, R. Albridge, S. T. Pantelides, N. H. Tolk, and G. Lukovsky, Phys. Rev. B **65**, 193103 (2002).

[29] M. I. Dyakonov and V. I. Perel, in: Optical Orientation, edited by F. Meyer and B. P. Zakharchenya (North-Holland, Amsterdam, 1984), p. 11.
G. E. Pikus and A. N. Titkov, ibid. p. 73.

[30] Yu. D. Glinka, T. V. Shahbazyan, I. E. Perakis, N. H. Tolk, X. Liu, Y. Sasaki, and J. K. Furdyna, Appl. Phys. Lett. **81**, 3717 (2002).

phys. stat. sol. (b) **240**, No. 3, 500–508 (2003) / **DOI** 10.1002/pssb.200303844

Strain induced optical anisotropies in zincblende semiconductors

L. F. Lastras-Martínez[*,1], **R. E. Balderas-Navarro**[1,2], **M. Chavira-Rodríguez**[**,1], **J. M. Flores-Camacho**[**,1], and **A. Lastras-Martínez**[1]

[1] Instituto de Investigación en Comunicación Optica, Universidad Autonóma de San Luis Potosí, Alvaro Obregón 64, San Luis Potosí, México
[2] Facultad de Ciencias, Universidad Autonóma de San Luis Potosí, Alvaro Obregón 64, San Luis Potosí, México

Received 15 November 2003, revised 30 November 2003, accepted 2 December 2003
Published online 25 November 2003

PACS 78.40.Fy, 78.66.Fd, 78.68+w

Reflectance Difference (RD/RAS) spectroscopy has emerged in the last two decades as a sensitive probe for the study of surface and surface-induced phenomena in zincblende semiconductors. This spectroscopy measures the difference in reflectivity between two mutually orthogonal polarizations and it is thus specific to semiconductor regions with symmetries lower than cubic. Photoreflectance-difference (PRD) spectroscopy is a technique closely related to RD spectroscopy that measures the difference between two photoreflectance spectra, one with linearly-polarized light and the other with non-polarized light. In contrast to RD spectroscopy, PRD spectroscopy is specific to reflectance anisotropies of electro-optic origin. In this paper we will discuss the application of both RD and PRD spectroscopies to the study of strain-induced optical anisotropies in zincblende semiconductors. We will present a theoretical line shape model for both RD and PRD spectra that describes with high accuracy the experimental spectra. Results show the high sensitivity of these techniques for the study of piezo-optical properties of zincblende semiconductors. Furthermore, results to be discussed should prove to be useful in the identification of strain-induced contributions to RD spectra that are know to comprise components with a number of physical origins.

1 Introduction

Reflectance-difference spectroscopy (RDS/RAS)) [1] has emerged in the last two decades as a sensitive tool for the characterization of surface and interface properties of zincblende semiconductors. RDS measures the difference in reflectivity between two mutually-perpendicular polarization states and it is thus sensitive to perturbations such as surface electric fields or applied uniaxial stresses, that change the crystal symmetry from cubic (T_d) to orthorhombic (C_{2v}). RDS spectroscopy offers advantages over other techniques due to its high sensitivity and experimental simplicity. It can be used likewise on vacuum or air conditions besides being non-destructive.

RDS have been found to comprise components of different physical origins, both bulk and surface. Among these components we can mention those associated to surface reconstruction (surface dim-

---

[*] Corresponding author: e-mail: lflm@cactus.iico.uaslp.mx, Phone: 52 (444) 825 0183, Fax: 52 (444) 825 0198
[**] CONACyT Fellow

mers) [2–6], dislocations [7–9], local field effects [10] and surface electric fields [11]. Applications of RDS include the *in situ* monitoring of the epitaxial growth of zincblende semiconductors and the determination of piezo-optical properties of semiconductors [12–18]. Nevertheless, despite its sensitivity and flexibility, due to the multiplicity of mechanisms that may contribute to RDS intensity, the present theoretical understanding of RDS line shapes is limited and have prevented the widespread use of RDS as an optical probe in a range of applications including epitaxial growth monitoring.

Photoreflectance-difference (PRD) spectroscopy is a technique closely related to RDS that measures the difference between two photoreflectance (PR) spectra, one taken with linearly polarized light and the other with unpolarized light [19]. Contrary to RDS, which responds to any cubic symmetry breaking mechanism, PRD spectroscopy is specific to the symmetry breakdown associated to surface and interface electric fields [19]. For a (001) zincblende surface, the PRD signal is associated to a piezoelectric strain that renders the semiconductor orthorhombic. In an unstressed crystal the PRD line shape comprise only a linear electro optic (LEO) component, as the quadratic electro optic (QEO) term, that dominates the PR spectrum, is isotropic for cubic symmetries [19]. In contrast, the PRD spectrum of a zincblende (001) crystal under a uniaxial stress along [110] comprises both LEO and QEO components [14].

In this paper we discuss theoretical models to describe both RD and PRD line shapes for critical points of Γ and Λ symmetry. We obtain theoretical line shape expressions that include perturbations due to a surface electric field and to an externally applied uniaxial stress. We present a number of experimental RD (GaAs and GaSb) and PRD (GaAs) spectra and show that the presented theoretical model describes them accurately. We conclude that the theoretical understanding of both RD and PRD line shapes allows us to use these spectroscopies in the determination of deformation potentials and in the characterization of surface and interface electric fields in zincblende semiconductors. Furthermore, results presented in this paper should prove to be useful for the identification of strain-induced components in the RD spectrum of zincblende semiconductors, both at atmospheric pressure and under ultra high vacuum conditions.

The rest of the paper is organized as follows. In Sec. 2 we present the theory for the PRD line shape for points of Λ symmetry, for semiconductor under a uniaxial stress along [110]. In Sec. 2 we also discuss the theory for the RD line shape for points of Γ and Λ symmetries, for a semiconductor with a uniaxial stress applied along [110]. In Sec. 3 we present experimental RD and PRD line shapes and discuss their fitting to the theoretical line shape presented in Sec. 2. Finally, conclusions are given in Sec. 4.

2 Theory

Let us consider a zincblende semiconductor with a (001) surface. We will consider the perturbation of the bands by two mechanisms: a piezo-electric stress induced by a surface electric field along [001] and an applied uniaxial stress along [110]. Under these conditions the non-zero elements of the strain tensor referred to the crystal 100 axes are [14]

$$e_{xx} = e_{yy} = \frac{S_{11} + S_{12}}{2} X, \qquad e_{zz} = S_{12} X, \qquad e_{xy} = e_{yx} = d_{14} F + \frac{S_{44}}{4} X, \tag{1}$$

where F and X are the strengths of the surface electric field and the applied [110] stress respectively, d_{14} is the piezo-electric modulus and S_{11}, S_{12} and S_{44} are the elastic compliance modulii. We take $F > 0$ for a n type semiconductor and $X < 0$ for a compressive stress. The effects of the strain tensor (1) on the electronic energy levels are described by the Pikus–Bir Hamiltonian [20]

$$H^{(i)} = -a^{(i)}(e_{xx} + e_{yy} + e_{zz}) - 3b^{(i)}\left[\left(L_x^2 - \frac{1}{3}L^2\right)e_{xx} + \left(L_y^2 - \frac{1}{3}L^2\right)e_{yy} + \left(L_z^2 - \frac{1}{3}L^2\right)e_{zz}\right]$$
$$- \frac{\sqrt{3}}{4}d^{(i)}(L_xL_y + L_yL_x)e_{xy}, \tag{2}$$

where $a^{(i)}$, $b^{(i)}$ and $d^{(i)}$ are the hydrostatic, tetragonal and orthorombic deformation potentials for the Γ symmetry bands, respectively, and i is the band index. L_x, L_y and L_z are the components of the angular momentum.

The Hamiltonian (2) splits and shifts the energy of critical points of Λ and Γ symmetry [14, 15, 20]. Furthermore, the squares of the interband transition matrix elements become polarization dependent [14, 15]. These facts induce an anisotropy on the dielectric function. We relate the anisotropy in the reflectivity spectrum with the anisotropy on the dielectric function through the relation [21]

$$\frac{\Delta R}{R} = \mathrm{Re}\left[(\alpha - i\beta)\,\Delta\varepsilon\right], \tag{3}$$

where α and β are Seraphin coefficients. For PRD spectroscopy, $\Delta\varepsilon$ is defined as the difference between the dielectric functions for light polarized along [110] and for non-polarized light. For RDS, $\Delta\varepsilon$ is defined as the difference between the dielectric functions for light polarized along [110] and for light polarized along [1$\bar{1}$0]. In what follows we will discuss the line shape of the components of PRD and RD spectra, induced by tensor (1) for critical points of Λ and Γ symmetries.

2.1 Uniaxial stress applied along [110] in transitions of Λ and Γ symmetries: E_1, $E_1 + \Delta_1$ and E_0'−triplet transitions

Using the energy shifts and the matrix elements for a stress along [110] reported elsewhere [14], the change in dielectric function for E_1 and $E_1 + \Delta_1$ interband transitions is given by:

$$\Delta\varepsilon' = \frac{1}{E^2}\frac{\partial E^2 \varepsilon'(E, E_0 - r\Delta_1/2 + \Delta E_h)}{\partial E}\frac{D_1^5 S_{44}}{4\sqrt{3}}X + \varepsilon'(E, E_0 - r\Delta_1/2 + \Delta E_h)\frac{2rD_5 S_{44}}{\sqrt{6}\Delta_1}X, \tag{4}$$

where ε' stands for the contribution of E_1 and $E_1 + \Delta_1$ to the overall dielectric function ε, Δ_1 is the spin orbit splitting energy for valence band, $r = +1(-1)$ refer to E_1 ($E_1 + \Delta_1$), E is the energy of the incident light and E_0 is the interband transition energy in the absence of perturbations. D_1^5 is the interband orthorrombic deformation potentials for transitions of Λ symmetry [22]. D_5 is the orthorrombic deformation potential for the valence band and it is related to d through $d = D_5/\sqrt{2}$ [22]. ΔE_h is the hydrostatic energy shift induced by X and is given by

$$\Delta E_h = \frac{D_1^1(S_{11} + 2S_{12})}{\sqrt{3}}X. \tag{5}$$

Parameter D_1^1 is the interband hydrostatic deformation potential for transitions of Λ symmetry [22].

The interband transitions corresponding to points of Γ symmetry are the components of the triplet E_0', $E_0' + \Delta_0'$, and $E_0' + \Delta_0' + \Delta_0$, corresponding to transitions $\Gamma_8^v \to \Gamma_7^c$, $\Gamma_8^v \to \Gamma_8^c$, and $\Gamma_7^v \to \Gamma_8^c$, respectively [15]. By using Eqs. (1–2) and the wave functions for critical points of Γ symmetry, the change in dielectric function for the E_0'−triplet is given by [15]:

$$\Delta\varepsilon'' = \frac{3}{8E^2}\frac{\mathrm{d}E^2\varepsilon''}{\mathrm{d}E}(\delta E_{0,1}^2 + 3\,\delta E_{1,1}^2)^{1/2} + \frac{3}{2}\frac{\delta E_{1,2}}{\Delta_0'}\varepsilon'', \tag{6}$$

$$\Delta\varepsilon''' = \left[\frac{2 - \sqrt{6}}{8}\left[\frac{\delta E_{0,1}}{\Delta_0} + \frac{\delta E_{0,2}}{\Delta_0'}\right] - 3\frac{2 + \sqrt{6}}{8}\left[\frac{\delta E_{1,1}}{\Delta_0} + \frac{\delta E_{1,2}}{\Delta_0'}\right]\right]\varepsilon''', \tag{7}$$

$$\Delta\varepsilon'''' = -\frac{3}{8E^2}\frac{\mathrm{d}E^2\varepsilon''''}{\mathrm{d}E}(\delta E_{0,2}^2 + 3\,\delta E_{1,2}^2)^{1/2} + \frac{3}{2}\frac{\delta E_{1,1}}{\Delta_0}\varepsilon'''', \tag{8}$$

where ε'', ε''' and ε'''' stand, respectively, for the contribution of E_0', $E_0' + \Delta_0'$ and $E_0' + \Delta_0' + \Delta_0$ to the overall dielectric function ε. Parameters in Eqs. (7)–(10) are defined as: $\delta E_{0,1} = 2b(S_{11} - S_{12})X$, $\delta E_{1,1} = dS_{44}X/\sqrt{3}$ and $\delta_1 = \Delta_0$ for valence band and $\delta E_{0,2} = 2b^c(S_{11} - S_{12})X$, $\delta E_{1,2} = d^c S_{44}X/\sqrt{3}$ and $\delta_2 = \Delta_0'$ for conduction band, where b and b^c are the tetragonal deformation potentials for valence and conduction bands, respectively, and d and d^c are the orthorhombic deformation potentials for valence and conduction bands, respectively. Δ_0 and Δ_0' are the spin orbit splitting energies for valence and for conduction bands, respectively. The RD line shape is obtained by using Eq. (3) with $\Delta\varepsilon = \Delta\varepsilon' + \Delta\varepsilon'' + \Delta\varepsilon''' + \Delta\varepsilon''''$.

2.2 Piezo electric strain in transitions of Λ symmetry: E_1 and $E_1 + \Delta_1$ transitions

Under a stress along [110] the transitions E_1 and $E_1 + \Delta_1$ split apart into two sets of 4 equivalent points, a first set containing points along [111], [$\bar{1}\bar{1}1$], [11$\bar{1}$] and [$\bar{1}11$] and a second set containing points along [$\bar{1}11$], [1$\bar{1}\bar{1}$], [1$\bar{1}1$] and [$\bar{1}1\bar{1}$] [14]. To obtain an expression for the PRD line shape, rather that first calculating the change in dielectric function, we will directly obtain an expression for the change in reflectivity. This procedure is supported by the relationship between R and ε reported elsewhere [19]. By using Eqs. (1)–(2) and the wave functions for critical points of Λ symmetry, the PRD line shape is given by [14]

$$\frac{\Delta R}{R} = r\frac{4D_5 d_{14} F}{\sqrt{6}\Delta_1} R(E, E_0 - r\Delta_1/2 + \Delta E_h) + \frac{1}{2R}\frac{\partial R(E, E_0 - r\Delta_1/2 + \Delta E_h)}{\partial E}\frac{D_1^5 d_{14} F}{\sqrt{3}}$$
$$+ 2r\frac{D_5 S_{44} X}{2\sqrt{6}\Delta_1} L(E, E_0 - r\Delta_1/2 + \Delta E_h) + \frac{1}{2}\frac{\partial L(E, E_0 - r\Delta_1/2 + \Delta E_h)}{\partial E}\frac{D_1^5 S_{44} X}{4\sqrt{3}},\qquad(9)$$

where R is the reflectivity spectrum obtained experimentally for unpolarized light. L is the unpolarized light photoreflectance spectrum of Fig. 1a. In the low-field limit, this spectrum has a quadratic electro-optic origin and is given by [23]

$$L = \frac{e^2\hbar^2}{24}\mu_{\parallel}^{-1} F^2 \operatorname{Re}\left[(\alpha - i\beta)\frac{1}{E^2}\frac{\partial^3 E^2 \varepsilon}{\partial E^3}\right],\qquad(10)$$

where $\mu_{\parallel}$ is the reduced mass in the field direction.

The first two terms in the right hand side of Eq. (9) correspond to the LEO component while, in the low-field limit, the last two terms represent the QEO component [14]. Equation (9) will be used in the next section to calculate the PRD spectrum.

3 Experimental results and discussion

Measurements were carried out on n-type GaAs (001) crystals with a carrier concentration of $n = 5.6 \times 10^{16}$ cm^{-3}, and p-type GaSb (001) crystals with a carrier concentration of $p = 1 \times 10^{16}$ cm^{-3}. Samples were cut out from commercial wafers in pieces $3.0 \times 3.0 \times 0.5$ mm in size. Crystal direction were determined by etching in KOH [24]. Directions determined in this way coincide with those specified by the wafer supplier. RD measurements were carried out by applying a stress along either [110] or [1$\bar{1}$0] by using a calibrated spring. The experimental set up allowed us to apply stresses up to -5.0×10^9 dyn/cm^2.

Spectra were obtained in the energy range from 2.5   5.5 eV by using setups described in the literature for RD [25] and PRD [19]. In both spectrometers a 75-Watt Xe lamp was employed as the light source and a silicon diode as the photodetector.

In order to model RD and PRD spectra we have determined the GaAs and GaSb dielectric function by spectroscopic ellipsometry (SE), by using a rotating analyzer ellipsometer [26]. The measured pseudo dielectric function $\langle\varepsilon\rangle$ was corrected for the presence of a oxide layer. The oxide thickness was determined by applying a three-phase (bulk-oxide-ambient) model [27], assuming sharp interfaces. By using literature data for the dielectric responses of both GaAs [28] and GaSb [29] oxides, we found thicknesses of 40 Å   and 80 Å for the oxide layer of GaAs and GaSb respectively. Results obtained were checked for consistency with literature data [30, 31].

3.1 Photoreflectance difference spectra of GaAs: Λ and Γ symmetry transitions

Figure 1 shows the measured PR spectra for GaAs (001) crystals under [110] stress in the energy range from 2.6–3.6 eV. Solid and dashed lines correspond to polarized light along [110] and to unpolarized light, respectively. Applied stress strengths are (a) $X = 0.0$, (b) $X = -1.0 \times 10^9$ dyn/cm^2 and (c) $X = -4.4 \times 10^9$ dyn/cm^2. The optical structure of the PR spectra of Fig. 1 is related to GaAs E_1

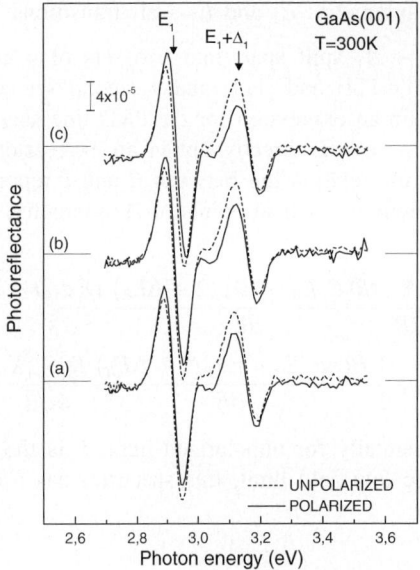

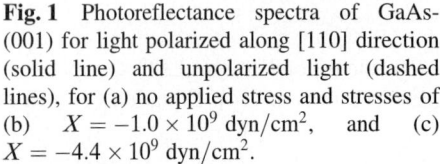

Fig. 1 Photoreflectance spectra of GaAs-(001) for light polarized along [110] direction (solid line) and unpolarized light (dashed lines), for (a) no applied stress and stresses of (b) $X = -1.0 \times 10^9$ dyn/cm^2, and (c) $X = -4.4 \times 10^9$ dyn/cm^2.

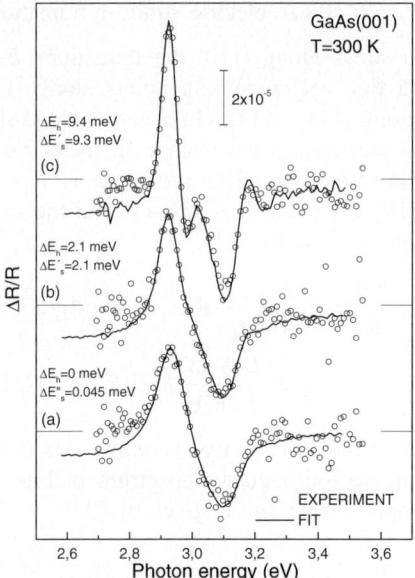

Fig. 2 Photoreflectance difference spectra of GaAs-(001) (open circles), for (a) no applied stress and stresses of (b) $X = -1.0 \times 10^9$ dyn/cm^2 and (c) $X = -4.4 \times 10^9$ dyn/cm^2. The solid lines are the fits obtained using Eq. (9).

and $E_1 + \Delta_1$ transitions whose energies are indicated by arrows. Note that PR line shapes in Fig. 1, that are dominated by the QEO component, are polarization-dependent [32]. PR line shapes show also a dependence on applied stress that is, however, somewhat difficult to follow. We may see, for instance, that the structure related to the E_1 transition is shifted slightly to higher energies when the applied stress is increased.

PRD spectra obtained by subtracting polarized from unpolarized PR spectra are shown in Fig. 2. We note that the stress dependence of the PRD spectra is evident, and in fact is quite complex, contrasting with the PR spectra behavior. This demonstrates the high sensitivity of PRD spectroscopy over PR spectroscopy for the characterization of stress-induced phenomena.

The evolution of the PRD spectra of Fig. 2 with applied stress can be described by using the line shape given by Eq. (9). As pointed out above, the first two terms of the right hand side of Eq. (9) correspond to the LEO component. To obtain the theoretical LEO contribution we fitted the reflectance spectrum of the non-stressed GaAs sample by using two Lorentzian line shapes, one for E_1 and the other for $E_1 + \Delta_1$. The QEO term was obtained using the unpolarized PR spectrum of Fig. 1a. Due to the fact that the contribution of E_1 and $E_1 + \Delta_1$ transitions to the PR spectrum are well resolved, we were able to deconvolute these contributions to generate the first term of the QEO line shape. For the second term of the QEO component we have used the first energy-derivative PR spectrum of Fig. 1a. A detailed discussion on the procedure employed has been reported previously [14].

To calculate the PRD line shapes we employed the following parameter values: $S_{11} = 0.117562 \times 10^{-11}$ cm^2/dyn, $S_{12} = -0.0365132 \times 10^{-11}$ cm^2/dyn and $S_{44} = 0.16835 \times 10^{-11}$ cm^2/dyn for the elastic compliance constants [33], $d_{14} = -2.7 \times 10^{-10}$ cm/V for the piezo-electric modulus [34] and $\Delta_1 = 0.21$ eV for the spin orbit splitting energy [20].

PRD fits obtained as described above are shown by solid lines in Fig. 2. From them we have obtained the values $D_1^1 = -8.40$ eV, $D_5 = -7.6$ eV and $D_1^5 = 8.8$ eV for the deformation potentials. These potentials agree quite well with the measured and theoretical values reported in the literature [14]. From the fit to the non-stressed spectrum of Fig. 2a we estimate a surface electric field strength of

$F \approx 3.4 \times 10^4 \, \text{V/cm}$. Note that the PRD spectrum of Fig. 2a is well described solely by the LEO component.

The line shape of the PRD spectra change when the stress is applied. The QEO component that is proportional to X becomes important when the stress is increased. For a stress of $X = -1.0 \times 10^9 \, \text{dyn/cm}^2$ a small shoulder can be noted around 3.0 eV in the spectrum of Fig. 2b. Note that this small structure is well described by our model (solid line). For a stress of $X = -4.4 \times 10^9 \, \text{dyn/cm}^2$ the change in line shape is evident. In this case the QEO component is larger than the LEO component. The structure around 3.0 eV is well resolved and the peaks of the spectrum are narrower. Note the excellent agreement between the PRD spectrum of Fig. 2c and our model.

From the fitting to the PRD spectra we obtain the hydrostatic energy shifts: $\Delta E_h = 2.1 \, \text{meV}$ and $\Delta E_h = 9.4 \, \text{meV}$, for the spectra of Figs. 3b and 3c respectively. Additionally, we obtain from the fittings the splitting energy shifts: $\Delta E_s'' = 0.04 \, \text{meV}$, $\Delta E_s' = 2.1 \, \text{meV}$ and $\Delta E_s' = 9.3 \, \text{meV}$ for the spectra of Figs. 2a, 2b and 2c respectively.

3.2 Reflectance difference spectra of GaAs: Λ and Γ symmetry transitions

We show in Fig. 3 RD spectra for GaAs under uniaxial stress along either [110] or [1$\bar{1}$0]. Note that RD spectra reverse sign when rotating the applied stress from [110] to [1$\bar{1}$0], indicating their linear-strain origin. In order to supress the background spectrum (non-stressed spectrum of Fig. 3d we have subtracted spectrum (Fig. 3g) from spectrum (Fig. 3a). This corrected spectrum is shown by open circles in Fig. (4).

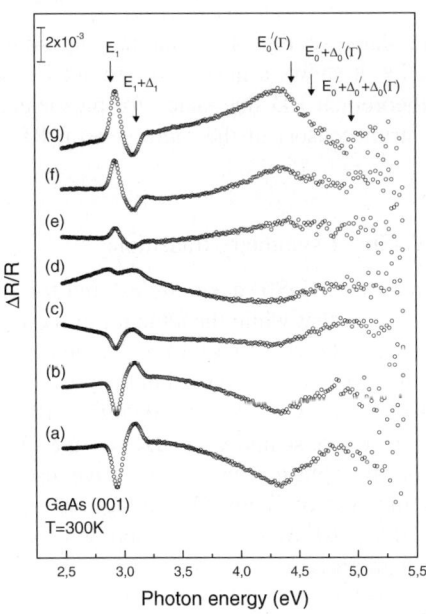

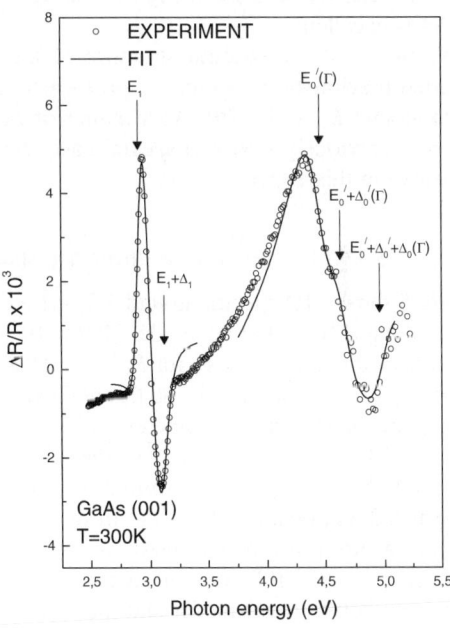

Fig. 3 GaAs RD spectra at room temperature for: [110] uniaxial stress with magnitudes (a) $-4.2 \times 10^9 \, \text{dyn/cm}^2$, (b) $-2.8 \times 0^9 \, \text{dyn/cm}^2$, (c) $-1.4 \times 10^9 \, \text{dyn/cm}^2$, (d) with not applied stress and [1$\bar{1}$0] uniaxial stresses with magnitudes (e) $-0.8 \times 10^9 \, \text{dyn/cm}^2$, (f) $-1.7 \times 10^9 \, \text{dyn/cm}^2$, (g) $-3.3 \times 10^9 \, \text{dyn/cm}^2$. Arrows indicate the energy position for the interband bulk transitions.

Fig. 4 RDS spectrum obtained by subtracting spectrum (g) from spectrum (a) of Fig. 3. The solid lines correspond to the calculated spectrum obtained using Eqs. (4), (6)–(8) and (3).

To calculate the Λ component of the RD spectrum we have deconvoluted the contributions of critical points E_1 and $E_1 + \Delta_1$ to the overall dielectric function. The deconvolution was carried out by fitting Lorentzian line shapes to the real and imaginary parts of the experimental dielectric function spectra, following the procedure described elsewhere [19]. With these fits and the energy-derivative of the dielectric function, Eqs. (3) and (4) allowed us to calculate the solid line spectrum in Fig. 4 around E_1 and $E_1 + \Delta_1$ energies. The strength of the stress was taken as a fitting parameter obtaining the value $X = -3.45 \times 10^9$ dyn/cm^2, in good agreement with the value $X = -3.7 \times 10^9$ dyn/cm^2 determined from the calibrated spring.

Carrying out a similar deconvolution for the contributions of the E_0'-triplet to the overall dielectric function poses some difficulties due to the fact that the E_0' and $E_0' + \Delta_0'$ transitions of Γ symmetry are separated for only 50 meV from their Δ-symmetry partners. [35] Furthermore, the $E_0' + \Delta_0' + \Delta_0$ transition is only 120 meV away from the dominating E_2 band and appears only as a small shoulder [35, 36]. To overcome this problem we have instead modeled the contributions ε'', ε''', and ε'''' of the $E_0'(\Gamma)$-triplet with three, excitonic-type, Lorentzian line shapes located at the reported critical point energies: $E_0'(\Gamma) = 4.45$ eV, $E_0' + \Delta_0'(\Gamma) = 4.62$ eV, and $E_0' + \Delta_0' + \Delta_0 = 4.95$ eV [35]. We have further used the known $4:2:1$ ratio for the amplitudes of the components of the $E_0'(\Gamma)$-triplet (i.e. the amplitude of the $E_0' + \Delta_0' + \Delta_0$ transition is fourtimes weaker than that of E_0') [35, 37].

By using the Lorenzian line shapes for ε'', ε''' and ε'''' and Eqs. (6)–(8) we obtained the fitted RD line shape given by the solid line around 4.5 eV in Fig. 4. We have used phases and broadening energies of the Lorenzian line shapes as fitting parameters. We have further used the following literature parameter values: $a' = -2.5$ eV [38] for the hydrostatic deformation potential, $b = -2.2$ eV, $d = -5.4$ eV [33] for the valence band and $b^c = 1.6$ eV and $d^c = -5.5$ eV [38] for the conduction band. As can be seen from Fig. 4, the agreement of the experimental spectrum with the theoretical model is excellent.

We note that the spectral structure of the RD line shape of Fig. 4 is qualitatively similar to that reported previously for strains of piezoelectric origin, i.e. it shows a maximum around E_1 and a minimum around $E_1 + \Delta_1$ [19]. As a matter of fact, the theoretical RD line shape for piezoelectric strains reported previously [19] is a special case (zero-trace strain tensor) of the more general RD line shape presented in this paper.

3.3 Photoreflectance difference spectra of GaSb: Λ symmetry transitions

Figure 5 shows RD spectra around $1.7 - 2.9$ eV for unstressed GaSb (open circles) and for GaSb under stress along [110] with magnitude -0.9×10^9 dyn/cm^2. Note that while the strength of the E_1 transition increases with applied stress, for $E_1 + \Delta_1$ decreases. Figure 6 shows the spectrum obtained by subtracting spectra of Fig. 5. Solid line is the fit obtained following the same approach used in the previous section on GaAs, by using the parameters: $S_{11} = 0.158 \times 10^{-11}$ cm^2/dyn, $S_{12} = -0.0495 \times 10^{-11}$cm^2/dyn, $S_{44} = 0.231 \times 10^{-11}$ cm^2/dyn for the elastic compliance constants [33], $D_5 = -6.8$ eV [33] and $D_1^5 = 7.4$ eV [39], for the deformation potentials. For the split-off energy we have used the value $\Delta_1 = 0.435$ eV, obtained from the fit of the dielectric function of GaSb. The strength of the stress was taken as a fitting parameter obtaining the value $X = -0.8 \times 10^9$ dyn/cm^2, in good agreement with the value $X = -0.9 \times 10^9$ dyn/cm^2 determined from the calibrated spring. As in the case of GaAs, the agreement between our model and the experiments is excellent.

4 Conclusions

We have investigated by means of PRD and RD spectroscopy, both theoretically and experimentally, the anisotropies induced in zincblende semiconductors by, both, surface electric field and external applied stresses. RD is sensitive to any cubic-symmetry breakdown mechanism, while PRD spectroscopy responds only to anisotropies associated to surface or interface electric fields. We have developed theoretical models that give accurate description of both RD and PRD spectra at critical points

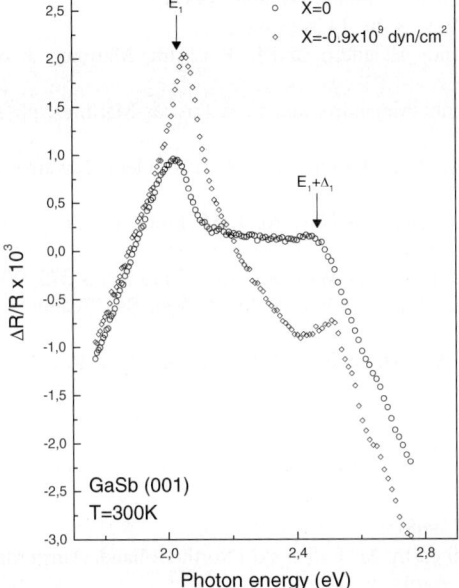

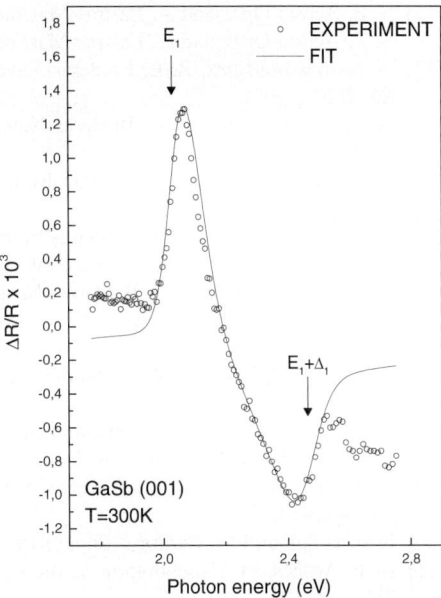

Fig. 5 GaSb RD spectra at room temperature for: [110] uniaxial stress with magnitud -0.8×10^9 dyn/cm² (open circle) and with not applied stress (open diamond). Arrows indicate the energy position for the interband bulk transitions.

Fig. 6 RDS spectrum obtained by subtracting the stressed spectrum (open circles) from the unstressed spectrum (open diamonds) of Fig. 5. The solid lines correspond to the calculated spectrum obtained using (3) and Eq. (4).

of Γ and Λ symmetries. Taking advantages of the theoretical understanding of RD and PRD spectra, both techniques can be used in a complementary way, to determine deformation potentials and surface and interface electric fields strength. Besides, results presented in this paper should prove to be useful for the identification of strain-induced components in the RD spectrum of zincblende semiconductors, both at atmosferic pressure and under ultra high vacuum conditions.

Acknowledgements We would like to thank E. Ontiveros and J. Ramírez for technical assistance. This work was supported by Consejo Nacional de Ciencia y Tecnología through grands 32147 E and 485100-5-33976E.

References

[1] D. E. Aspnes and A. A. Studna, Phys. Rev. Lett. **54**, 1956 (1985).

[2] Itaru Kamiya, D. E. Aspnes, L.T. Florez, and J. P. Harbison Phys. Rev. B **46**, 15894 (1992).

[3] M. J. Begarney, L. Li, C. H. Li, D. C. Law, Q. Fuand, and R. F. Hicks, Phys. Rev. B **62**, 8092 (2000).

[4] W. G. Schmidt, F. Bechstedt, K. Fleischer, C. Cobet, N. Esser, W. Richter, J. Bernholc, and G. Onida, phys. stat. sol. (a) **188**, 1401 (2001).

[5] S. Visbeck, T. Hannappel, M. Zorn, J.-T. Zettler, and F. Willing, Phys. Rev. B **63**, 245303 (2001).
K. Hingler, R. E. Balderas-Navarro, W. Hilber, A. Bonanni, and D. Stiffter, Phys. Rev. B **62** 13048 (2000).

[6] L. F. Lastras-Martínez, D. Rönnow, P. V. Santos, M. Cardona, and K. Eberl, Phys. Rev. B **64**, 245303 (2001).

[7] L. F. Lastras-Martínez and A. Lastras-Martínez, Solid State Comun. **98**, 479 (1996).
L. F. Lastras-Martínez and A. Lastras-Martínez, Phys. Rev. B **54**, 10726 (1996).

[8] L. F. Lastras-Martínez and A. Lastras-Martínez, Phys. Rev. B **64**, 085309 (2001).

[9] R. E. Balderas-Navarro, K. Hingerl, A. Bonanni, H. Sitter, and D. Stifter, Appl. Phys. Lett. **78**, 3615 (2001).

[10] W. L. Mochán and R. G. Barrera, Phys. Rev. Lett. **55**, 1192 (1985).

[11] S. E. Acosta-Ortíz and A. Lastras-Martínez, Solid State Commun. **64**, 809 (1987).
      S. E. Acosta-Ortíz and A. Lastras-Martínez, Phys. Rev. B **40**, 1426 (1989).
[12] A. Lastras-Martínez, R. E. Balderas-Navarro, P. Cantú-Alejandro, and L. F. Lastras-Martínez, J. Appl. Phys.
      86, 2062 (1999).
[13] A. Lastras-Martínez, R. E. Balderas-Navarro, P. Cantú-Alejandro, and L. F. Lastras-Martínez, phys. stat. sol.
      (a) **175**, 45 (1999).
[14] L. F. Lastras-Martínez, M. Chavira-Rodríguez, A. Lastras-Martínez, and R. E. Balderas-Navarro, Phys. Rev.
      B **66** 075315 (2002).
[15] L. F. Lastras-Martínez, M. Chavira-Rodríguez, R. E. Balderas-Navarro, J. M. Flores-Camacho, and A. Las-
      tras-Martínez, submitted to, Phys. Rev. B (2003).
[16] A. Lastras-Martínez, R. E. Balderas-Navarro, and L. F. Lastras-Martínez, Thin Solid Films **373**, 207 (2000).
[17] D. E. Aspnes, J. P. Harbison, A. A. Studna, L. T. Florez, and K. Kally, J. Vac. Sci. Technol. B **6**, 1127
      (1988).
      D. E. Aspnes, E. Colas, A. A. Studna, R. Bhat, M. A. Koza, and V. G. Keramidas, Phys. Rev. Lett. **61**, 2782
      (1988).
[18] T. Kita, O. Wada, T. Nakayama, and M. Murayama, Phys. Rev. B **66**, 195312 (2002).
      M. Pristovsek, S. Tsukamoto, B. Han, J.-T. Zettler, and W. Richter, J. Cryst. Growth **248**, 254 (2003).
[19] A. Lastras-Martínez, R. E. Balderas-Navarro, L. F. Lastras-Martínez, and M. A. Vidal, Phys. Rev. B **59**,
      10234 (1999).
[20] F. H. Pollak and M. Cardona, Phys. Rev. **172**, 816 (1968).
[21] D. E. Aspnes, in: Handbook on Semiconductors, edited by M. Balkanski (North-Holland, Amsterdam, 1980),
      Vol. 2, p. 121.
[22] F. H. Pollak, in: Strained-Layer Supperlattices: Physics, edited by T. P. Pearsall, Semiconductors and Semi-
      metals Vol. 32 (Academic Press, New York, 1990), p. 17.
[23] D. E. Aspnes, in: Handbook on Semiconductors, edited by M. Balkanski (North-Holland, Amsterdam, 1980),
      Vol. 2, p. 145.
[24] K. Sangwal, in: Defects in Solids, edited by S. Amelinckx and J. Nihoul (North-Holland, Amsterdam, 1987),
      Vol. 15, p. 427.
[25] L. F. Lastras-Martínez, A. Lastras-Martínez, and R. E. Balderas-Navarro, Rev. Sci. Instrum. **64**, 2147 (1993).
[26] D. E. Aspnes and A. A. Studna, Appl. Opt. **14**, 220 (1975).
[27] R. M. A. Azzam and N. M. Bashara, Ellipsometry and Polarized Light (North-Holland, Amsterdam, 1997).
[28] D. E. Aspnes, G. P. Schwartz, G. J. Gualtieri, A. A. Studna, and B. Schwartz, J. Electrochem. Soc. **128**, 590
      (1981).
[29] D. E. Aspnes, B. Schwartz, A. A. Studna, L. Derick, and L. A. Koszi, J. Appl. Phys. **48**, 3510 (1977).
[30] G. E. Jellison, Jr., Opt. Mater. **1**, 151 (1992).
[31] D. E. Aspnes and A. A. Studna, Phys. Rev. B **27**, 985 (1983).
[32] For a cubic crystal the QEO component is isotropic as discussed in Ref. [19]. We note however that under
      the applied unaxial stress the crystal becomes orthorhombic and thus the QEO component is not longer
      isotropic.
[33] Elements and III–V Compounds, edited by O. Mandelung, M. Schultz, and H. Weiss, Landolt-Börnstein,
      New Series, Vol. III/17a (Springer-Verlag, Berlin, 1982).
[34] Sadao Adachi, J. Appl. Phys. **58**, R1 (1985).
[35] D. E. Aspnes and A. A. Studna, Phys. Rev. B **7**, 4605 (1973).
[36] P. Lautenschlager, M. Garriga, S. Logothetidis, and M. Cardona, Phys. Rev. B **35**, 9174 (1987).
[37] D. E. Aspnes, Phys. Rev. Lett. **28**, 913 (1972).
[38] A. Blacha, H. Presting, and M. Cardona, phys. stat. sol. (b) **126**, 11 (1984).
[39] T. Tuomi, M. Cardona, and F. H. Pollak, phys. stat. sol. **40**, 227 (1970).

phys. stat. sol. (b) **240**, No. 3, 509–517 (2003) / **DOI** 10.1002/pssb.200303825

Simplified bond-hyperpolarizability model of second- and fourth-harmonic generation: application to Si–SiO₂ interfaces

D. E. Aspnes[*, 1], **J.-K. Hansen**[1, 2], **H.-J. Peng**[1], **G. D. Powell**[1, 3], and **J.-F. T. Wang**[1, 4]

[1] Department of Physics, North Carolina State University, Raleigh, NC 27695-8202, USA
[2] Now at Rabben 43, NO-5108, Hordvik, Norway
[3] Now at Magma Design Automation, Inc., Cupertino, CA 95014, USA
[4] Now at Department of Physics and Astronomy, Vanderbilt University, Nashville, TN 37235, USA

Received 30 May 2003, revised 4 August 2003, accepted 11 August 2003
Published online 25 November 2003

PACS 42.65.An, 42.65.Ky, 78.68.+m

We recently developed a simplified bond-hyperpolarizability model (SBHM) to describe the variation of second- and fourth-harmonic-generation (SHG, FHG) intensities as a function of sample azimuth angle, and applied it to various Si-dielectric interfaces. The approach provides an efficient representation of these data with fewer parameters than required by Fourier or tensorial representations. In addition, these parameters have a direct physical meaning in microscopic terms. The model is simple enough to allow analytic expressions to be obtained for SHG and FHG intensities for (001) and (111) interfaces. SHG absorption is shown to result in easily recognized features in anisotropy data. Relative amplitudes of FHG intensities of the (001)Si–SiO₂ interface among the three nonvanishing polarization combinations are shown to be consistent to a factor of about 2.

1 Introduction

The development of the femtosecond (fs) laser has generated a significant resurgence of interest in the use of nonlinear optics (NLO) for studying bulk material, thin films, interfaces, and surfaces [1, 2]. NLO phenomena possess richer selection rules than their linear-optic equivalent, and are therefore potentially more powerful as diagnostic tools [1–3]. In addition, the symmetry of surfaces and interfaces is lower than that of bulk material, thereby providing a means of isolating their contributions from those of the bulk. For example, with the exception of higher-order processes that can be neglected here, both second- (SHG) and fourth-harmonic-generation (FHG) are forbidden in the bulk of centrosymmetric crystals such as Si or randomly order materials such as amorphous SiO₂, and hence SHG and FHG signals of oxidized or nitrided Si wafers originate from a region no more than several atomic layers thick, where there is a regular or nearly regular repeating geometry of asymmetric bonds.

However, until recently a major difficulty has been the description and interpretation of SHG and FHG data. First-principles theoretical calculation are arcane, and being based on one-electron band theory, are not suitable for providing direct physical insight at the level of interface bonding [4, 5]. Bond approaches have been investigated more recently, but these have also proven somewhat difficult to relate to experiment [6–9]. As a result data are typically reported as dependences of SHG and FHG intensities on sample azimuths at specific wavelengths, representing these dependences as polar plots or tables of Fourier or tensor coefficients consistent with group-theoretical constraints [10, 11]. While these ap-

* Corresponding author: e-mail: aspnes@unity.ncsu.edu, Phone: +001-919-515-4261, Fax: +001-919-515-1333

proaches are consistent with general phenomenological treatments, they do not provide insight at the level of the atomic bonds.

To interpret our SHG data we found it necessary to take a different approach, one that is based more closely on the microscopic physics of SHG [12, 13]. We later showed that the same model could also be used to describe FHG data as well [14]. In short, we describe SHG and FHG intensities in terms of the coherent superposition of fields radiated by bonds driven anharmonically by the laser field, with several simplifying assumptions: (1) the directions of the interface bonds are the same as those of the bulk material; (2) only the motion along the bond axis is relevant; and (3) the observed intensity is the square of the coherent superposition of fields radiated by the individual bonds, with the radiation calculated in the dipole approximation. The first assumption recognizes the basic symmetry of the bulk material. The second is equivalent to assuming that the bonds are rotationally symmetric, and reduces the calculation of the driven dipoles to the solution of a one-dimensional anharmonic force model as discussed for example by Shen [3]. The third is an adaptation of the approach used to derive the Ewald-Oseen extinction theorem of linear optics [15, 16]. This simplified bond-hyperpolarizability model (SBHM) [12–14] has the necessary mathematical breadth and at the same time is mathematically much more efficient than previous approaches. Despite its simplicity it provides a very good representation of both SHG and FHG anisotropy data for Si-dielectric interfaces.

Further details, including a summary of previous calculations based on bond models and how the SBHM differs, are given in [12–14]. In the present work we provide the basic details and discuss the application of the SBHM to previously published SHG[17] and FHG[18] data for Si–SiO$_2$ interfaces.

2   Model

The approach and main results can be summarized as follows. Let the dipole components p_{2j} and p_{4j} be the dipole components associated with the second- and fourth-order longitudinal hyperpolarizabilities α_{2j} and α_{4j} of the jth bond, respectively. These dipole components can be written

$$p_{2j} = q_j \, \Delta x_2 \, \hat{b}_j = \alpha_{2j} (\hat{b}_j \cdot \boldsymbol{E})^2 \, \hat{b}_j; \qquad p_{4j} = q_j \, \Delta x_4 \, \hat{b}_j = \alpha_{4j} (\hat{b}_j \cdot \boldsymbol{E})^4 \, \hat{b}_j. \tag{1}$$

The observed intensities $I_{2\omega}$ and $I_{4\omega}$ far from the sample are proportional to the absolute squares $|\boldsymbol{E}_{2\omega}^{\text{far}}|^2$ and $|\boldsymbol{E}_{4\omega}^{\text{far}}|^2$ of the fields $\boldsymbol{E}_{2\omega}^{\text{far}}$ and $\boldsymbol{E}_{4\omega}^{\text{far}}$ that are the coherent sums of the fields radiated by the different bonds. These fields can be written

$$\boldsymbol{E}_{2\omega}^{\text{far}} \propto \frac{e^{ikr}}{r} \, [\underline{1} - \widehat{kk}] \bullet (\Sigma \hat{b}_j \hat{b}_j \hat{b}_j) \bullet \bullet \hat{E}\hat{E}, \tag{2a}$$

$$\boldsymbol{E}_{4\omega}^{\text{far}} \propto \frac{e^{ikr}}{r} \, [\underline{1} - \widehat{kk}] \bullet (\Sigma \hat{b}_j \hat{b}_j \hat{b}_j \hat{b}_j \hat{b}_j) \bullet \bullet \bullet \bullet \hat{E}\hat{E}\hat{E}\hat{E}, \tag{2b}$$

where $\hat{k}$ is the unit vector in the direction of the observer. These equations have the form of external projection operators acting on the products of the $\hat{b}_j$, here the cubic $\sum_j \hat{b}_j \hat{b}_j \hat{b}_j$ or the quintic $\sum_j \hat{b}_j \hat{b}_j \hat{b}_j \hat{b}_j \hat{b}_j$ for the SHG and FHG cases, respectively. These operations provide a recipe for generating a subset of the complete set of functions (for example xxx, xxy, etc. for SHG) that is consistent with the symmetry of the interface but is more restricted than that allowed by interface symmetry alone. In so doing we are clearly making assumptions, but as shown below these appear to work rather well.

For the (111) interface we take the b_j as follows:

$$\hat{b}_1 = \hat{z}; \qquad \hat{b}_2 = \frac{\sqrt{8}}{3}\hat{x} - \frac{1}{3}\hat{z}; \qquad \hat{b}_{3,4} = -\frac{\sqrt{2}}{3}\hat{x} \pm \frac{\sqrt{6}}{3}\hat{y} - \frac{1}{3}\hat{z}. \tag{3a}$$

For the (001)Si–SiO$_2$ interface we use a more general representation:

$$\hat{b}_{1,2} = \pm\hat{x}\sin\beta_4 + \hat{z}\cos\beta_4; \qquad \hat{b}_{3,4} = \pm\hat{y}\sin\beta_4 - \hat{z}\cos\beta_4, \tag{3b}$$

where in the ideal case $\beta_4 = \cos^{-1}(1/\sqrt{3}) = 54.7356°$ is the angle between the bonds and the z axis.

On a macroscopic scale the (001)Si surface consists of two types of domains differing by a 90° rotation of the bonds, since the two sublattices of the diamond structure are chemically identical. Hence any interface generated by purely statistical means will contain nominally equal areas of both. Adding to the set Eq. (3b) a second set of vectors rotated azimuthally by 90° from the original group yields a set of 8 bonds. If the α_{4j} were all equal, the contributions from the upper and lower bonds would cancel. This is clearly not the case, so we replace the original α_{4j} with their difference $\Delta\alpha_{4j} = \alpha_{4j,\text{up}} - \alpha_{4j,\text{down}}$ and sum only over the four bonds for which $\cos\beta_4$ is positive. Finally, assuming that the angles of incidence and observation are both equal to θ and considering incoming and outgoing s- and p-polarized light separately, we can write the incoming fields as $\boldsymbol{E}_s = E_s\hat{y}$ and $\hat{E}_p = E_p(-\cos\theta\,\hat{x} + \sin\theta\,\hat{z})$, the outgoing wave vector as $\hat{k} = (-\sin\theta\,\hat{x} + \cos\theta\,\hat{z})$, and the outgoing s- and p-polarized beams as the components $\hat{y}$ and $\hat{p} = (\cos\theta\,\hat{x} + \sin\theta\,\hat{z})$, respectively, of the total outgoing field. Summing over all four bonds we find the SHG fields in the radiation zone for the (111) interface to be:

$$E_{2\omega,\text{pp}}^{\text{far}} = \hat{p}\,[\alpha_u\sin^3\theta + \alpha_d(\cos^3\beta\sin^3\theta - \tfrac{3}{4}\sin\beta\sin2\beta\cos^2\theta\sin\theta + \tfrac{3}{4}\sin^3\beta\cos^3\theta\cos3\phi)], \tag{4a}$$

$$E_{2\omega,\text{ps}}^{\text{far}} = \hat{y}\,[\tfrac{3}{4}\alpha_d\sin^3\beta\cos^2\theta\sin3\phi], \tag{4b}$$

$$E_{2\omega,\text{sp}}^{\text{far}} = \hat{p}\,[\tfrac{3}{4}\alpha_d\sin\beta\,(\sin2\beta\sin\theta - \sin^2\beta\cos\theta\cos3\phi], \tag{4c}$$

$$E_{2\omega,\text{ss}}^{\text{far}} = -\hat{y}\,[\tfrac{3}{4}\alpha_d\sin^3\beta\sin3\phi]; \tag{4d}$$

and the FHG fields for the (001) interface to be:

$$\boldsymbol{E}_{4\omega,\text{pp}}^{\text{far}} = \tfrac{1}{2}\Delta\alpha_4\sin\theta\cos\beta_4(-9\cos^4\theta\sin^4\beta_4 + 8\sin^2\theta\cos^2\theta\sin^2\beta_4\cos^2\beta_4$$
$$+ 8\sin^4\theta\cos^4\beta_4 - 3\cos^4\theta\sin^4\beta_4\cos4\phi)\,\hat{p}, \tag{5a}$$

$$\boldsymbol{E}_{4\omega,\text{ps}}^{\text{far}} = \tfrac{1}{2}\Delta\alpha_4\sin\theta\cos\beta_4(4\cos^3\theta\sin^4\beta_4\sin4\phi)\,\hat{y}, \tag{5b}$$

$$\boldsymbol{E}_{4\omega,\text{xp}}^{\text{far}} = \tfrac{1}{2}\Delta\alpha_4\sin\theta\cos\beta_4(\sin^4\beta_4(3+\cos4\phi))\,\hat{p}, \tag{5c}$$

$$\boldsymbol{E}_{4\omega,\text{ss}}^{\text{far}} = 0. \tag{5d}$$

The above results are easily generalized to lower-symmetry situations, i.e., to rough interfaces where the distribution of bonds can be considered random. Summing over such bond sets suppresses all but the isotropic term, as is easily shown. For example, the simplest representation of a completely random orientation of equivalent bonds is obtained by adding four more bonds at initial angles of 45°, 135°, 225°, and 315° to the (001) expressions above. These additional bonds contribute the same constant value but a $\cos4\phi$ contribution of opposite sign, thereby canceling the $\cos4\phi$ contribution of the original four bonds.

3 Application

3.1 SHG

We assess the validity of the model by applying it to the SHG data of Lüpke, Bottomley, and van Driel [17] obtained on an oxidized vicinal (111) Si wafer. These data are appropriate because all four combinations of polarizations are given. However, the amplitudes of the measured SHG anisotropies are normalized to the 3ϕ Fourier component, which means that the information that insprinciple is available from

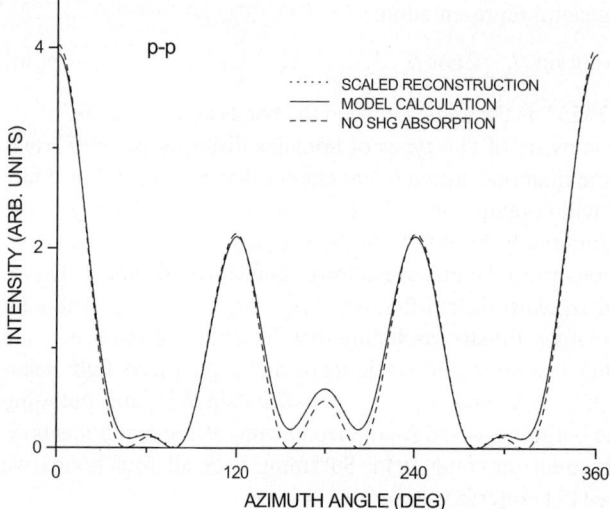

Fig. 1 Points: Reconstructed pp SHG data of Lüpke et al. [17] for a 5° vicinal (111)Si−SiO₂ interface. Solid line: SBHM result using the best-fit parameters given in the text. The data and calculation are essentially indistinguishable on the scale of the figure. Dashed line: best-fit SBHM result with SHG absorption assumed to be zero (after [13]).

the relative amplitudes of the four configurations is lost, and that comparisons must be done on the basis of lineshapes alone. For these data the pump wavelength was 765 nm, the nominal miscut angle 5° toward the [11$\bar{2}$] direction, and the nominal angles of incidence and observation 45°. Setting aside the normalization coefficients the data are presented as a table of seven real and two complex Fourier coefficients (Table 6 of [17]), which are also given apart from normalization as seven complex coefficients of χ_2 (Table 7 of [17]). Using their Fourier coefficients we have reconstructed their data as intensities, normalizing them in our case to the 6ϕ coefficients.

We first determine the complex second-order bond hyperpolarizabilities by least-squares fitting the azimuthal variation of the SHG intensity for the pp configuration. The result, shown in Fig. 1, is an essentially exact representation of the reconstructed data. The least-squares values obtained are 29.5° for the angles of incidence and observation, 1.12° for the vicinal angle, and 1.78 + i0.17, 2.76 + i1.43, 2.19 + i0.00, and 2.19 for the hyperpolarizabilities of the up, step, and two back bonds, respectively. The imaginary part of the hyperpolarizibility of the final back bond is arbitrarily set equal to zero, since the absolute phase cannot be determined in an intensity measurement. The values for the back bonds are equivalent, as expected. We find therefore that, again apart from normalization, we can represent the pp lineshape essentially with three parameters, one less than the number of Fourier coefficients needed to describe the same data in [17]. Thus the SBHM not only has the mathematical breadth necessary to represent SHG anisotropies, but also does it in a more efficient and physically more meaningful manner.

The existence of SHG absorption can be demonstrated unequivocally by repeating the least-squares fit with the imaginary parts of the hyperpolarizabilities of the first two bonds set equal to zero. The result is the dashed line in Fig. 1, which is obtained with the least-squares values 1.743, 3.158, 2.230, and 2.230. More to the point, in the absence of SHG absorption the SHG intensity reaches zero at the local minima near 30, 155, 205, and 280°. Thus, having identified how its phase-shift characteristics appear in the data, SHG absorption is relatively easy to recognize.

Taking the best-fit pp parameters we now calculate the expected intensity anisotropies for the ps, sp, and ss configurations. The reconstructed data and least-squares fits are shown in Fig. 2. As can be seen, the SBHM provides a very good representation of these data, provided that the predicted lineshapes are normalized by the least-squares-determined values 1.40, 1.93, and 2.26, respectively. These normalization factors are needed because the data shown in Fig. 2 were scaled to the 6ϕ Fourier component. Without going into details, from the best-fit value of θ we can conclude that, for this oxidized (111)Si interface, the SHG signal originates essentially from the SiO₂ side of the interface. The SBHM therefore reconstructs the ps, sp, and ss anisotropies with the same three parameters determined for the pp configura-

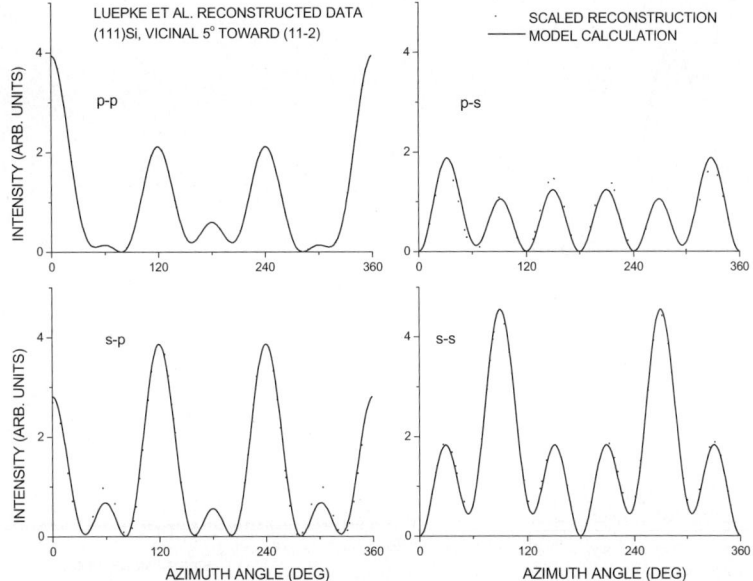

Fig. 2 Points: Reconstructed SHG data of Lüpke et al. [17] for all four polarization combinations, including that of Fig. 1. Solid lines: predictions of the SBHM using the best-fit parameters for pp. Normalization factors of 1.40, 1.93, and 2.26 were used for the ps, sp, and ss dependences (after [13]).

tion, in contrast to the seven additional Fourier coefficients and fourteen overall tensor coefficients needed in [17].

As an additional assessment we performed a least-squares fit to all four polarization combinations simultaneously. The lowest residual, 78% of that which results from the pp-determined parameters, is obtained at angles of incidence and observation of 25°. The fits are not noticeably different from those shown in Fig. 2, which is not surprising since the residuals are nearly the same. However, in the latter case the hyperpolarizability of the "up" bond is nearly twice as large, which is also not surprising given the functional dependence $\sin^3 \theta$ of the associated coefficient in Eq. (4a). The interesting aspect here is that this sensitivity should allow a better determination of the angles of incidence and observation to be made when theoretical calculations become sufficiently accurate.

For the oxidized vicinal (001) Si interface similar good agreement is obtained, as shown by our data in Fig. 3. This sample was prepared by thermally growing a 10 nm sacrificial oxide at 900°, stripping the oxide with HF, then thermally regrowing a 3 nm oxide at 750° by annealing at 900°. For these data the angles of incidence and observation were 45° and the wavelength was 830 nm. The ps data were multiplied by a factor of 4 to compensate for reflectance effects and different pulse durations and intensities that resulted as a consequence from the need to use different experimental conditions with respect to the p-incident configuration. The least-squares fitting parameters were determined to be 12° for the angles of incidence and observation, 7.28° for the vicinal angle, and 105 + i30, 197 + i55, 133 + i0, and 131 for the hyperpolarizabilities of the four bonds, again in arbitrary units and with the phase of the hyperpolarizability of the final bond arbitrarily set equal to zero. The "step" bond again shows the highest polarizability and absorption, although the essentially threefold pattern of Fig. 2 shows that most of the signal is coming from the vicinity of the steps. The ps anisotropy is found to be particularly sensitive to the angles of incidence and observation, and we estimate that the 12° value is determined to within 3°. This angle is consistent with Snell's Law for a refractive index of 3.4, which indicates that the SHG signal here is originating predominantly from the Si side of the interface.

© 2003 WILEY-VCH Verlag GmbH & Co. KGaA, Weinheim

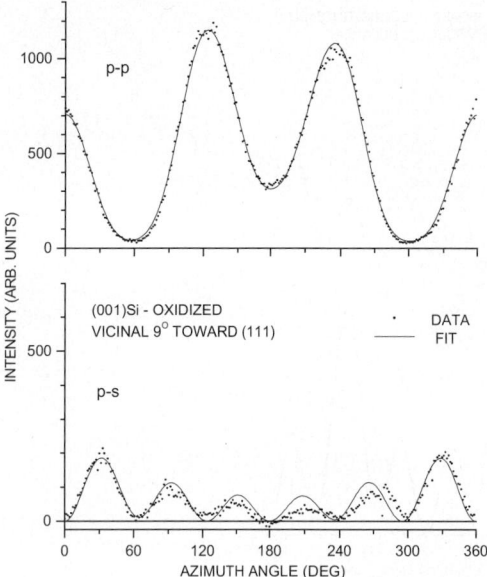

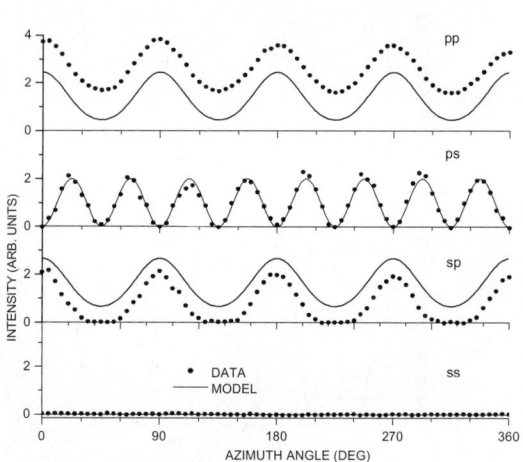

Fig. 3 Points: SHG anisotropy data for a 9° vicinal Si–SiO₂ interface. Solid lines: best-fit SBHM results (after [13]).

Fig. 4 Points: Normalized FHG data of Lee and Downer [18] for a (001)Si–SiO₂ interface for all four polarization combinations. Solid lines: variations predicted by the SBHM (after [14]).

3.2 FHG

Lee and Downer [18] reported FHG anisotropy data of (001)Si–SiO₂ interfaces measured at angles of incidence and observation of 45° to which we can compare the above calculations. As discussed previously [5], our calculations are done with $\theta = 29.1°$ to incorporate the effect of refraction at the air-SiO₂ interface. The data of [18] are shown in Fig. 4, along with the results of SBHM calculations. For the FHG investigation we normalized both data and calculations to unity amplitude of the dominant Fourier component, except for the null case ss where the data were multiplied by the average of the scaling factors used for sp and ps. As can be seen the SBHM describes the I_{ps} and I_{ss} data almost perfectly. For I_{pp} and I_{sp} the lineshapes and phases are predicted correctly, but the calculated dependences exhibit an offset (constant background) that does not agree with the data. Since the calculations are done with no adjustable parameters except for normalizations, the observed agreement for all four combinations of polarizations is clearly a consequence of the geometric and symmetry properties of the lattice of covalent bonds. However, to explain the offset discrepancy for I_{pp} and I_{sp} we must consider additional effects.

The key to understanding the offset discrepancy is the data of Lee and Downer [18] as a function of etching time, as reproduced in Fig. 5. These data exhibit the anisotropy changes that result when the (001)Si surface is etched in a 7:1 NH₄F:HF solution for different durations. This etch roughens this surface, and therefore the interface between the (001)Si substrate and the SiO₂ overlayer that grows after the sample is removed from the etch [19]. The relative magnitudes of the isotropic and azimuthally dependent contributions depend on how much the surface has been roughened.

We incorporate the effect of surface roughness in the SBHM by assuming that a fraction f of the interface bonds are statistically distributed in azimuth with a different hyperpolarizability $\Delta\alpha_r$ and mean angle β_r with respect to the z axis. Our expectation that roughness contributes only to the constant terms of Eqs. (5) is supported by the data: we predict offset discrepancies only for pp and sp, since constant terms do not appear for ps and ss.

The need for two parameters to describe roughness follows from Fig. 4 and Eqs. (5). First, the predicted offset for I_{sp} is larger than that observed, whereas that for I_{pp} is smaller. The only way the offset of

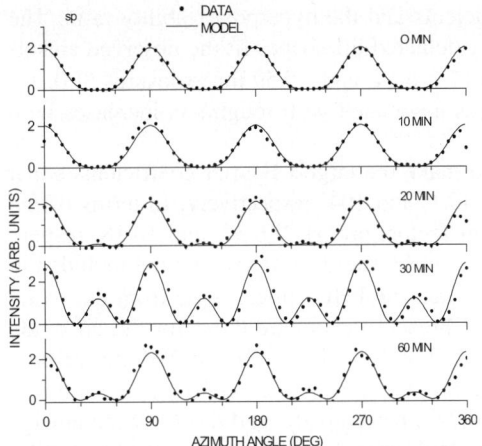

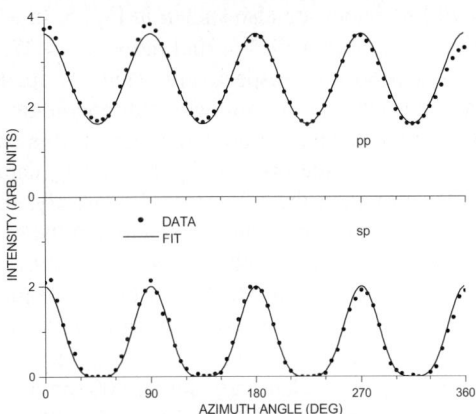

Fig. 5 Points: as for Fig. 4 following exposure of the sample to a buffered oxide etch for durations as indicated. Solid lines: SBHM calculations including interface roughness (after [14]).

Fig. 6 Points: Normalized FHG from Fig. 4 for the polarization combinations pp and sp. Solid lines: SBHM calculations including interface roughness (after [14]).

I_{sp} can be reduced is if $\Delta\alpha_4$ and $\Delta\alpha_r$ have opposite signs. However, if $\beta_r = \beta_4$ then the I_{pp} offset will also be reduced, leading to a greater discrepancy for the pp configuration. However, if β_r is less than $\beta_{r0} = 34.2°$ for $\theta = 29.1°$ the constant term will have the opposite sign and the I_{pp} discrepancy will also be reduced. Thus microscopic roughness can provide the explanation for the offset discrepancies in I_{pp} and I_{sp}, but only if $\Delta\alpha_4$ and $\Delta\alpha_r$ have opposite signs and β_r is less than 34.2°. The nearly systematic reduction in the offset with increased etching seen in Fig. 5 can therefore be attributed to a nearly systematic increase in the fraction f of random bonds. Since this is likely to be substantially less than the fraction $(1 - f)$ of oriented bonds, the data of Fig. 5 imply that $\Delta\alpha_r$ is considerably larger than $\Delta\alpha_4$.

We make the above results quantitative by performing a simultaneous least-squares fit of the SBHM to the I_{pp}, I_{ps}, and I_{sp} data of Fig. 4 The results of the fitting procedure for the two cases where offset discrepancies occur, pp and sp, are shown in Fig. 6. The values of the parameters are $\Delta\alpha_4 = 2.895$, $c_{pp} = 0.7903$, and $c_{sp} = 0.3583$, where c_{pp} and c_{sp} are the additional dc offsets needed to bring the pp and sp lineshapes in agreement with experiment. By using these values in Eqs. (5) we find $\beta_r = 27.2°$ and $\Delta\alpha_r = -12.63$ referenced to $\Delta\alpha_4 = 2.90$, which yields $\Delta\alpha_r/\Delta\alpha_4 = -4.36$. Since the surface density of bonds associated with roughness is expected to be much less than the intrinsic bond density of 2 per surface atom, the fourth-order longitudinal hyperpolarizability of the former must be much larger than that of the latter. For example if 10% of the interface bonds are associated with roughness, the hyperpolarizability associated with roughness is about 50 times larger. Theoretical calculations will be necessary to understand this relatively large difference in detail, but the fact that disordered bonds exhibit more anharmonic average bonding may not be unreasonable.

With β_r established and unlikely to change significantly with etching, we can now determine the $\Delta\alpha_r/\Delta\alpha_4$ ratio for the data of Fig. 5. We write $E_{sp} = a + b\cos 4\phi$, in which case

$$I_{sp} = a^2 + b^2/2 + 2ab\cos 4\phi + (b^2/2)\cos 8\phi, \tag{6a}$$

$$= c_0 + c_4\cos 4\phi + c_8\cos 8\phi, \tag{6b}$$

where c_0, c_4, and c_8 are the Fourier (best-fit) coefficients for the anisotropies. After some algebra we find

$$\frac{\Delta\alpha_r}{\Delta\alpha_4} = \left(\frac{a}{3b} - 1\right)\frac{\cos\beta_4\sin^4\beta_4}{\cos\beta_r\sin^4\beta_r}, \tag{7}$$

which is the desired connection between the Fourier coefficients and the hyperpolarizability ratio. The best-fit lineshapes are also shown in Fig. 5. In all cases the calculated lineshapes fit the observed anisotropies remarkably well. We find ratios of -4.37, -4.23, -5.17, -6.14, and -5.59 for exposures of 0, 10, 20, 30, and 60 min, respectively. Thus the fraction of bonds associated with roughness increases with increasing etching time, although not monotonically.

We now consider the normalization factors explicitly. To make the largest Fourier coefficients equal to 1 we divided the original I_{pp}, I_{ps}, and I_{sp} data by 1690, 40.2, and 104, respectively. In terms of the fields, with which the SBHM works directly, the equivalent factors are 41.2, 6.34, and 10.18, respectively. However, these values also contain the contributions from Fresnel factors that are not included in the model. Consistent with the use of $\theta = 29.1°$ we suppose that the FHG signals arise from the oxide side of the Si–SiO$_2$ interface, in which case the appropriate Fresnel factors are those for the air-oxide interface. Given the dielectric function values of SiO$_2$ of 2.112 and 2.408 at 800 and 200 nm, respectively, and using 45° for the angles of incidence and observation, the transmission coefficients for the p- and s-polarized incident radiation at 800 nm are 0.767 and 0.747, respectively, and those for the emerging radiation at 200 nm are 1.346 and 1.297, respectively. The Fresnel contributions are obtained by raising the appropriate first value to the fourth power and multiplying the result by the second. For the pp, ps, sp, and ss combinations we obtain 0.467, 0.450, 0.420, and 0.404, respectively. For completeness, the corresponding factors assuming the signal originates at the Si side of the Si-SiO$_2$ interface, using dielectric functions of $13.59 + i0.04$ and $-6.87 + i5.47$ for Si at 800 and 200 nm, respectively, are 0.0669, 0.0435, 0.0306, and 0.0199.

Removing the Fresnel contributions, we find that our actual scaling factors are 88.1, 14.1, and 24.2, respectively. We can now assess the results of the least-squares calculation, which are obtained relative to pp. The scaling-factor values show that we fit ps and sp fields that were too large by $88.1/14.1 = 6.25$ and $88.1/24.2 = 3.63$, respectively. The least-squares equivalents are 2.95 and 3.89. Thus the SBHM results for the ps and sp fields must be multiplied by 2.12 and 0.93, respectively, to match the data. Had we used the values for the Si-air interface the corresponding multipliers are 0.71 and 2.07, respectively. Some adjustments are possible by varying θ and β, but this would violate the basic assumptions of the model. Thus we can conclude that the SBHM, while not perfect, gives reasonable agreement with respect to relative intensities as well as correctly predicting symmetries, and hence as with the SHG case also describes the major fraction of these FHG data with a minimum set of physically meaningful parameters.

4   Conclusion

Despite its simplicity the SBHM provides a remarkably good representation of both SHG and FHG data with a minimal number of parameters with a direct physical meaning on the atomic scale. This should make the task of interpreting SHG and FHG data simpler. Although we treat these parameters as phenomenological, it should be relatively straightforward to calculate their equivalents from first principles. The capability of representing these data in a few parameters also makes this approach an attractive interface between theory and experiment. As with linear optics, the most useful information is expected to be obtained by spectroscopy, and we hope that the results given here will encourage more spectroscopic nonlinear-optical work in the future.

Acknowledgements   We gratefully acknowledge the generosity of Lee and Downer for sharing their data files with us, and the Office of Naval Research for providing the funding by which this work was accomplished. One of us (JKH), is supported by the Norwegian Research Council (NFR).

References

[1] J. F. McGilp, Prog. Surf. Sci. **49**, 1 (1995).
[2] G. Lüpke, Surf. Sci. Rep. **35**, 75 (1999).
[3] Y. R. Shen, The Principles of Nonlinear Optics (Wiley, New York, 1984).
[4] See, e.g., M. Cini, Phys. Rev. B **43**, 4792 (1991).

[5] V. I. Gavrilenko, R. Q. Wu, M. C. Downer, J. G. Ekerdt, D. Lim, and P. Parkinson, Phys. Rev. B **63**, 165325 (2001).

[6] C. H. Patterson, D. Weaire, and J. F. McGilp, J. Phys.: Condens. Matter **4**, 4017 (1992).

[7] C. M. J. Wijers, P. L. de Boeij, C. W. van Hasselt, and Th. Rasing, Solid State Commun. **93**, 17 (1995).

[8] B. S. Mendoza and W. L. Mochán, Phys. Rev. B **55**, 2489 (1997).

[9] N. Arzate and B. S. Mendoza, Phys. Rev. B **63**, 113303 (2001).

[10] J. E. Sipe, D. J. Moss, and H. M. van Driel, Phys. Rev. B **35**, 1129 (1987).

[11] J. E. Sipe and A. I. Shkrebtii, Phys. Rev. B **61**, 5337 (2000).

[12] G. D. Powell, J.-F. T. Wang, and D. E. Aspnes, Phys. Rev. B **65**, 205320 (2002).

[13] J.-F. T. Wang, G. D. Powell, R. S. Johnson, G. Lucovsky, and D. E. Aspnes, J. Vac. Sci. Technol. B **20**, 1699, (2002).

[14] J.-K. Hansen, H.-J. Peng, and D. E. Aspnes, J. Vac. Sci. Technol. B **21**, 1798 (2003).

[15] P. P. Ewald, Dissertation, Munich, 1912; Ann. Phys. (Leipzig) **49**, 1 (1916).

[16] C. W. Oseen, Ann. Phys. (Leipzig) **48**, 1 (1915).

[17] G. Lüpke, D. J. Bottomley, and H. M. van Driel, J. Opt. Soc. Am. B **11**, 33 (1994).

[18] Y.-S. Lee and M. C. Downer, Opt. Lett. **23**, 918 (1998).

[19] K. Sawara, T. Yasaka, S. Miyazaki, and M. Hirose, J. Appl. Phys. Jpn. **31**, L931 (1992).

phys. stat. sol. (b) **240**, No. 3, 518–526 (2003) / **DOI** 10.1002/pssb.200303855

Second harmonic generation from a collection of nanoparticles

Vera L. Brudny[*,1], **W. L. Mochán**[2], **Jesús A. Maytorena**[2], and **Bernardo S. Mendoza**[3]

[1] Departamento de Física, Facultad de Ciencias Exactas y Naturales, Universidad de Buenos Aires, Ciudad Universitaria Pab. I, 1428 Buenos Aires, Argentina
[2] Centro de Ciencias Físicas, Universidad Nacional Autónoma de México, Apartado Postal 48-3, 62251 Cuernavaca, Morelos, México
[3] Centro de Investigaciones en Optica, Apartado Postal 948-1, 37000 León, Guanajuato, México

Received 30 May 2003, revised 30 August 2003, accepted 1 September 2003
Published online 28 November 2003

PACS 03.50.De, 42.65.−Ky, 73.20.Mf, 78.66.−w

In this paper we discuss the SH radiation produced by a composite film made up of nanosized spherical centrosymmetric homogeneous particles. Although each individual sphere is unable to radiate in the forward direction when illuminated by a plane wave, our results show that for a thin composite film illuminated by a focused gaussian linearly polarized beam the radiation pattern presents two narrow lobes displaced along the polarization direction by a small angle of the order of the diffraction-induced angular divergence of the linear far field. The SH radiation produced by a non-homogeneous film, on the other hand presents non-trivial patterns which depend on the fundamental frequency, the waist of the illuminating beam, the sharpness of the edge, the density profile and both the incoming and outgoing polarizations. We hope that these results can be extended to develop characterization techniques for composite films based on surface SHG or SFG.

1 Introduction

Optical second harmonic generation (SHG) in centrosymmetric systems has proved to be a useful spectroscopic probe of surfaces, as the bulk contribution is strongly suppressed due to symmetry [1]. As opposed to other surface techniques, SHG permits observation of surfaces out of UHV and in many different ambients. For instance, it may be employed to study buried interfaces. Although most of the theoretical and experimental work has been performed on flat surfaces [2, 3], other shapes have also been explored, most notably, spherical particles [4]. Recently, SHG was employed to explore a thin composite layer made up of Si spherical nanocrystallites embedded within a SiO_2 matrix [5, 6]. It was found that the signal came mainly from the nanoparticles, not from the matrix, thus demonstrating the sensitivity of SHG to the condition of the surface of the particles and its usefulness for studying novel electronic devices [7]. However, experiment showed that the transmitted intensity displayed a peak along the forward direction and with a very narrow angular aperture. This result was unexpected, as previous theory [8] indicated that the radiation produced by a single sphere illuminated by a plane wave vanishes identically along the forward direction, as has been corroborated through experiments with suspended colloidal particles [9, 10].

The theory of Dadap et al. [8] mentioned above was developed for spheres illuminated by a plane electromagnetic wave. However, the nonlinear response of the particles actually depends on the nature of the exciting field, as shown by Brudny et al. [11] which obtained alternative results for the case of

---

[*] Corresponding author: e-mail: vera@df.uba.ar, Phone: +54 11 4576 3353, Fax: +54 11 4576 3357

a longitudinal exciting field. In general, the polarizing field is neither of a pure longitudinal character nor is it a simple plane wave. For example, within a composite, the electric field has longitudinal short range spatial fluctuations [12] besides the transverse macroscopic field, and the latter is
not necessarily a plane wave. A general theory for the quadratic nonlinear response of a single
small nonmagnetic centrosymmetric sphere illuminated by an arbitrary non-homogeneous electromagnetic field was developed by Mochán et al. in [13], from where the results in both [8, 11] can
be derived.

In this paper we use the hyperpolarizabilities obtained in [13] to calculate the macroscopic second
order susceptibilities of a composite made up of a random distribution of these spheres. We calculate
the second-harmonic radiation from a thin homogeneous slab of this composite illuminated by a
focused gaussian beam, and from an inhomogeneous composite with a density gradient. We show that
a finite sized beam produces a non-trivial radiation pattern around the forward direction when it illuminates a homogeneous composite film. The density gradient at the edge of the sample produces a
signal that may be larger than that coming from the interior and has different patterns and polarizations, depending on the direction of the incoming electric field with respect to the edge.

2 Nonlinear response of a single sphere

In this section we review the main results presented in Ref. [13] for the nonlinear response of a single
sphere of nanoscopic dimensions made up of a centrosymmetric material. Although the centrosymmetry is locally lost close to the surface of each sphere, opposite sides acquire different polarizations
according to their orientation with respect to the applied field. As the particle is globally centrosymmetric, its total quadratic dipole moment is null when illuminated by an homogeneous field. However,
a non-uniform field can produce a finite dipole moment $p^{(2)}$, that arises from the field variations
across the particle (see Fig. 1).

Assuming that the particles are small with respect to the scale of variation of the applied inhomogeneous field, each individual particle can be characterized by a few hyperpolarizabilities γ^e, γ^m, γ^q and
$\gamma^{\tilde{q}}$ which yield the quadratic dipole $p^{(2)}$ and quadrupole $\mathbf{Q}^{(2)}$ moments induced within the particle by
an inhomogeneous electromagnetic field through

$$p^{(2)} = \gamma^e \mathbf{E} \cdot \nabla \mathbf{E} + \gamma^m \mathbf{E} \times (\nabla \times \mathbf{E}),\tag{1}$$

$$\mathbf{Q}^{(2)} = \gamma^q (\mathbf{E}\mathbf{E} - \tfrac{1}{3}E^2 \mathbf{1}),\tag{2}$$

and

$$\tilde{Q}^{(2)} = \gamma^{\tilde{q}} E^2,\tag{3}$$

where $\tilde{Q}^{(2)}$ is the scalar second moment of the nonlinear induced charge distribution.

Assuming that the radius of the particle, R, is much larger than ℓ, the depth of the selvedge region
from where the surface response of the material originates, we may safely consider the sphere to be
locally flat. We shall also consider R to be smaller than the spatial variation of the field, so in particular, it should be smaller than the optical wavelength λ. As ℓ is of the order of the screening depth,
typically about an atomic distance, while λ is commonly two to three orders of magnitude larger,

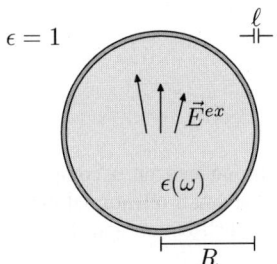

Fig. 1 Nonmagnetic sphere of radius R with dielectric function $\epsilon(\omega)$ within vacuum, being acted by an inhomogeneous electric field $\mathbf{E}^{\text{ex}}$. The selvedge of width ℓ is shown as a dark shell.

there is a range of sizes R for which the conditions $\ell \ll R \ll \lambda$ may be obeyed, such as for nanosized particles.

Under these assumptions expressions for the nonlinear polarizabilities γ^ν, $\nu = e, m, q, \tilde{q}$ of a single sphere may be obtained [13] in terms of the dipolarly allowed surface response functions (namely, the dimensionless functions $a(\omega)$, $b(\omega)$ and $f(\omega)$ which are commonly used to parametrize the response of a flat, homogeneous surface [14, 15]), and the quadrupolarly allowed bulk response ($\gamma(\omega)$ and $\delta'(\omega)$) of a flat semiinfinite homogeneous isotropic material [16]. Thus, we may take advandtage of various models and measurements of those intrinsic response functions to obtain the non linear susceptibility of small particles.

3   Nonlinear response of a composite

We now proceed to analyze the nonlinear response of a composite formed by nanoscopic spheres made up of a centrosymmetric homogeneous material enbedded in an also homogeneous matrix. Eqs. (1) and (2) imply that each particle would acquire a dipole moment along the propagation direction when illuminated by a plane monochromatic wave. Furthermore, it would acquire an axially symmetric quadrupole moment along the field. Thus, neiher the dipole moment nor the quadrupole moment nonlinearly induced by a plane wave are able to radiate along the forward direction, in apparent contradiction to experiment [5].

In Ref. [13] it was proposed that the finite extension of the wavefront could be the key to understanding the experimental result. Indeed, the electromagnetic field within a finite beam has spatial changes not only along the nominal propagation direction but also along the wavefront plane (see Fig. 2). The later variations might induce a dipole and a quadrupole moment which are not aligned with the wavevector and which therefore can radiate along the forward direction. Particles lying symmetrically at opposite sides of the incoming beam are expected to radiate fields with exactly opposite phases along the exact forward direction, and therefore should cancel each other; thus, no SH radiation is to be expected exactly along the forward direction. However, this cancellation would not be exact at small finite angles to the normal direction, so that a double lobed intensity profile within a narrow cone with an angular aperture similar to that of the diffracted fundamental beam is to be expected.

To illustrate this possibility, in this section we calculate the coherent SH radiated by a composite illuminated by a finite beam. We consider thus a system made up of a very narrow disordered array of small spheres distributed with a density $n_s(\boldsymbol{r})$ and illuminated by a finite beam. The place of the external field acting on a given sphere should be taken by the actual external field plus the field produced by other spheres. For simplicity, we further ignore the local field effect and identify $\boldsymbol{E}^{\text{loc}}$ with the macroscopic field $\boldsymbol{E}$.

The macroscopic nonlinear polarization of the composite medium may be obtained from the nonlinear dipole and quadrupole of each of its component spheres [17],

$$\boldsymbol{P}^{nl} = n_s \boldsymbol{p}^{(2)} - \tfrac{1}{6} \nabla \cdot n_s \mathbf{Q}^{(2)} - \tfrac{1}{6} \nabla n_s \tilde{Q}^{(2)} , \tag{4}$$

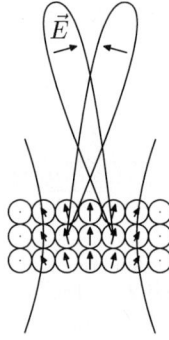

Fig. 2   Schematic representation of the SH dipole moments induced on a composite system illuminated by a focused beam. The SH field radiated at small off-normal directions is indicated, as well as the expected two lobed intensity distribution and the contributions of some dipoles to those lobes.

where we introduce the scalar $\tilde{Q}$ as $\mathbf{Q}$ is traceless. Within a homogeneous composite, $n_s(r)$ has a constant value n_b, and we employ Eqs. (1)–(3) to write

$$\mathbf{P}^{nl} = n_b \gamma^e \mathbf{E} \cdot \nabla \mathbf{E} + n_b \gamma^m \mathbf{E} \times (\nabla \times \mathbf{E}) - n_b \frac{\gamma^q}{6} \mathbf{E} \cdot \nabla \mathbf{E} + n_b \frac{\gamma^q - 3\tilde{\gamma}^q}{18} \nabla E^2 , \tag{5}$$

which becomes

$$\mathbf{P}^{nl} = \Gamma \nabla E^2 + \Delta' \mathbf{E} \cdot \nabla \mathbf{E} , \tag{6}$$

after expanding the curl terms and eliminating the field divergence, where we have introduced the response functions

$$\Gamma = \frac{n_b}{18} \left(9\gamma^m + \gamma^q - 3\tilde{\gamma}^q \right) \tag{7}$$

and

$$\Delta' \equiv n_b (\gamma^e - \gamma^m - \gamma^q/6) . \tag{8}$$

Now, we consider a thin sample of width l lying on the $z = 0$ plane illuminated by a beam with a finite waist w_0, as in Ref. [5]. The unscreened SH vector potential $\mathbf{A}$ obeys

$$\nabla^2 \mathbf{A}^T + (2q)^2 \mathbf{A}^T = -\frac{4\pi}{c} \mathbf{j}^T , \tag{9}$$

where the superscript T denotes the transverse projection of a vector field and $\mathbf{j}$ is the SH current

$$\mathbf{j} = \frac{\partial}{\partial t} \mathbf{P}^{nl} = -2i\omega \mathbf{P}^{nl} , \tag{10}$$

and solve Eq. (9) to obtain

$$\mathbf{A}^T(\mathbf{r}) = -2iq \int \mathrm{d}^3 r' \frac{e^{2iq|\mathbf{r}-\mathbf{r}'|}}{|\mathbf{r}-\mathbf{r}'|} [\mathbf{P}^{nl}(\mathbf{r}')]^T . \tag{11}$$

At large distances r from the illuminated spot we may perform the usual approximations $1/|\mathbf{r}-\mathbf{r}'| \approx 1/r$ and $e^{2iq|\mathbf{r}-\mathbf{r}'|} \approx e^{2iqr} e^{-2iq\hat{n}\cdot\mathbf{r}'}$ with $\hat{n} = (\sin\theta\cos\varphi, \sin\theta\sin\varphi, \cos\theta)$ a unit vector in the direction of $\mathbf{r}$, to obtain

$$\mathbf{A}^T(\mathbf{r}) = -2iql \frac{e^{2iqr}}{r} (\mathbf{P}^{nl}_K)^T , \tag{12}$$

where we assumed the film is much narrower than the wavelength, $ql \ll 1$, and we introduced the 2D Fourier transform

$$\mathbf{P}^{nl}_K = \int \mathrm{d}^2 r_{\parallel} \mathbf{P}^{nl}(\mathbf{r}_{\parallel}, z = 0) \, e^{-i\mathbf{K}\cdot\mathbf{r}_{\parallel}} \tag{13}$$

with wavevector $\mathbf{K} = 2q\hat{n}_{\parallel}$. The first term in Eq. (6) is purely longitudinal, so it cannot contribute to the radiated field. Thus, substituting into Eq. (12), we obtain the transverse potential

$$\mathbf{A}^T(\mathbf{r}) = -2iql \frac{e^{2iqr}}{r} \Delta' (\mathbf{E} \cdot \nabla \mathbf{E})^T_K . \tag{14}$$

Using Eq. (14), we have reduced the problem of SH radiation from a thin composite film illuminated by a beam of finite cross section to that of the calculation of the in-plane Fourier transform of the nonlinear driving term $\mathbf{E} \cdot \nabla \mathbf{E}$.

To proceed, we assume that at $z = 0$ the driving field $\mathbf{E}$ is given by the waist of a Gaussian beam and we assume that its width w_0 is much larger than the wavelength, so that the beam is paraxial around the nominal propagation direction z. Thus, we can ignore the small variations of order K/q in its polarization direction, and we can take a linearly polarized beam of the form

$$\mathbf{E}(\mathbf{r}_{\parallel}, 0) = E_0 \, e^{-r_{\parallel}^2/w_0^2} \, \hat{x} , \tag{15}$$

© 2003 WILEY-VCH Verlag GmbH & Co. KGaA, Weinheim

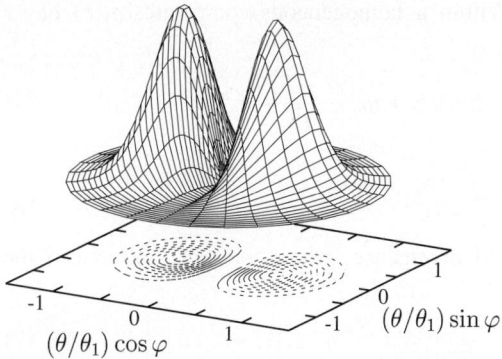

Fig. 3 SH intensity vs. angular position of the detector for a disordered array of spheres illuminated by a gaussian beam of width w_0 and frequency $\omega = qc$.

$(\theta/\theta_1) \cos \varphi$

$(\theta/\theta_1) \sin \varphi$

where $\hat{x}$ is the unit vector in the x direction. Since we can consider x to be a transverse direction, normal to the nominal propagation axis z, we approximate $(E \cdot \nabla E)^T \approx E \cdot \nabla E$, and the vector potential turns out to be

$$A^T(r) = \pi q^2 w_0^2 l \theta \cos \varphi \, e^{-q^2 w_0^2 \theta^2/2} \frac{e^{2iqr}}{r} \Delta' E_0^2 \hat{x}, \tag{16}$$

where we wrote the wavevector in terms of the scattering direction and further assumed $\sin \theta \approx \theta$.

The radiated electric field $E^r = 2iqA^T$ is immediatelly obtained from the potential (16), and from it we obtain Poynting's vector $S = c|E^r|^2 \hat{n}/8\pi$, the SH radiated intensity per unit solid angle into direction $\hat{n}$, $dI(\hat{n})/d\Omega = r^2 S(r\hat{n}) \cdot \hat{n}$ and the differential efficiency of SH generation [13],

$$\frac{d\mathcal{E}}{d\Omega} \equiv \frac{1}{\mathcal{P}^2} \frac{dI}{d\Omega} . \tag{17}$$

In Fig. 3 we show the radiation pattern obtained from Eq. (16). Notice that there is no radiation along the forward direction, but there is a strong nonlinear signal along two lobes very close to the forward direction, displaced along the incoming polarization direction $\varphi = 0$ by an angle $\pm\theta_2 = \pm 1/(qw_0)$. The angular displacement $\theta_2 = \theta_1/2 = \theta_2^*/\sqrt{2}$ where θ_1 and θ_2^* are the beam divergence half-angles of the fundamental beam and of the SH field that would have been generated by a homogeneous non-centrosymmetric material, such as the quartz plates usually employed as a reference in SH experiments. We remark that the characteristic angles θ_1 and θ_2^* are defined in terms of the angular decay of the fundamental and SH far field *amplitudes* [18]. More directly observable are the corresponding quantities defined in terms of the angular decay of the fundamental and SH *intensities*, or equivalently, by the second moment of the angular distribution of the intensities, which are $\sqrt{2}$ times smaller. Indeed, a recent repetition of the measurement in [5] showed a resolved, two lobed structed as the one predicted in [13], instead of the one peaked structure previously reported [19].

We should remark that the randomness within composites yields necessarily some SH incoherent radiation and the efficiency of both contributions may be comparable in size [13]. However, the incoherent radiation is distributed over a wide solid angle, according to a superposition of dipolar and quadrupolar radiation patterns, while the coherent radiation has a very narrow distribution close to the forward direction (Fig. 3) and thus, has a much larger intensity.

4 Nonlinear response at the edge of the composite

Another observation was reported in Ref. [5] regarding the intensity of the SH signal as the fundamental beam was scanned laterally through the sample. The observed SH intensity as the beam crossed the edge of the composite was about an order of magnitude larger than the signal from well within the composite, instead of interpolating smoothly between the signal from the interior and the practically inexistent signal from the exterior, as could have been naively expected. Furthermore, the signal displayed strong oscillations close to the edge. Thus, in this section we concentrate our attention on the calculation of the SH radiation from the lateral edge of a thin composite made of small spheres, such as that considered in the previous section (see Fig. 4). Accounting for the variation of the density $n_s \equiv n_b \xi(x, y)$ across the illuminated spot, from Eq. (4) we obtain

$$\boldsymbol{P}^{nl} = \Delta' \xi \boldsymbol{E} \cdot \nabla \boldsymbol{E} + \Gamma \nabla(\xi E^2) - n_b \frac{\gamma^m}{2} E^2 \nabla \xi - n_b \frac{\gamma^q}{6} \boldsymbol{E} \boldsymbol{E} \cdot \nabla \xi , \qquad (18)$$

instead of Eq. (6), where ξ is a function which varies from 0 outside the composite to its bulk value 1 within. Without loss of generality we take the edge along the y axis, so that $\xi = \xi(x)$, and we write $\nabla n_s = n_b(\mathrm{d}\xi/\mathrm{d}x)\hat{x} \equiv n_b \xi'(x)\,\hat{x}$. Then, Eq. (18) reduces to

$$\boldsymbol{P}^{nl} = \Delta' \xi \boldsymbol{E} \cdot \nabla \boldsymbol{E} + \Gamma \nabla(\xi E^2) + \Upsilon_\nu \xi'(x)\, \hat{x} E^2 , \qquad (19)$$

where Υ_ν depends on whether the external field $\boldsymbol{E}$ points in the direction parallel ($\nu = \|$) or perpendicular ($\nu = \perp$) to the gradient of n_s,

$$\Upsilon_\| = -n_b \left(\frac{\gamma^m}{2} + \frac{\gamma^q}{6} \right), \qquad \Upsilon_\perp = -n_b \frac{\gamma^m}{2} . \qquad (20)$$

The far field is now given by

$$\boldsymbol{A}^T(\boldsymbol{r}) = -2iql \frac{e^{2iqr}}{r} \left(\Delta' \xi \boldsymbol{E} \cdot \nabla \boldsymbol{E} + \Upsilon_\nu \xi' \hat{x} E^2 \right)_K^T . \qquad (21)$$

To proceed, we need to provide a model for the density fluctuation of the nanocrystals as well as for their nonlinear response function. It can then be observed that the differential and the total SHG efficiencies depend on both the incoming and the outgoing polarizations. To illustrate some features of the nonlinear response from the edge of the composite, in Fig. 5 we show the SH radiation patterns obtained for a linear density variation of the inclusions and for different values of the ratio between the width of the beam and the size of the edge of the composite film. We have also assumed the nanospheres to be made of Si. We have calculated the surface and bulk nonlinear response function of Si from its linear dielectric function, using the continuous dipolium model [20, 21]. The incoming energy $\hbar\omega = 1.55$ eV and polarization are the same as in the experiment of Jiang et al. [5]. We assumed the nominal beam focused at the middle of the edge, with a nominal density $n_s = 0.5 n_b$. We notice that for a wide edge, or equivalently, for a thin beam, the differential efficiency is the same as the two lobbed radiation pattern of Fig. 3 for a homogeneous film (left panel). However, for a wider beam or a thinner edge a new contribution coming from the gradient ∇n_s of the density appears, filling the minimum between the two lobes (middle panel). Furthermore, the interference between the bulk-like and edge signals produces an asymmetry between both lobes; the lobe that leans towards ∇n_s is smaller. For $w_0/L \approx 0.1$ the gradient contribution dominates and the two-lobbed structure is almost lost (right panel), although a slight asymmetry is still visible. In this case the peak is about twice as high as those for the homogeneous film.

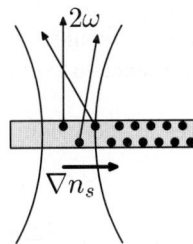

Fig. 4 A Gaussian beam is focused on the edge of a thin composite producing SH radiation (thin arrows). The gradient of the density of spheres ∇n_s close to the edge is indicated (thick arrow).

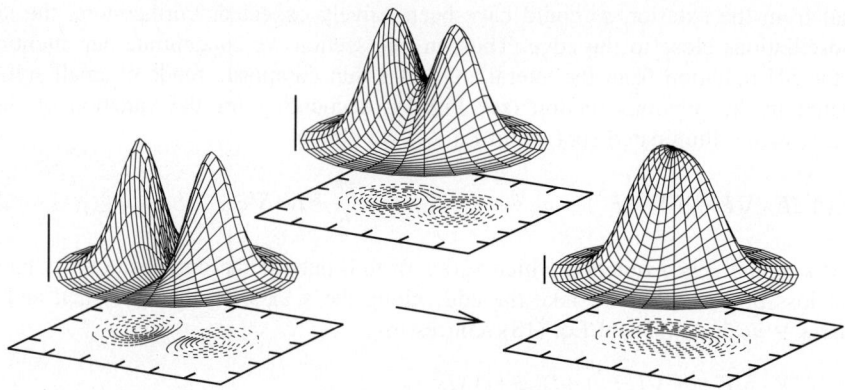

Fig. 5 SH radiation patterns $d\mathcal{E}_{\parallel}/d\Omega$ from the edge of a composite thin film made up of Si nanospheres for various values of the beam width w_0 and the edge size L, $w_0/L = 0.01, 0.05,$ and 0.1. The energy of the incoming photons is $\hbar\omega = 1.55$ eV and we chose the nominal density at half the bulk value $n_s = 0.5n_b$. For reference, small vertical bars corresponding to the same fixed height $0.5\mathcal{E}_b$ are shown in each panel. The arrow indicates the direction of the density gradient.

5   Some remarks on the intensity of the SH signal

In the following we will use qualitative and dimensional arguments to derive the efficiency of SHG from a thin film when illuminated by a finite beam. As the macroscopic polarization has the same units as the polarizing electric field, then the linear susceptibility $\chi^{(1)}$ in cgs units is a dimensionless quantity. Similarly, the dipolar quadratic susceptibility in a non-centrosymmetrc system [20] has units of inverse electric field, and thus $\chi^{(2)} \sim 1/E_0$, where E_0 is a quantity with units of electric field that depends on the intrinsic properties of the material. Naturally, E_0 must be of the order of the field that acts on an electron as it moves along the material, namely, $E_0 \sim e/a_B^2$, where e is the electronic charge and a_B is Bohr's radius, i.e., $E_0 \sim 1$ in atomic units. Due to the dipolar selection rules, within a centrosymmetric material $\chi^{(2)} = 0$ and the response of the system depends on the field variation, $P^{(nl)} \propto a/dE^2$, where $d \equiv |\nabla E|/|E|$ is the spatial scale of variation of the field and a is the size of the microscopic polarizable units of the system. Naturally, $a \sim a_B$ so that Γ and Δ' in Eq. (6) are expected to be of order a_B^3/e.

The far field produced by a single dipole p is of order $p/r\lambda^2$. Thus, the coherent SH far field produced by a thin film of width l illuminated by a beam of finite width w_0 would be of order $E^{(2)} \sim lw_0^2 P^{(nl)}/r\lambda^2$. The corresponding SH intensity $I^{(2)} \approx c(E^{(2)})^2$ would be $l^2 w_0^4 (P^{(nl)})^2 c/r^2\lambda^4$. The total radiated power would be $\mathcal{P} = Ir^2\Omega$ with Ω the solid angle illuminated by the outgoing radiation. The field produced by all non-linear dipoles would add up in phase only within a cone with an angular aperture $\approx \lambda^2/w_0^2$ Thus, $\mathcal{P} = l^2 w_0^2 P^2 c/\lambda^2$. In the case of SHG, based on dimensional reasons, the nonlinear polarization is $P^{(2)} \approx (a/E_0)E^2/w_0$ where $a \approx a_B$ is of the order of an atomic distance and $E_0 \approx e^2/a_B$ is about an atomic field. Thus, $\mathcal{P}^{(2)} \approx (a_B^6/ce^4)(l^2/\lambda^2)\mathcal{P}^2/w_0^4$, where $\mathcal{P}$ is the power of a fundamental pulse. Notice that the SHG signal falls as $1/w_0^4$ instead of depending on the inverse illuminated area of the sample, $1/w_0^2$. Thus, the total SH radiated power is proportional to the square of the intensity of the incoming pulses and NOT proportional to the square of their powers. This means that, contrary to what could naively be expected, increasing the incoming power would not necessarily result in an increased signal, if it is necessary to keep the beam intensity constant to avoid sample damage.

Let us now consider the possibility of employing two linearly polarized plane waves as the incoming field, so that

$$E^{in} = E_1^0 \, e^{i(k_1 \cdot r)} + E_2^0 \, e^{i(k_2 \cdot r)}, \qquad (22)$$

with $i = 1, 2$. The relevant terms in Eq. (6) would then be of the form

$$P^{(2)} \propto \boldsymbol{E}_1 \cdot \boldsymbol{k}_1 \boldsymbol{E}_1 \, \mathrm{e}^{2i\boldsymbol{k}_1 \cdot \boldsymbol{r}} + \boldsymbol{E}_1 \cdot \boldsymbol{k}_2 \boldsymbol{E}_2 \, \mathrm{e}^{i(\boldsymbol{k}_1+\boldsymbol{k}_2) \cdot \boldsymbol{r}} + \boldsymbol{E}_2 \cdot \boldsymbol{k}_1 \boldsymbol{E}_1 \, \mathrm{e}^{i(\boldsymbol{k}_1+\boldsymbol{k}_2) \cdot \boldsymbol{r}} + \boldsymbol{E}_2 \cdot \boldsymbol{k}_2 \boldsymbol{E}_2 \, \mathrm{e}^{2i\boldsymbol{k}_2 \cdot \boldsymbol{r}} \qquad (23)$$

The first term would yield the same SHG as a single plane wave in a homogeneous medium, namely, no SHG as the fundamental field is a transverse wave. The same could be said about the last term. However the second term could produce, in principle, a sizable SH polarization of the order of $(a_B^3/e^2) E_1 E_2/\lambda$ if $\boldsymbol{E}_1$ were p-polarized, and, if the beam were finite, it would produce a SH power $\mathcal{P}^{(2)} \approx (l^2 a_B^6/ce^2\lambda^2) (\mathcal{P}_1 \mathcal{P}_2/w_0^2\lambda^2)$, where $\mathcal{P}_i$ are the powers associated to the fundamental beams. When compared with the SH power yielded by a single finite beam, we find an increase in the radiated power by a factor of $(w_0/\lambda)^2$, which may be several orders of magnitude.

Care must be taken, however. If $\boldsymbol{E}_2$ is also p-polarized, then the third term in Eq. (23) would also radiate SH, and the corresonding field would be exactly opposite to the second term, yielding destructive interference and thus no enhanced SHG. Therefore, it could seem that the interference pattern does not enhance SHG. Similarly, if both beams were s-polarized, both terms in Eq. (1) would be null and would yield no enhancement. However, if $\boldsymbol{E}_1$ were p-polarized and $\boldsymbol{E}_2$ were s-polarized, then the second term would yield an enhanced SHG while the third term would be null, leading to a net enhancement of SHG. As $\boldsymbol{E}_1 \cdot \nabla \boldsymbol{E}_2$ depends on the angle α between both beams as $\sin \alpha$, the enhancement would also be proportional to $\sin^2 \alpha$. This could be thought of as a 'polarization direction' grating. If we assume that both incoming waves make opposite angles to say, the z direction, then the polarization of the total field would rotate as we move along the (x, y) plane. According to this analysis, if two p-ploarized or two s-polarized beams were made to interfere within a thin sample, there would be no sizable SHG enhancement, but if one s-polarized beam and one p-polarized beam were to cross each other at an angle α within a thin sample, the SHG signal would be enhanced by a factor $\approx \sin^2 \alpha w_0^2/\lambda^2$, where w_0 is the width of the incoming beams and λ their wavelength. The enhanced SHG radiation should appear along the bisector of both beams. Smaller two lobed signals should also be observable along the directions of the original beams. Therefore, it would look like the use of two crossed, opposite polarized beams may provide an alternative procedure to enhance the SH signal.

6 Discussion and conclusions

In this paper we discussed the SH radiation produced by a composite film made up of nanosized spherical particles, starting from the results obtained for the hyperpolarizabilities of a single centrosymmetric isotropic sphere [13].

As has been discussed previously [8] each individual sphere is unable to radiate in the forward direction when illuminated by a plane wave. Our results show that for a thin composite film illuminated by a focused gaussian linearly polarized beam no radiation is expected exactly in the forward direction either. However, the radiation pattern shows two narrow lobes displaced along the polarization direction by a small angle of the order of the diffraction-induced angular divergence of the linear far field. The forward peak reported experimentally for an array of Si nanospheres [5] appears to be these unresolved lobes [19].

Experiment also showed that the signal may be enhanced at the lateral edge of finite films. Thus, we calculated the SH radiation produced at a non-homogeneous film. This radiation presents nontrivial patterns which may be modified by changing the frequency, the waist of the illuminating beam, the sharpness of the edge, the density profile and both the incoming and outgoing polarization. By analyzing the polarization of the outgoing beams gradient and bulk-like contributions may be separated, facilitating the analysis of the response functions. Calculations for a Si nanosphere composite within a simple dipolium model [20] agree qualitatively with the experimental observations [5].

The solid-angle resolved SH patterns radiated by the bulk and edge we have calculated have not been explored experimentally yet. The bulk radiation of the composite depends on a single parameter Δ', which in turn depends, through the hyperpolarizabilities γ^v, on the bulk and surface response functions δ', γ, a, b, and f, of an individual sphere. As the radiation from the edge involves different combinations Υ_ν of the above response functions, proper measurements might provide additional information to partially disentangle the separate contributions.

Finally, we showed that as the SH field depends on the variations of the fundamental field and not only on its size, it turns out that the SH power from a thin film made up of small particles is not proportional to the square of the incident power, but to the square of its intensity. We propose an alternative illumating configuration that may enhance the SH power by several orders of magnitude.

We hope that this work will stimulate further theoretical and experimental research on the subject that may lead to characterization techniques for composite films.

Acknowledgements   We are grateful to Michael Downer for providing us his experimental results prior to publication and to him and Tony Heinz for useful discussions. We acknowledge support from DGAPA-UNAM under grant IN110999 (WLM and JM), from Conacyt under grant 36033-E (BM) and from Fundación Antorchas (VLB). VLB is also with CONICET, Argentina.

References

[1] H. W. K. Tom, T. F. Heinz, and Y. R. Shen, Phys. Rev. Lett **51**, 1983 (1983).
   J. E. Sipe, D. J. Moss, and H. M. van Driel, Phys. Rev. B **35**, 1129 (1987).
   P. Guyot-Sionnest and Y. R. Shen, Phys. Rev. B **38**, 7985 (1988).
   S. Janz and H. M. van Driel, Int. J. Nonlinear Opt. Phys. **2**, 1 (1993).
[2] G. A. Reider and T. F. Heinz, in: Photonic Probes of Surfaces, edited by P. Halevi (Elsevier, Amsterdam, 1995), Chap. 9.
   J. E. McGilp, Surf. Rev. Lett. **6**, 529 (1999).
   G. Lüpke, Surf. Sci. Rep. **35**, 75 (1999).
   N. Bloembergen, Appl. Phys. B, **68**, 289 (1999).
   M. C. Downer, B. S. Mendoza, and V. I. Gavrilenko, Surf. Interface Anal. **31**, 966 (2001).
[3] T. F. Heinz, in: Nonlinear Surface Electromagnetic Phenomena, edited by H.-E. Ponath and G. I. Stegeman (Elsevier, Amsterdam, 1991), p. 354.
[4] G. S. Agarwal and S. S. Jha, Solid State Commun. **41**, 499 (1982).
   X. M. Hua and J. I. Gersten, Phys. Rev. B **33**, 3756 (1986).
   D. Östling, P. Stampfli, and K. H. Bennemann, Z. Phys. D **28**, 169 (1993).
   J. Martorell, R. Vilaseca, and R. Corbalán, Phys. Rev. A **55**, 4520 (1997).
[5] Y. Jiang, P. T. Wilson, M. C. Downer, C. W. White, and S. P. Withrow, Appl. Phys. Lett. **78**, 766 (2001).
[6] Y. Jiang, L. Sun, M. C. Downer, Appl. Phys. Lett. **81**, 3034 (2002).
[7] L. Pavesi, L. Dal Negro, C. Mazzoleni, G. Franzò, F. Priolo, Nature **408**, 440 (2000).
[8] J. I. Dadap, J. Shan, K. B. Eisenthal, T. F. Heinz, Phys. Rev. Lett. **83**, 4045 (1999).
[9] J. I. Dadap, J. Shan, and T. F. Heinz, in: Book of Abstracts, Adriatico Research Conference on Lasers in Surface Science (International Centre for Theoretical Physics, Trieste, Italy, 11–15 September, 2000), p. 38.
[10] N. Yang, W. E. Angerer, and A. G. Yodh, Phys. Rev. Lett. **87**, 103902 (2001).
[11] Vera L. Brudny, Bernardo S. Mendoza, and W. Luis Mochán, Phys. Rev. B **62**, 11152 (2000).
[12] W. Luis Mochán and Rubén G. Barrera, Phys. Rev. B **32**, 4984 (1985); **32**, 4989 (1985).
[13] W. L. Mochán, J. A. Maytorena, B. S. Mendoza, and V. L. Brudny, Phys. Rev. B **68**(3), 85318, 2003.
[14] J. Rudnick and E. A. Stern, Phys. Rev. B **4**, 4274 (1971).
[15] P. Guyot-Sionnest, A. Tadjeddine, and A. Liebsch, Phys. Rev. Lett. **64**, 1678 (1990).
[16] N. Bloembergen, R. K. Chang, S. S. Jha, and C. H. Lee, Phys. Rev. **174**, 813 (1968); **178**, 1528(E) (1969).
[17] G. Russakoff, Am. J. Phys. **38**, 1188 (1970).
[18] H. Kogelnik and T. Li, Appl. Opt. **5**, 1550 (1966).
[19] M. Downer, private communication.
[20] Bernardo S. Mendoza and W. Luis Mochán, Phys. Rev. B **53**, 4999 (1996).
[21] D. E. Aspnes and A. A. Studna, Phys. Rev. B **27**, 985 (1983).

phys. stat. sol. (b) **240**, No. 3, 527–536 (2003) / **DOI** 10.1002/pssb.200303860

Depth resolved nonlinear optical nanoscopy

W. Luis Mochán[*,1], **Catalina López-Bastidas**[2], **Jesús A. Maytorena**[1],
Bernardo S. Mendoza[3], and **Vera L. Brudny**[4]

[1] Centro de Ciencias Físicas, Universidad Nacional Autónoma de México, Apartado Postal 48-3,
62251 Cuernavaca, Morelos, México

[2] Centro de Ciencias de la Materia Condensada, Universidad Nacional Autónoma de México, Apartado
Postal 2681, Ensenada, Baja California, 22800, Mexico

[3] Centro de Investigaciones en Optica, León, Guanajuato, México

[4] Departamento de Física, Facultad de Ciencias Exactas y Naturales, Universidad de Buenos Aires,
Ciudad Universitaria Pab. I, 1428 Buenos Aires, Argentina

Received 30 May 2003, revised 4 August 2003, accepted 11 August 2003
Published online 28 November 2003

PACS 07.79.Fc, 42.65.Ky, 78.67.−n, 78.68.+m

An electromagnetic field forced to vary along a plane with a spatial scale d much smaller than its free
space wavelength λ decays exponentially along its normal with a decay length $\sim d$. This decay, similar
to that of the wavefunction of tunneling electrons, has allowed the development of scanning near-field
optical microscopes (SNOMs), reminiscent of scanning tunneling and atomic force microscopes, which
have been able to resolve structures in the nanometer scale. However, existing SNOMs are unable to
determine the depth below the surface from which the optical signals arise due to the monotonic decay
of the optical evanescent probe fields. In this paper we study the optical second harmonic generation
(SHG) produced by mixing of the evanescent fields produced by a SNOM tip. We show that employing
an appropriately spatially-patterned tip, a non-monotonic non-linear probing field may be produced
which has a maximum at a given distance beyond the tip, yielding a novel microscopy which may
attain depth resolution with nanometric lengthscales. We estimate the size of the optical signal and we
compare it with that arising in the usual SHG-based surface spectroscopy of centrosymmetric materials.

1 Introduction

The light collected by the imaging system of a conventional optical microscope has wavevector
projections Q parallel to the image plane which are necessarily within the light cone
$-n\omega/c \leq Q \leq n\omega/c$, where ω is the frequency of the light and n the index of refraction of the ambient. With that finite range of wavevectors, the spatial resolution of the microscope is limited to spatial
structures not much smaller than the wavelength λ [1]. Nevertheless, light constitutes a very attractive
probe of matter, as it can be non-invasive, non-destructive, it is very finely tunable and many linear
and nonlinear, elastic and inelastic spectroscopies have been built on it. Therefore, a large effort has
been made in the last two decades to circumvent the Abbe resolution limit. In order to increment the
resolving power, larger wavevectors have to be involved in the imaging processes. Writing the wave
equation $\nabla^2 \phi = (n^2 \omega^2/c^2)\,\phi$ for any component ϕ of the electromagnetic field as
$\partial_z^2 \phi = [n^2(\omega^2/c^2) - Q^2]\,\phi$, where we assumed that the field oscillates along the xy plane, we see that
$Q > n\omega/c$ implies an exponential decay along the normal z direction. Thus, for a fixed wavelength,

---

[*] Corresponding author: e-mail: mochan@fis.unam.mx, Phone: +52 777 3291734, Fax: 3291775

better resolution may be attained by employing the evanescent near-field [2]. The exponential decay of the near-field is reminiscent of the exponential decay of the wavefunction of electrons that tunnel through classically inaccesible regions. This suggested the development [3] of scanning near-field optical microscopes (SNOMs) using a technology similar to that employed in scanning tunneling microscopes (STMs).

The first SNOMs employed apertures in metallic thin films [4, 5] and later optical fibre waveguides with a tapered tip covered with a metal layer with a small aperture through which the evanescent field could leak out towards a nearby surface. In one of its operation modes, the intensity I of the light reflected through the same waveguide is monitored as the tip is scanned along the xy directions while the height z above the nominal surface is kept fixed. Similarly, the reflected intensity could be kept fixed by raising or lowering the tip of the microscope through a feedback loop as it is scanned along the surface. The intensity $I(x, y)$ or the height $z(x, y)$ may be employed to draw three dimensional images of the surface which, if properly interpreted, can yield information on its composition and geometry. The horizontal resolving distance is of the order of the width of the aperture and the vertical decay length is about an order of magnitude smaller; both may be much smaller than the wavelength, to which they are quite insensitive.

There are disadvantages to the so called *aperture* SNOM described above. For instance, light can penetrate a finite distance into a metal. Thus, the field at the tip occupies an area somewhat larger than the actual size of the aperture, reducing the resolution. On the other hand, the signal is typically very small. Increasing the power to increase the signal is frequently not feasible, as the electromagnetic energy absorbed by the metallic screen heats the tip and may disrupt the sample under study [6]. To realize the highest resolutions the tip should be very close to the sample and thus topographical artifacts become entangled with variations in optical properties [2, 7–10].

There have been many alternative configurations designed to overcome the limitations of the aperture SNOM. In apertureless SNOM one can eliminate the waveguide altogether by focusing the field propagating from a far away source onto the tip of a very sharp metallic needle. The strong near-field induced by the needle in the vicinity of its tip may be scattered by a nearby surface and the intensity and/or phase of the corresponding far field may be monitored as the tip is scanned over the surface. Another alternative is to use uncoated optical fibers with sharp dielectric tips used for both SNOM and atomic force microscopy (AFM) in the shear mode. The AFM allows the identification of topographic features and their separation from localized chemical and physical features [11]. Advantages of tip and aperture type nanoscopes have been combined by growing a tip within an aperture [12]. The signal may be further enhanced over the background by illuminating the tip with the evanescent fields that accompany totally reflected light [13, 14].

The resolution attained with scattering SNOMs has gone below 10 nm [15, 16]. Field modulations with an even faster scale are present in the surface local field at crystalline surfaces. Thus, atomic resolution employing the crystalline near-field is feasible. This has been illustrated in an analysis of the optical spectra of adsorbate covered surfaces (though without a scanning instrument). As the field scattered by an adsorbate physisorbed over a crystalline surface is proportional to the square of the local field that polarizes it, incorporation of the near-field in the theoretical interpretation has enabled the extraction of geometrical information with a sub Å resolution [17].

SNOMs have been succesfully employed to observe many kinds of samples. Interferometric measurements [18] have yielded not only the amplitude but also the phase of the plasmon resonances of metallic nanoparticles optically excited with carbon nanotubes [19]. Beyond the relatively simple solid conducting and insulating surfaces, SNOMs have also been employed to investigate biological tissues in liquid ambients [20]. SNOMs may take advantage of the great variety of optical spectroscopies available. Thus, besides employing the intensity, phase and polarization [21, 22] of the linearly and elastically scattered near-fields, it can also use inelastical scattering to make images in the nanometer scale. For example, single-wall carbon nanotubes have been imaged with the Raman scattered fields [23]. Fluorescence and photobleaching have been used to study single adsorbed molecules [24]. Additional information may be obtained by combining other spectroscopies with a SNOM. For example, elemental and molecular analysis has been achieved by employing time-of-flight mass spectrometry on laser-desorbed molecules, where the laser light is delivered by a SNOM [25].

Non-linear optical processes have also been used both to image and to modify surfaces using near-fields. Metallic rough surfaces illuminated by a broad beam have been observed with a 100 nm resolution by scanning a tappered optical waveguide to collect the generated second harmonic (SH) near-field [26]. The enhancement of the SH at a defect may be an order of magnitude larger than the enhancement of the fundamental field due to the lighting rod effect [27]. The near SH field generated by nanopads of nonlinear materials on a substrate have been shown to be more localized than the [28] fundamental fields. This allows a high contrast which allows the study of mesoscopic domains in ferroelectric films [29]. The theory of SHG in nonlinear SNOM is complicated by the fact that both the fundamental and SH fields scattered by the object under study, the SNOM probe, and the substrate, have to be obtained self-consistently [30, 31]. Besides producing SH by the system under study, the SH light may be generated at the tip taking advantage of the strong field enhancement [32], and afterwards used as a very confined light source whose near-field may be scattered by an underlying surface over which it may be scanned. The field enhancement at a metallic tip is larger when the fundamental field has a component along the tip axis, as when the tip is at the focus of a high order Hermite-Gaussian laser mode [33]. Other useful non-linear effects include two-photon litography. Using the near-field produced by the SNOM tip [34], patterns with a scale < 100 nm have been produced.

One problem in surface science is the observation of buried structures. The usual surface probes (low energy electrons, ions, atoms, etc.) have a small penetration into matter and so they are sensitive mostly to the outermost layers. Thus, to study properties along the normal to the plane it is common to make a crater so that buried structures are brought to the surface where they can be conventionally examined. For example, secondary ion mass spectrometry (SIMS) is commonly used to study composition depth profiles as a surface is evaporated away [35]. The main disadvantage of this kind of techniques is that they destroy the sample. Optical fields may probe the interior of matter non-destructively, but their penetration depth is very large on an atomic scale. Nevertheless, the reduction in symmetry at interfaces has allowed the optical observation of buried interfaces [36] employing nonlinear spectroscopy. Non planar interfaces such as that between Si nanospheres and the SiO_2 matrix within which they were produced have also been observed with SHG [37, 38]. However, to obtain depth resolution it is not enough to reach the buried interface and identify its contribution to the signal but the depth itself has to be measured. Use of two photon optical processes such as SHG or two-photon optical-beam-induced current [39, 40] has permitted 3D imaging of circuit components in an integrated circuit with 1 μm lateral and 100 nm axial resolution. Differential confocal microscopy [41] has reached a depth resolution of 2 nm. Much better lateral resolutions have been obtained in SNOMs. Thus, the purpose of the present paper is to propose a scheme that would allow the observation of structures buried below a surface employing the near-field produced by a SNOM tip permitting the extraction of depth information with a resolution in the nanometer scale.

The paper is organized as follows. In Sec. 2 we present a scheme that employs sum frequency generation to obtain depth resolved information. In Sec. 3 we propose a simpler system based on SHG. A more realistic realization and calculation is developed in Sec. 4 and we devote Sec. 5 to conclusions.

2   Scheme

One drawback of all the techniques associated so far with SNOM is that the sensitivity is a monotonically decaying function of distance. Thus, it is not possible to distinguish the near-field produced by a strong scatterer located relatively deep below a surface from that produced by a weak scatterer at the surface or at a very shallow position. A similar problem was the topographical masking of optical data at high resolutions [2, 7–10]. Topographic artifacts have been minimized by monitoring the height above the surface with an AFM integrated with the SNOM [11]. However, this scheme is not useful for buried structures.

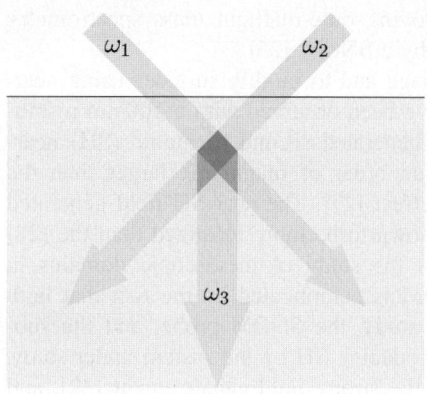

Fig. 1 Sum frequency generation. Photons with frequency $\omega_3 = \omega_1 + \omega_2$ are produced only in the region (dark gray) within the sample (light gray) where the beams with frequency ω_1 and ω_2 overlap. In principle, this region may be scanned over the 3D sample.

A solution to the problem of depth resolved nanoscopy may be found if we first look at a similar problem regarding the far field. Consider a three wave non-linear optical mixing effect such as sum frequency generation (SFG). In order to produce a photon of frequency $\omega_3 = \omega_1 + \omega_2$ by mixing two photons with fundamental frequencies ω_1 and ω_2, both photons must arrive simultaneously at the same place. Thus, SFG can only take place within a volume where the two beams with frequencies ω_1 and ω_2 overlap (Fig. 1). In principle, this volume can be scanned over the interior of a sample in order to obtain a 3D image.

A more practical idea based in a similar same principle has been to look at the SHG, two photon fluorescence, two photon induced current [39, 40, 42] or any two photon process which takes place mainly close to the focus of a tightly focused beam, where the fundamental intensity is largest.

Consider then a SNOM tip made up of, or coated with, a periodic array of alternating layers of materials such that one of them is perfectly transparent to light of frequency ω_1 and opaque to light of frequency ω_2 and the other one has the complementary behavior, as illustrated in Fig. 2.

If the tip is illuminated simultaneously by two beams with frequencies ω_a ($a = 1, 2$), we expect the corresponding field E_a to attain some finite value at positions immediately below the layers that are transparent to that frequency, and to be almost null immediately below the layers that are opaque to that frequency. Oversimplifying to some extent, the fields might look somewhat like the periodic step

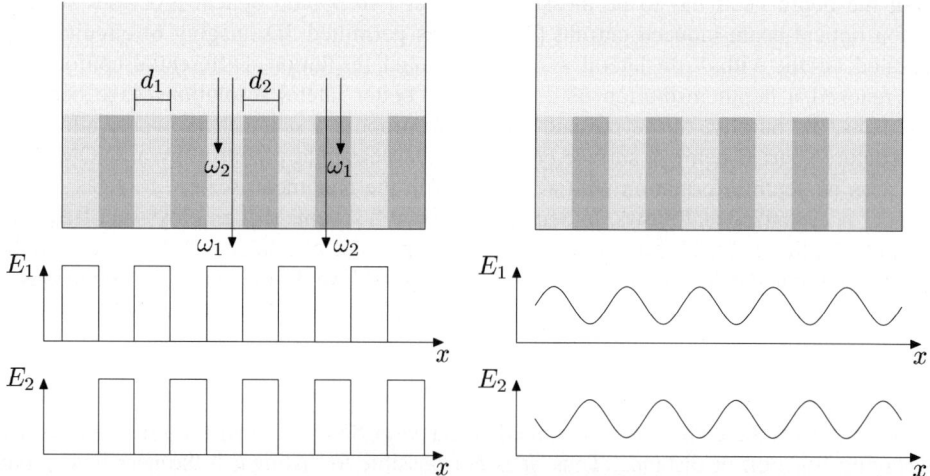

Fig. 2 SNOM tip (light gray) coated with a periodic array of alternating layers of widths d_1 and d_2 of materials which are either perfectly transparent at frequency ω_1 and opaque at ω_2 or the other way around. The frequency components of the field are either absorbed (short arrows) or transmitted (long arrows) through each layer, giving rise to the transmitted fields E_a at ω_a ($a = 1, 2$) plotted immediately below the tip (left panel) and farther away from the tip (right panel).

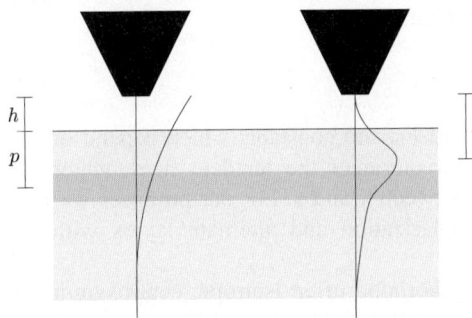

Fig. 3 A typical SNOM with a sensitivity that decays monotonically from the tip (left) and a SNOM similar to that in Fig. 2 (right) with a sensitivity that has a maximum a distance ℓ below the tip. For the latter, the signal produced by a layer (shaded) buried a distance p below the surface is maximized when the tip is at a height $h = \ell - p$ above the surface.

functions displayed in the left panel of Fig. 2. Although both E_1 and E_2 are relatively large close to the tip, one of them is almost null at those positions where the other is maximum, so that the product $E_1 E_2$ is essentially zero. The tip acts for each field as a grating, inducing a diffracted field with diffraction wave-vectors G which are integer multiples of $2\pi/d$ with $d = d_1 + d_2$, where d_a is the width of a layer of type a. However, if we choose d in the nanometer scale, all the diffracted fields are evanescent and decay along the normal direction with a decay length $2\pi/|G|$. Thus, field components with larger G decay faster and the field variations along the surface smoothen out as we move away from the tip. Thus, the overlap between E_1 and E_2, i.e., the product $\langle E_1 E_2 \rangle$ averaged along the surface, may be an increasing function of distance. Finally, far enough from the tip, most Fourier components would decay so that $\langle E_1 E_2 \rangle$ would again be zero. The main result of this qualitative discussion is that $\langle E_1 E_2 \rangle$ *is not monotonic and we expect it to exhibit a maximum at a finite distance* $\ell \approx d$. If the tip in Fig. 2 is brought close to the surface of a material, the evanescent fields E_1 and E_2 may mix to produce a propagating field E_3 with frequency $\omega_3 = \omega_1 + \omega_2$. However, unlike the usual SNOM signals, this instrument would be most sensitive at a distance ℓ from the tip and not at the tip itself. Thus, scanning the tip vertically and detecting the maximum in the SFG intensity we may infer the depth where the nonlinear signal is produced (Fig. 3).

3 SHG

In order to avoid the need of illuminating the tip of the SNOM with two different beams to produce SFG, we consider now the possibility of generating SHG. Instead of separating at the tip fields E_1 and E_2 of different frequencies, we consider separating fields of different polarizations but with the same frequency. Thus, we assume that the layers in Fig. 2 are now perfect polarizers that either absorb light polarized along the x direction while letting through light polarized along the y direction or vice versa. As the field is scattered along the x direction we identify the xz plane as the plane of propagation, $a = 1 \rightarrow s$ layers as s-polarizers and $a = 2 \rightarrow p$ layers with p polarizers. Therefore, we identify E_1 with E_s and E_2 with E_p and write the fields below the tip as

$$E_{sy} = \sum_G E_{sG} e^{iGx} e^{-|G|z} \,, \tag{1}$$

and

$$E_{px} = \sum_G E_{pG} e^{iGx} e^{-|G|z} \,, \tag{2}$$

which satisfy the wave equation for $z > 0$ in the limit of a small period d, i.e., the non-retarded Laplace equation, and we assumed periodicity along x, ignoring the edge effects at the boundary of the tip. We assume that the tip is at $z \le 0$ and the system under study is somewhere at $z > 0$, z points downwards in Fig. 2, and for simplicity, we calculate the fundamental fields as if in vacuum. The z component of $\boldsymbol{E}_p$ may be obtained from Eq. (2) by applying Gauss law $\nabla \cdot \boldsymbol{E}_p = 0$. The periodic step functions shown in Fig. 2 immediately below the tip yield the amplitudes

$$E_{aG} = 2 \frac{E_{0a}}{Gd} e^{iGx_a} \sin \frac{Gd_a}{2} , \tag{3}$$

where x_a denotes the center of a layer of type a. Here E_{0s} and E_{0p} are constants which depend on the absorbance and transmittance of each layer and on the polarization of the incident field, which we assumed to be polarized along some intermediate direction between x and y. For our purposes we may take $E_{0s} = E_{0p} = E_0$, which corresponds to $45°$ incident polarization and alternate layers with the same transmittance.

We assume now that the nanoscope is brought to the neighborhood of an isotropic centrosymmetric material with nonlinear susceptibilities γ and δ', where the s and p fields (1) and (2) are mixed and produce a nonlinear polarization $\boldsymbol{P}$ at the second harmonic frequency [43]

$$\boldsymbol{P} = \gamma \nabla E^2 + \delta' \boldsymbol{E} \cdot \nabla \boldsymbol{E} . \tag{4}$$

By filtering out the p component of the induced polarization we are left with only crossed terms of the form

$$\boldsymbol{P}_s = \delta' \boldsymbol{E}_p \cdot \nabla \boldsymbol{E}_s . \tag{5}$$

Thus, as in the previous section, we expect no signal from the immediate vicinity of the tip, where $\boldsymbol{E}_s$ and $\boldsymbol{E}_p$ are non-overlapping, and no signal originated far from the tip, where there are no evanescent fields left. Thus, the situation seems qualitatively similar to that illustrated in Fig. 3, and we expect a distance where the nonlinear polarization acquires a maximum. Unfortunately, a calculation of $\langle \boldsymbol{P}_s \rangle$ employing Eqs. (1), (2), and (5), yields a null result.

The reason for the null polarization is that our tip is symmetric under the inversion $x \leftrightarrow -x$. Thus, the s SH polarization induced by mixing an s fundamental field scattered with any wavevector $G_s = G$ with a p polarized fundamental field with the wavevector $G_p = -G$ has the same size and the opposite phase to the polarization induced by an s field with $G_s = -G$ mixed with a p field with $G_p = G$. We notice that only fields with $G_p = -G_s$ contribute to $\langle \boldsymbol{P}_s \rangle$ and thus to a far field. Combinations with $G_p \neq -G_s$ produce polarizations which oscilate along the x direction with a large wavevector $G_p + G_s$ and therefore produce only evanescent SH fields. Although they may be scattered into far fields by small structures, if not null, $\langle \boldsymbol{P}_s \rangle$ should dominate.

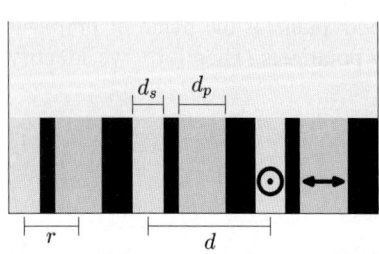

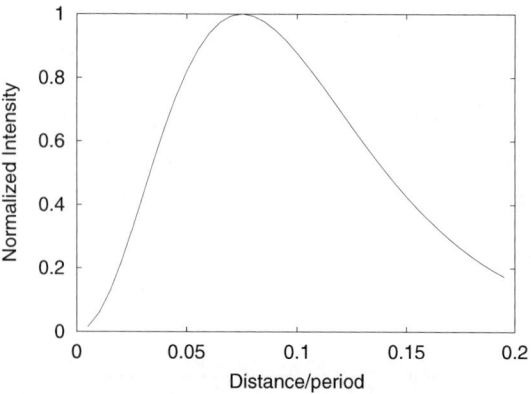

Fig. 4 Tip coated with a periodic array of polarizing layers of widths d_s and d_p (shaded) separated by perfectly absorbing spacers (black). The period d, the distance between the centers of two layers r and the transmitted polarizations of each kind of layers are indicated.

Fig. 5 Normalized intensity of the SHG generated by a nonlinear layer buried beneath a surface and excited by the near-fields produced by the tip of Fig. 4 with $d_s = d_p = 0.2d$ and $r = 0.65d$ as a function of its distance to the tip normalized to the superlattice period d.

In order to produce a finite $\langle P_s \rangle$ we have to break the inversion symmetry of the surface. This may be done by asymmetrically introducing perfectly absorbing spacers between the polarizers, as indicated in Fig. 4.

We have calculated the intensity of the SH light generated by a nonlinear layer buried beneath a surface as a function of the distance to the tip of the nanoscope. As exemplified by Fig. 5, there is a peak at a distance $\ell \approx 0.08d$, very close to the decay length $d/4\pi$ of the intensity of an evanescent wave with wavelength d equal to the period of the polarizer array.

4   Realization

Having shown that it is possible to produce a maximum in the sensitivity of an SHG nonlinear SNOM at a given distance ℓ from the tip, in this section we investigate a more realistic realization of the nanoscope tip. We consider a monolayer consisting of anisotropic molecules layed down in a periodic pattern such as that shown in Fig. 6.

When illuminated by fundamental light polarized along an oblique direction with both x and y components, both kinds of molecules become polarized and produce fields with rapid spatial variations and with a polarization that alternates between the x and y directions. Therefore, we can expect that the SH field induced in a sample below the tip might have a maximum analogous to that in the previous section.

Before calculating the field produced by a polarized array of molecules as that in Fig. 6 we realize that even though the period d may be very small, a fully nonretarded calculation would be inadequate. The reason is that, according to Eq. (4), the spatially averaged nonlinear polarization produced by a field E derived from a potential ϕ is

$$\langle P \rangle = \gamma \langle \nabla (\nabla \phi)^2 \rangle + \delta' \langle \nabla \phi \cdot \nabla \nabla \phi \rangle \tag{6}$$

Integrating by parts over one unit cell of the structure to calculate the averages we obtain $\langle P \rangle = 0$ as the integral over a period of a gradient is identically zero. Thus, we have to incorporate magnetic effects in our calculation. For simplicity, we will assume that s is much smaller that d, so that we approximate the system as if it were continuous along the y direction and take account of the spatial variations along x only, as in the previous section. Then, we can view the system as a collection of long flat *capacitors* with electrodes separated along the x direction alternating with *wires* oriented along the y direction. To lowest order in the retardation, the former produce an electric field

$$E_p = -2\pi \frac{p_p}{sd} \sum_G (|G|, 0, iG) \, e^{iGx} e^{-|G|z} , \tag{7}$$

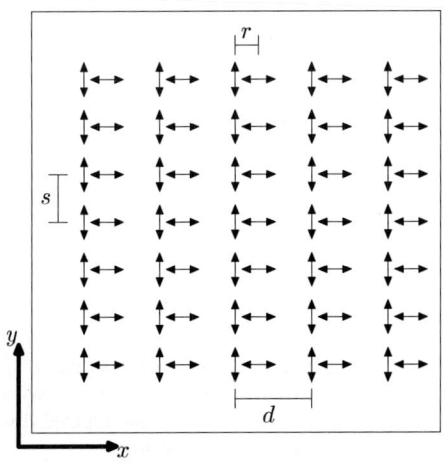

Fig. 6 Bottom view of a SNOM tip on whose surface a periodic array consisting of alternating rows of anisotropic polarizable molecules with mutually perpendicular orientations. The principal direction with the largest linear polarizability is indicated for each molecule by double arrows. The period d along the alternation direction, the period s along the perpendicular direction and the distance r between successive rows are indicated, as well as the coordinate axes.

while the latter produce a magnetic field

$$\boldsymbol{B}_s = -2\pi i q \frac{p_s}{sd} \sum_G (1, 0, i\, \mathrm{sgn}\,(G))\, e^{iG(x-r)} e^{-|G|z} \,, \tag{8}$$

where p_p and p_s are the amplitudes of the dipoles induced in the molecules which are aligned with the x and y axes respectively, $q = \omega/c$ and ω is the fundamental frequency. We write Eq. (5) as

$$\boldsymbol{P}_s = -\delta' \boldsymbol{E}_p \times (\nabla \times \boldsymbol{E}_s)\,, \tag{9}$$

we use $\nabla \times \boldsymbol{E}_s = iq\boldsymbol{B}_s$ and we substitute Eqs. (7) and (8) to calculate the intensity of SHG induced by the tip of Fig. 6 on a nonlinear material a distance z below. The nonlinear polarization (9) depends on the nonlinear response δ' of the material and the driving term, which we write as

$$\boldsymbol{E}_p \times (\nabla \times \boldsymbol{E}_s) = \zeta(qs)^2\, (s/d)^2\, P_1^2/d\,, \tag{10}$$

where ζ is a dimensionless number which characterizes the potentiality of the near-field to generate an SH signal and P_1 is the volumetric linear polarization $P_1 = p/s^3$ induced in the dipolar overlayer at the tip of the SNOM (we chose $p_s = p_p \equiv p$). In Fig. 7 we show ζ as a function of the asymmetry parameter r/d and of the normalized distance z/d between the tip and the nonlinear material. As expected from the previous section, there is no SHG for an inversion invariant tip $r = .5d$ but ζ may acquire quite large values of the order of 10^3 for small r. The results are symmetrical around $r = .5d$.

An analysis of Fig. 7 shows that as r varies from $0.45d$ to $0.1d$, both the position ℓ of the peak and its width move from $d/10$ to about $d/40$.

The size of the SH signal may be estimated as follows: A thin region of width ℓ with an SH polarization P_s produces an SH far field $E_s \approx q\ell P_s$. For our arrangement, $P_s \approx (a_B^3/e)\, \boldsymbol{E} \times (\nabla \times \boldsymbol{E}) \approx \zeta(a_B^3/e)\, (qs)^2(s/d)^2\, P_1^2/d$, where we approximated the nonlinear quadrupolar response δ' by its typical value a_B^3/e, i.e., the inverse of a typical internal field divided by an atomic distance. We found that $\ell \approx 0.1d$ and we may reasonably take $s \approx 10a_B$ and $d \approx 10s$. We finally write $P_1 = \chi E$. On the other hand, the typical polarization for the usual dipolar surface SHG is $P_s \approx (a_B^2/e)E^2$ and the polarized selvedge has a size $\ell \approx a_B$. Putting all of these quantities together we obtain $E(\mathrm{SNOM})/E(\mathrm{surf}) \approx 10^{-1}(qa_B)^2 \chi^2 \zeta$. Fig. 7 shows that ζ may be as large as 10^3 and at resonance χ might easily reach 10 or 100, so that the signal from the nonlinear nanoscope is expected to be comparable to that of the usual surface SHG.

The calculations above have been performed for a perfectly periodic system, which is necessarily extended along the xy plane and therefore provides no lateral resolution. Preliminar calculations show

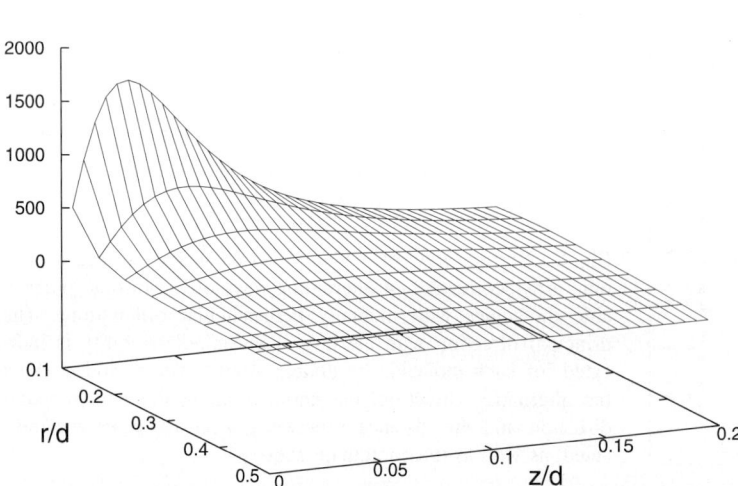

Fig. 7 SHG source strength $|\zeta|$ for different asymmetries r/d as a function of the normalized distance z/d to the tip.

that qualitatively similar results hold for finite tips, even when they have only one period, although the sensitivity would fall off as a power instead of exponentially. In this case the SH intensity would be smaller than the estimate above, but as in other SNOMs, it could be increased by tuning the fundamental field to a strong resonance of the tip.

5 Conclusions

We have shown that three wave mixing may be employed to construct an optical near-field surface probe capable of observing buried structures with a depth resolution of the order of nanometers. By coating the tip with a small-period patterned material, evanescent fundamental fields may be produced which are spatially separated at the tip. Separated s and p fields may be produced by alternating linear chains of oriented anisotropic molecules. The arrangement has to lack inversion symmetry for this scheme to work. The source for the SHG has a peak at a distance close to a tenth of the modulation period and the generated signal is expected to be not smaller than that for ordinary surface SHG. Thus, we believe that the instrument we propose here is viable and worth of further research.

Acknowledgements We acknowledge the support from DGAPA-UNAM under grant IN117402 (WLM, CLB and JM), from Conacyt under grants 36033-E (BM) and C01-4113 (CLB), and from Fundación Antorchas (VLB). VLB is also with CONICET, Argentina.

References

[1] M. Born and E. Wolf, Principles of Optics, 7th edition (Cambridge Univ. Press, Cambridge, 1999), sec. 8.6.
[2] L. Novotny, B. Hecht, and D. W. Pohl, Ultramicroscopy **71**, 341 (1998).
[3] G. A. Massey, Appl. Opt. **23**, 658 (1984).
[4] D. W. Pohl, W. Denk, and M. Lanz, Appl. Phys. Lett. **44**, 651 (1984).
[5] U. Ch. Fischer, J. Vac. Sci. and Technol. B **3**, 386 (1985).
[6] A. H. La Rosa, B. I. Yakobson, and H. D. Hallen, Appl. Phys. Lett. **67**, 2597 (1995).
[7] B. Knoll and F. Keilmann, Nature 399, 134 (1999).
[8] R. Hillenbrand and F. Keilmann, Phys. Rev. Lett. **85**, 3029 (2000).
[9] R. Hillenbrand and F. Keilmann, Appl. Phys. B **73**, 239 (2001).
[10] R. Hillenbrand, T. Taubner, and F. Keilmann, Nature 418, 159 (2002).
[11] G. Kaupp, A. Herrmann, J. Phys. Org. Chem. **10**, 675 (1997).
[12] Heinrich G. Frey, Fritz Keilmann, Armin Kriele, and Reinhard Guckenberger, Appl. Phys. Lett. **81**, 5030 (2002).
[13] R. C. Reddick, R. J. Warmack, and T. L. Ferrell, Phys. Rev. B **39**, 767 (1989).
[14] M. Spajer, D. Courjon, K. Sarayeddine, A. Jalocha, and J.-M. Vigoureux, J. Phys. III (France) **1**, 1 (1991).
[15] F. Zenhausern, Y. Martin, and H. K. Wickramasinghe, Science **269**, 1083 (1995).
[16] R. Hillenbrand and F. Keilmann, Appl. Phys. Lett. **80**, 25 (2002).
[17] W. Luis Mochán and Rubén G. Barrera, Phys. Rev. Lett. **56**, 2221 (1986).
[18] T. Taubner, R. Hillenbrand, and F. Keilmann, J. Microsc. **210** 311 (2003).
[19] R. Hillenbrand, F. Keilmann, P. Hanarp, D. S. Sutherland, and J. Aizpurua, Appl. Phys. Lett. **83**, 368 (2003).
[20] A. Kramer, A. Wintergalen, M. Sieber, H. J. Galla, M. Amrein, and R. Guckenberger, Biophys. J. **78**, 458 (2000).
[21] Kunio Nakajima, Yasuyuki Mitsuoka, Norio Chiba, Hiroshi Muramatsu, Tatsuaki Ataka, Katsuaki Sato, and Masamichi Fujihira, Ultramicroscopy **71**, 257 (1995).
[22] Georg Eggers, Andreas Rosenberger, Nicole Held, Ansgar Münnemann, Gernot Güntherodt, and Paul Fumagalli, Ultramicroscopy **71**, 249 (1998).
[23] Achim Hartschuh, Erik J. Sánchez, X. Sunney Xie, and Lukas Novotny, Phys. Rev. Lett. **90**, 095503 (2003)
[24] W. Patrick Ambrose, Peter M. Goodwin, John C. Martin, and Richard A. Keller, Phys. Rev. Lett. **72**, 160 (1994).
[25] D. A. Kossakovski, S. D. O'Connor, M. Widmer, J. D. Baldeschwieler, and J. L. Beauchamp, Ultramicroscopy **71**, 111 (1998).
[26] Igor I. Smolyaninov, Anatoly V. Zayats, and Christopher C. Davis, Phys. Rev. B **56**, 9290 (1997).

[27] Anatoly V. Zayats, Thomas Kalkbrenner, Vahid Sandoghdar, and Jürgen Mlynek, Phys. Rev. B **61**, 4545 (2000).
[28] Zhi-Yuan Li, Ben-Yuan Gu, and Guo-Zhen Yang, Phys. Rev. B **59**, 12622 (1999).
[29] Ai-Fang Xie, Ben-Yuan Gu, Guo-Zhen Yand, and Ze-Bo Zhang, Phys. Rev. B **63**, 054104 (2001).
[30] Sergey I. Bozhevolnyi and Valeri Z. Lozovski, Phys. Rev. B **61**, 11139 (2000).
[31] Sergey I. Bozhevolnyi and Valeri Z. Lozovski, Phys. Rev. B **65**, 235420 (2002).
[32] Satoshi Takahashi and Anatoly V. Zayats, Appl. Phys. Lett. **80**, 3479 (2002).
[33] A. Bouhelier, M. Beversluis, A. Hartschuh, and L. Novotny, Phys. Rev. Lett. **90**, 013903 (2003).
[34] Xiaobo Yin, Nicholas Fang, Xiang Zhang, Ignacio B. Martini, and Benjamin J. Schwartz, Appl. Phys. Lett. **81**, 3663 (2002).
[35] M. V. Ramana Murty, Surf. Sci. **500** 523 (2002).
[36] Christopher T. Williams and David A. Beattie, Surf. Sci. **500**, 545 (2002).
[37] Y. Jiang, P. T. Wilson, M. C. Downer, C. W. White, and S. P. Withrow, Appl. Phys. Lett. **78**, 766 (2001).
[38] W. Luis Mochán, Jesús A. Maytorena, Bernardo S. Mendoza, and Vera L. Brudny, Phys. Rev. B **68**, 085318 (2003).
[39] E. Ramsay, D. T. Reid, and K. Weilsher, Appl. Phys. Lett. **81**, 7 (2002).
[40] D. T. Reid, E. Ramsay, and K. Weilsher, Optics and Photonic News **13**(12), 17 (2002).
[41] Chau-Hwang Lee and Jyhpyng Wang, Opt. Commun. **135**, 233 (1997).
[42] Changhuei Yang and Jerome Mertz, in: Proceedings of the SPIE **4963**, Multiphoton Microscopy in the Biomedical Sciences III, edited by Ammasi Periasamy, and Peter T. C. So (SPIE, Bellingham, Washington, 2003), p. 52.
[43] N. Bloembergen, R. K. Chang, S. S. Jha, and C. H. Lee, Phys. Rev. **174**, 813 (1968); **178**, 1528(E) (1969).

physica

p s s **status solidi** **c**

www.physica-status-solidi.com

conferences and critical reviews

phys. stat. sol. (c) **0**, No. 8, 2921–2925 (2003) / **DOI** 10.1002/pssc.200303826

Surface-induced broadening and shift of exciton ground-state resonance in quantum wells

N. Atenco-Analco[1], **N. M. Makarov**[2], and **F. Pérez-Rodríguez**[*, 1]

[1] Instituto de Física, Universidad Autónoma de Puebla, Apartado Postal J-48, Puebla, Pue. 72570, México
[2] Instituto de Ciencias, Universidad Autónoma de Puebla, Priv. 17 Norte No 3417, Puebla, Pue. 72050, México

Received 30 May 2003, revised 4 August 2003, accepted 11 August 2003
Published online 10 November 2003

PACS 71.35.–y, 73.20.Mf, 78.67.De

Basing on the Green's function method for the study of disordered systems, we propose a simple model that describes the shift and broadening of excitonic resonance in interface-corrugated quantum wells. This model employs the original microscopic excitonic Hamiltonian and is applicable in a wide region of varying parameters from the sharp resonance to the classical limit. The relaxation frequency ν and the shift $\delta\omega_0$ of exciton ground-state resonance are calculated analytically for realistic quantum wells with finite potential barriers and anisotropic effective masses of the electron and hole. It is shown that the type of the excitonic resonance and its line-shape are mainly controlled by the competition between the correlation properties of surface disorder and finite height of potential barriers.

1 Introduction   At present, confined excitonic semiconductor structures are intensively studied because of their intriguing optical and spectral properties and potential applications to optoelectronic devices. Such properties are determined by the strength of confinement and, consequently, are strongly subjected to the inherent interface roughness, which causes the inhomogeneous damping and shift of excitonic resonances (see, for example, Refs. [1-10]). There are two qualitatively different regimes: weak and strong confinement regimes.

In the regime of weak confinement (also known as thin film regime) the characteristic size d of the system is much larger than the exciton Bohr radius a_0 ($d \gg a_0$). In this case, the behavior of the electron-hole pair is as in the bulk system, but thanks to the confinement, the exciton center-of-mass motion is quantized, giving rise to resonances in optical spectra (absorption, transmission, specular and diffuse reflection).

In the regime of strong exciton confinement (or quantum well regime) for II-VI and III-V semiconductor structures, the thickness d of the quantum well (QW) is smaller than the exciton radius a_0 ($d < a_0$). Here, the Coulomb attraction between the electron and hole is suppressed in the growth direction because of the dominant effect of the confining potential. Therefore, in this direction the motion of the electron and hole is quantized separately. Due to the in-plane Coulomb interaction, the electron-hole pair in QW forms a quasi-2D exciton. Consequently, in analyzing surface scattering of exciton we should treat the individual electron-surface and hole-surface interaction.

In this contribution, basing on Green's function method, we present a unified theory for describing the exciton spectrum of surface-corrugated quantum wells from the broad to sharp resonances. The former is also known as classical limit (see, e.g., [4-6] and references therein). In the second case, the surface-induced broadening ν and shift $\delta\omega_0$ of exciton ground-state resonance are calculated within the self-consistent Born approximation [7-10], which takes into account the inherent action of the exciton-surface

[*] Corresponding author: e-mail: fperez@sirio.ifuap.buap.mx, Phone: +52 222 229 5500 ext. 2003, Fax: +52 222 2295611

scattering on itself and, therefore, is appropriate in studying excitonic resonances. Our calculations are carried out with the use of the original microscopic excitonic Hamiltonian for realistic quantum wells, i.e. with finite potential barriers and anisotropic effective masses of the electron and hole. The found expressions for v and $\delta\omega_0$ reveal clearly their dependences on the parameters of the quantum well, its rough interfaces and exciton-resonance detuning.

2 Problem formulation   To start with, one should define the excitonic model for QW with flat interfaces. We introduce the z-axis in the transverse to the well-plane direction so that the ideal QW is confined within the region $0 \le z \le d$. The unperturbed excitonic Hamiltonian can be suitably written as

$$\hat{H} = E_{gap} - \frac{\hbar^2}{2M}\frac{\partial^2}{\partial\mathbf{R}^2} - \frac{\hbar^2}{2\mu}\frac{\partial^2}{\partial\boldsymbol{\rho}^2} - \frac{e^2}{\varepsilon\rho} - \frac{\hbar^2}{2m_{ez}}\frac{\partial^2}{\partial z_e^2} - \frac{\hbar^2}{2m_{hz}}\frac{\partial^2}{\partial z_h^2} + U_e(z_e,\mathbf{r}_e) + U_h(z_h,\mathbf{r}_h) - i\hbar v_0. \quad (2.1)$$

Here, E_{gap} is the energy gap between the conduction and valence bands. The second term describes the kinetic energy of the exciton center-of-mass with in-plane total mass $M = m_{e\|} + m_{h\|}$ and in-plane radius vector $\mathbf{R}$. The third term specifies the kinetic energy of the relative electron-hole motion, which is characterized by the in-plane reduced mass $\mu = m_{e\|}m_{h\|}/M$ and the internal vector $\boldsymbol{\rho}$. Two-dimensional Coulomb potential is given by the fourth term with ε being the dielectric constant of the semiconductor. The fifth and sixth terms are responsible for the individual transverse motion of electron or hole, respectively. The quantities $U_e(z_e,\mathbf{r}_e)$ and $U_h(z_h,\mathbf{r}_h)$ are the QW confining potentials for the electron and the hole ($\mathbf{r}_{e,h}$ is the electron/hole in-plane position vector, $\mathbf{r}_{e,h} = \mathbf{R} \pm \mu\boldsymbol{\rho}/m_{e\|,h\|}$, $\boldsymbol{\rho} = \mathbf{r}_e - \mathbf{r}_h$). In the Hamiltonian (2.1) we have inserted a homogeneous exciton-bulk damping v_0 to take into account its effect on the exciton-surface scattering.

Now the lower interface is assumed to be of a corrugated relief $z = \xi(\mathbf{r})$ while the upper one $z = d$ remains, for simplicity, flat. The surface roughness is described by a random function $\xi(\mathbf{r})$ of the in-plane position vector $\mathbf{r}$ with probability density $P_\sigma(\xi)$ and standard statistical properties [11],

$$\overline{\xi(\mathbf{r})} = 0, \qquad\qquad \overline{\xi(\mathbf{r})\xi(\mathbf{r}')} = \sigma^2 W(|\mathbf{r} - \mathbf{r}'|). \qquad (2.2)$$

The upper line stands for statistical average over realizations of $\xi(\mathbf{r})$, σ is the root-mean-square (r.m.s.) roughness height. The correlator $W(|\mathbf{r}|)$ is normalized to its maximal value, $W(0) = 1$, and has a scale of decrease R_c (the correlation radius). Both the amplitude and gradient roughness of the corrugated interface is taken to be small, $\sigma \ll d$ and $\sigma \ll R_c$. These limitations are common in the theories of weak surface scattering that are based on an appropriate perturbative approach (see, e.g., Ref. [11]). Note that such an approach is applicable for any random process $\xi(\mathbf{r})$ (not only for Gaussian one), excluding the presence of extremely large, flat islands at the interface [12].

The electron-surface V_e and hole-surface V_h perturbation potentials that for surface-disordered QW should be added to the Hamiltonian (2.1), are introduced as the difference between the confining potentials of the corrugated $z = \xi(\mathbf{r})$ and ideally flat $z = 0$ interfaces,

$$V_{e,h} = U_{e,h}\left[\Theta(\xi(\mathbf{r}_{e,h}) - z_{e,h}) - \Theta(-z_{e,h})\right] = U_{e,h}\,\delta(z_{e,h})\xi(\mathbf{R} \pm \mu\boldsymbol{\rho}/m_{e\|,h\|}). \qquad (2.3)$$

Here $U_{e,h}$ is the finite height of the potential barriers for the electron (e) and the hole (h) and $\Theta(x)$ is the unit-step function. The delta-function $\delta(z_{e,h})$ emerges in Eq. (2.3) due to the condition $\sigma \ll d$ and evidences the surface nature of the scattering potentials $V_{e,h}$.

In what follows the Green's function G shall be derived. For the convenience of subsequent averaging, we employ the Dyson-type equation that relates G to the Green's function G_0 of the unperturbed Hamiltonian (2.1). Our consideration is restricted to the most important from the experimental point case of the exciton ground resonance. Therefore, we need to obtain only the Green's matrix element $g(k,k') = <k \mid G \mid k'>$ with respect to the unperturbed exciton-ground-states. Note, the in-plane wave vector k is attributed to the continuous exciton center-of-mass motion, the other (discrete) quantum numbers are not written for simplicity. In accordance with the general theory of Green's functions (see, e.g., Ref. [13]), it can be shown that $g(k,k')$ obeys the following Dyson-type equation:

$$g_0^{-1}(k)g(k,k') = (2\pi)^2 \delta(k - k') + \int_{-\infty}^{\infty} \frac{d^2 k_1}{(2\pi)^2} [\tilde{V}_e(k - k_1) + \tilde{V}_h(k - k_1)]g(k_1,k') \qquad (2.4)$$

with $g_0(k)$ and $\tilde{V}_{e,h}(k - k_1) = <k \mid V_{e,h} \mid k_1>$ being, respectively, the exciton-ground-state matrix elements of G_0 and surface-scattering potentials (2.3). Evidently, $g_0^{-1}(k) = \hbar[\omega - \omega_0 - (\hbar k^2 / 2M) + i v_0]$, where $\hbar\omega$ and $\hbar\omega_0$ are the excitonic energy and its eigenvalue. Now we should average Eq. (2.4). This can be performed by two different ways depending on the relation between variation scales of the functions entering the integral. Below, we present the main results derived by both methods.

3 Broad resonance (The classical limit) If the Fourier transform $\tilde{V}_{e,h}$ of the scattering potentials are much sharper than g (the correlation radius R_c is anomalously large), the latter can be taken outside the integral at $k_1 = k$. Therefore, the Green's function $g(k,k')$ turns out to be equal to the unperturbed one but with the energy shifted to the left by the value of the matrix elements of $V_e + V_h$,

$$g(k,k') = (2\pi)^2 \delta(k - k')g(k) , \qquad g^{-1}(k) = g_0^{-1}(k) - \left[U_e \mid \Psi_1^e(0) \mid^2 + U_h \mid \Psi_1^h(0) \mid^2 \right]\xi . \qquad (3.1)$$

According to the surface origin of the scattering potentials (2.3), their matrix elements [the second term in Eq. (3.1)], contain the ground-state confinement eigenfunctions $\Psi_1^{e,h}(z_{e,h})$ for the electron (e) and the hole (h) taken at the unperturbed well interface $z_{e,h} = 0$. It is noteworthy that for infinitely deep QW when $U_{e,h} = \infty$ and $\Psi^{e,h}(0) = 0$, the product $U_{e,h} \mid \Psi_1^{e,h}(0) \mid^2 = \pi^2 \hbar^2 / m_{ez,hz} d^3$ has finite value.

From Eq. (3.1) one can see that averaging of $g(k)$ reduces to its integration with the probability density $P_\sigma(\xi)$ of the random process ξ. After corresponding change of the integration variable the averaged Green's function $\overline{g}(k)$ is expressed by the convolution of the unperturbed Green's coefficient $g_0(k)$ with the function $P_{V_N}(\omega) = (\sigma / v_N)P_\sigma(\omega\sigma / v_N)$,

$$\hbar\overline{g}(k) = \hbar \int_{-\infty}^{\infty} d\xi P_\sigma(\xi)g(k) = \int_{-\infty}^{\infty} \frac{P_{V_N}(\omega - \omega')d\omega'}{\omega' - \omega_0 - (\hbar k^2 / 2M) + i v_0} . \qquad (3.2)$$

It is clear that function $P_{V_N}(\omega)$ plays the role of the probability density of the potential barrier fluctuations (in frequency units) and the frequency v_N is r.m.s. value of these fluctuations,

$$v_N = \frac{\sigma}{\hbar} \left[U_e \mid \Psi_1^e(0) \mid^2 + U_h \mid \Psi_1^h(0) \mid^2 \right]. \qquad (3.3)$$

This quantity controls the effect of finiteness of potential barriers and of transverse electron-hole quantization. The result (3.2) generalizes to the case of arbitrary distribution function $P_{v_N}(\omega)$, that is obtained for the Gaussian fluctuations with $P_{v_N}(\omega) = (2\pi v_N^2)^{-1/2} \exp(-\omega^2 / 2v_N^2)$ (see, e.g., Refs. [4-6]). Besides with the aid of Eqs. (2.3) and (3.3), we have explicitly associated the characteristics of the energetic and interface disorder.

As is known, the imaginary part of the average Green's function $\overline{g}(k)$ determines the optical absorptivity of QW. According to Eq. (3.2), for small enough values of the exciton-bulk damping $v_0 \ll v_N$ and normal incidence of light ($k = 0$), $\mathrm{Im}\,\overline{g}(k = 0)$ follows the distribution function $P_{v_N}(\omega - \omega_0)$. Therefore, the quantity v_N can be regarded as the exciton-surface scattering frequency for the classical limit. Note, in this case the resonance shift vanishes. Finally, as can be easily clear up from the method deriving Eq. (3.2), it is applicable within the region $v_N \gg \omega_W = \hbar R_c^{-2} / 2M$.

4   Sharp resonance

If the above mentioned inequality does not hold, i.e. $v_N < \omega_W$, one can average the exact equation (2.4) by making use of the perturbative diagrammatic method [11] or the technique proposed in Ref. [14]. As a result, within the self-consistent Born approximation we obtain

$$\hbar\overline{g}(k,k') = (2\pi)^2 \delta(k - k')\Big[\omega - \omega_0 - (\hbar k^2 / 2M) + iv_0 - \Sigma\Big]^{-1}. \qquad (4.1)$$

The quantity Σ enters into this equation as the self-energy. Its real and imaginary parts give, respectively, the surface-induced shift $\delta\omega_0 = \mathrm{Re}\,\Sigma$ and broadening $v = -\,\mathrm{Im}\,\Sigma$ of the excitonic resonance. Because of the self-consistency, Σ is governed by the equation

$$\Sigma = \frac{M}{\hbar} \int_0^\infty \frac{d\omega_t}{2\pi} \frac{Q(\omega_t)\,\widetilde{W}(\sqrt{2M\omega_t / \hbar})}{\omega - \omega_0 - \omega_t + iv_0 - \Sigma}. \qquad (4.2)$$

Here the Fourier transform $\widetilde{W}(k)$ of the roughness correlator $W(|r|)$ as a function of ω_t, has the scale of decrease ω_W and the characteristic value $\widetilde{W}(0) \sim R_c^2$. The function $Q(\omega_t)$ is defined by

$$Q(\omega_t) = \Big\{v_N^{(e)}\varphi_e(\omega_t) + v_N^{(h)}\varphi_h(\omega_t)\Big\}^2, \qquad \varphi_{e,h}(\omega_t) = \Big[1 + \omega_t / (\hbar a_{e,h}^{-2} / 2M)\Big]^{-3/2} \qquad (4.3)$$

with $v_N^{(e,h)}$ being the first (e) or second (h) terms from Eq. (3.3) and $a_{e,h} = \hbar^2 \varepsilon / 4e^2 m_{e\|,h\|}$.

The analysis of Eq. (4.2) shows that in the most relevant situation when $v \gg v_0$, the resonance lineshape in $\delta\omega_0$ and v is determined by the relation between the characteristic frequencies ω_W and v_N. If $v_N^2 \ll \omega_W^2$, the excitonic resonance is sharp and asymmetric. Here, at the resonance point $\omega = \omega_0 + \delta\omega_0$, the surface scattering frequency v is of the order of v_N^2 / ω_W and the resonance shift is negative, $\delta\omega_0 \sim -v \ln(\omega_W / v)$. As the frequencies ω_W and v_N approach each other the excitonic resonance in v is enhanced and the relaxation frequency tends to its maximal value equal to v_N. At the same time, the resonance shift $\delta\omega_0$ vanishes. Finally, when ω_W becomes negligible in comparison with v_N we arrive at the classical limit.

From the analysis performed above, we come to a very interesting and non-trivial conclusion: The type of the excitonic resonance and its line-shape are determined by the competition between the correlation properties of surface disorder and finiteness of potential barriers of QW. The increase of the correlation radius R_c (the decrease of ω_W) leads towards the broad resonance. Otherwise, the less height $U_{e,h}$ of potential barriers (the smaller v_N), the sharper and more asymmetric the resonance. It should be emphasized also that the described features of the excitonic resonance are specific predictions of the self-consistent approach, which takes into account the inherent action of exciton-surface scattering on itself.

It is noteworthy that Eqs. (4.1), (4.2), being applied to the region of the broad resonance $v_N \gg \omega_W$, lead to the result $\hbar \operatorname{Im} \overline{g}(\boldsymbol{k} = 0) = -v_N^{-1} \sqrt{1 - (\omega - \omega_0)^2/4v_N^2}$. It is in qualitative agreement with that obtained from Eq. (3.2). The only (but substantial) distinction is that the line-shape is ellipsoidal instead of coinciding with the line-shape of the roughness distribution function. So, if one is interested just in qualitative (rather than exact) description of the excitonic resonance, the self-consistent expressions (4.1), (4.2) can be used independently of the ratio between v_N and ω_W.

5 Summary We have studied theoretically the surface-induced broadening v and shift $\delta\omega_0$ of ground-state exciton in quasi-2D quantum wells with finite barriers and anisotropic effective masses of the electron and hole. Our theoretical approach is based on the Green's function method and describes both broad (classical) and sharp exciton resonances. In the latter case, we used the self-consistent Born approximation to understand the transition from one resonance to the other. We found out that such a transition is controlled by the competition of the correlation properties of surface disorder and the finite height of potential barriers. Analytical expressions, showing explicitly the dependencies of v and $\delta\omega_0$ on the parameters of the quantum well, its rough interfaces and exciton-resonance detuning, are exhibited.

Acknowledgements This work was partially supported by the Consejo Nacional de Ciencia y Tecnología (CO-NACYT, México) under Grant No. 36047-E and by VIEP-BUAP (México) under Grant II-104G02.

References

[1] R. Zimmermann and E. Runge, J. Lumin. **60&61**, 320 (1994).
[2] R. Zimmermann, F. Große, and E. Runge, Pure & Appl. Chem. **69**, 1179 (1997).
[3] E. Runge and R. Zimmermann, phys. stat. sol. (b) **206**, 167 (1998).
[4] M. Wilkinson, F. Yang, E. J. Austin, and K. P. O'Donnel, J. Phys.: Condens. Matter **4**, 8863 (1992).
[5] F. Yang, M. Wilkinson, E. J. Austin, and K. P. O'Donnel, Phys. Rev. Lett. **70**, 323 (1993).
[6] S. Glutsch and F. Bechstedt, Phys. Rev. B **50**, 7733 (1994).
[7] T. Stroucken, et al., Phys. Rev. Lett. **74**, 2391 (1995).
[8] T. Stroucken, et al., phys. stat. sol. (b) **188**, 539 (1995).
[9] A. Atenco-Analco, N. M. Makarov, and F. Pérez-Rodríguez, Solid State Commun. **119**, 163 (2001).
[10] A. Atenco-Analco, N. M. Makarov, and F. Pérez-Rodríguez, Microelectron. J. **33**, 375 (2002).
[11] F. G. Bass and I. M. Fuks, Wave Scattering from Statistically Rough Surfaces (Pergamon, New York, 1979).
[12] R. F. Kopf, E. F. Schubert, T. D. Harris, and R. S. Becker, Appl. Phys. Lett. **58**, 631 (1991).
[13] R. D. Mattuck, A guide to Feynman diagrams in the many-body problem (McGraw-Hill, New York, 1976).
[14] A. R. McGurn and A. A. Maradudin, Phys. Rev. B **30**, 3136 (1984).

phys. stat. sol. (c) **0**, No. 8, 2926–2930 (2003) / **DOI** 10.1002/pssc.200303856

Fine structure of reflectance spectra due to exciton quantization in near-surface quantum wells based on ZnS$_x$Se$_{1-x}$ ternary alloys

P. I. Kuznetsov[1], **J. Madrigal-Melchor**[2], **F. Pérez-Rodríguez**[*, 3], **S. O. Romanovsky**[4], **A. V. Sel'kin**[3, 4], and **G. G. Yakushcheva**[1]

[1] Institute of Radioengineering and Electronics, Russian Academy of Sciences, Vvedensky 1, Fryazino, Moscow Region, 141120, Russia

[2] Escuela de Física, Universidad Autónoma de Zacatecas, Apdo. Post. C-580, Zacatecas, Zac. 98060, México

[3] Instituto de Física, Universidad Autónoma de Puebla, Apdo. Post. J-48, Puebla, Pue. 72570, México

[4] A.F. Ioffe Physico-Technical Institute, Russian Academy of Sciences, St. Petersburg, Politekhnicheskaya 26, 194021, Russia

Received 30 May 2003, revised 30 August 2003, accepted 1 September 2003
Published online 10 November 2003

PACS 71.35.-y, 78.40.Fy, 78.66.Hf

ZnS$_x$Se$_{1-x}$ alloy films with and without near-surface quantum wells were deposited on the GaAs substrates in the photo-induced MOCVD growth process. The reflection spectra of the structures have been studied experimentally and theoretically. The spectra from the film containing a photo-induced near-surface quantum well show additional reflectance peaks as compared to the quantum well free film. The number and position of the additional peaks are governed by the lateral local composition x. A theoretical model explaining the principal exciton-mediated features of the experimental reflectivity spectra is developed. The resonant spectral structure of the reflectivity is shown to be due to the center-of mass quantization of both light- and heavy-hole excitons inside the quantum well and due to free excitons in the thick (barrier) part of the ZnS$_x$Se$_{1-x}$ film.

1 Introduction   During the last two decades, the optical properties of semiconductors with near-surface localized excitons have been intensively investigated (see, for example, [1-11] and references therein). Extrinsic near-surface potential wells, where the exciton center-of-mass motion is quantized, can be created by means of surface treatments such as electron and ion bombardment, heating, doping, illumination, and the application of an electric field. Also, sufficiently wide and deep extrinsic potential wells, having a large number of excitonic bound states, can be formed in semiconductor solid solutions by varying their composition in the near-surface region [1,6,7]. Near-surface localized excitons manifest themselves in optical spectra (specular [1-9,11] and diffuse [10] reflection, transmission [4]) as resonances at frequencies very close to their corresponding eigenvalues. For this reason, the excitonic optical spectra turn out to be very useful in characterizing both optical and structural properties of the near-surface region of solids.

In the present paper we investigate experimentally and theoretically the exciton quantization in near-surface potential wells of the ZnS$_x$Se$_{1-x}$ based planar structures. It is noteworthy that the ZnS$_x$Se$_{1-x}$ alloys are of much interest now because of their potential applications in optoelectronic devices in blue-near-ultraviolet spectral range. The behaviour of excitons in ZnS$_x$Se$_{1-x}$ quantum wells is especially important for understanding fundamental optical properties of the materials suitable for the design of light emitters, heterostructure lasers as well as other waveguiding devices utilizing these materials.

---

[*] Corresponding author: e-mail: fperez@sirio.ifuap.buap.mx, Phone: +52 222 229 5500 ext. 2003, Fax: +52 222 2295611

We present here experimental reflectance spectra for ZnS_xSe_{1-x} ternary solution films deposited on the GaAs substrates by the photo-induced MOCVD techniques. The films were fabricated with and without near-surface quantum wells (section 2). Furthermore, we develop a model (section 3) to explain the principal experimental features observed. The model describes the behaviour of the multi-mode polaritons associated with light- and heavy-hole free excitons, confined inside the photo-induced near-surface quantum well and extended in the thick ZnS_xSe_{1-x} film deposited on the GaAs substrate. In section 4, we compare the theoretical reflectivity spectra with the experimental ones.

2 Experiment In this work we used ZnS_xSe_{1-x} heterostructures grown by MOCVD on (100) GaAs substrates. The composition x was varied step-wise along the growth direction by photo-induced growth processes. Two different growing zones were formed on the same substrate. During all the growth time, one of the zones was permanently illuminated and the other one was constantly shaded. Depending on the growth conditions, the typical values of sulphur content for illuminated and dark regions were $x = 5 - 7$ % and $x = 11 - 13$ %, respectively.

In order to characterize optically the fabricated structures, their low temperature ($T = 77$ K) excitonic reflectance spectra were measured in polarized light at oblique incidence and at different points of the sample surface (Figs. 1 and 2). The usual Fabry-Perot interference patterns appear outside the exciton-resonance region. Inside this region, a very complicated fine structure of reflectance, depending on the light polarization and the angle of incidence, was observed. In contrast to the spectra taken from dark regions (Fig. 1), the spectra from illuminated ones (Fig. 2), i.e. the regions with a near-surface quantum well, show additional reflectance peaks, the number and position of which is governed by the lateral local composition x.

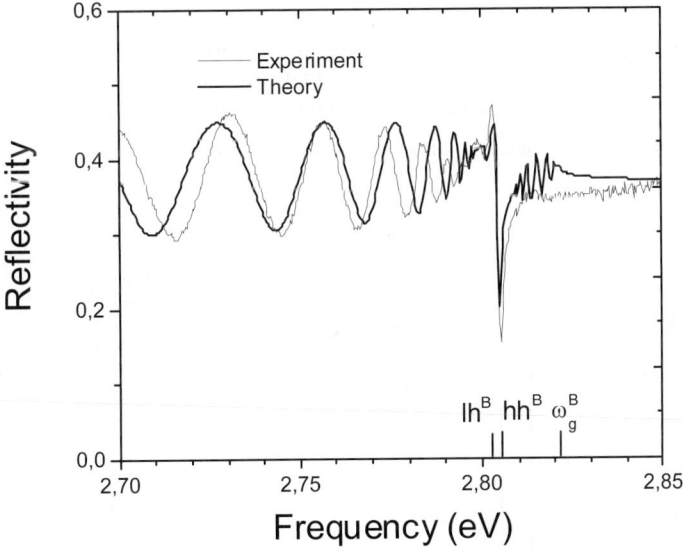

Fig. 1 Experimental reflection spectrum of ZnS_xSe_{1-x} on GaAs substrate for s-polarized light at the incidence $\theta = 45^0$, obtained from a sample region that was not illuminated during the growth process. The theoretical curve was calculated for a film-substrate system (without a near-surface potential well).

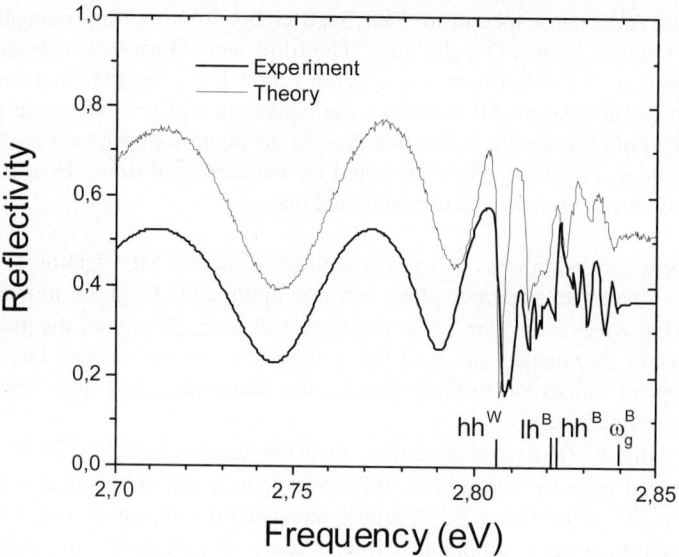

Fig. 2   Experimental reflection spectrum of ZnS_xSe_{1-x} on GaAs substrate for s-polarized light at the incidence $\theta = 45^0$, obtained from a sample region that was illuminated at the final stage of the growth process. The theoretical curve was calculated for a system formed by a near-surface potential-well, thick film, and substrate.

3   Theoretical model

To explain the reflection spectra for the ZnS_xSe_{1-x} heterostructures grown on GaAs substrate, we consider a multi-layer system schematically shown in Fig.3. There are a dead (exciton-free) layer with thickness L_D and dielectric constant ε_D, a near-surface potential well of thickness L_W, a thick film (barrier) of thickness L_B, and a semi-infinite exciton-free substrate with dielectric constant ε_S.

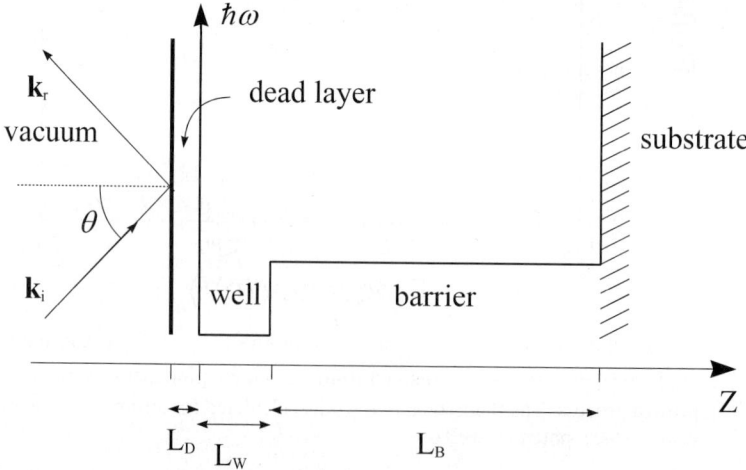

Fig. 3   Schematic diagram of the multi-layer system under study. k_i and k_r are the wave vectors of incident and reflected light, respectively, θ being the angle of incidence.

The non-local dielectric functions for the potential well and the thick film are given by Eq. (1) corresponding to two-oscillator model (heavy-hole (*hh*) and light-hole (*lh*) excitons):

$$\varepsilon(\boldsymbol{k},\omega) = \varepsilon_0(\omega) + \frac{\omega_{P,hh}^2}{\omega_{T,hh}^2 + D_{hh}k^2 - \omega^2 - i\omega\Gamma_{hh}} + \frac{\omega_{P,lh}^2}{\omega_{T,lh}^2 + D_{lh}k^2 - \omega^2 - i\omega\Gamma_{lh}}, \tag{1}$$

where $\boldsymbol{k}$ and ω are, respectively, the wavevector and frequency of the electromagnetic field, $\omega_{T,hh}$ ($\omega_{T,lh}$) is the *hh*-(*lh*-)exciton resonance frequency, $\omega_{P,hh}$ and $\omega_{P,lh}$ are measures of the oscillator strengths for the excitonic optical transitions. The coefficients D_{hh} and D_{lh} are parameters of the non-locality (spatial dispersion) given by the formulas:

$$D_{hh} = \hbar\omega_{T,hh} / M_{hh}, \qquad D_{lh} = \hbar\omega_{T,lh} / M_{lh}, \tag{2}$$

where M_{hh} and M_{lh} are the exciton translational masses. The frequency-dependent background contribution, $\varepsilon_0(\omega)$, to the dielectric function $\varepsilon(\boldsymbol{k},\omega)$ is modeled here as

$$\varepsilon_0(\omega) = \varepsilon_{00} + \frac{\hbar\omega_P^2}{4R_y\omega_g} \ln\left(\frac{\omega_g + \omega + i\Gamma_g/2}{\omega_g - \omega - i\Gamma_g/2}\right), \tag{3}$$

where ε_{00} is the term independent of frequency, ω_P^2 denotes the sum of the oscillator strengths ($\omega_P^2 = \omega_{P,hh}^2 + \omega_{P,lh}^2$), $\hbar\omega_g$ is the gap energy, R_y is the exciton Rydberg in the bulk. Γ_{hh}, Γ_{lh} and Γ_g in Eqs. (1) and (3) are the damping parameters . The formula (3) describes the step-like change of the imaginary part of $\varepsilon_0(\omega)$ at the gap frequency ($\omega = \omega_g$) due to the electron-hole continuum (open orbit excitons [12]). This formula is derived with an account of the fact that the density of states is continuous function of frequency at its gap value.

The reflectivity spectra for the model system are calculated by applying Maxwell boundary conditions for the electromagnetic fields (i.e. continuity of the tangential components of the electric and magnetic fields) and additional boundary conditions for the *hh*- and *lh*-exciton polarizations. The last means that i) both polarizations vanish at the inner boundary of the dead layer and at the film-substrate interface, ii) the *hh*- and *lh*-exciton polarizations and their derivatives with respect to z-coordinate (see Fig.3) are continuous at any other inner interface of the heterostructure.

4 Numerical results and comparison with experiment

The theoretical reflectivity spectra in Figs 1 and 2 were calculated for a model system without ("dark" region, $L_W = 0$ nm) and with ("light" region, $L_W = 40$ nm) a near-surface potential well. The parameter values are given in Tables 1-3.

Table 1 Common parameters (m_0 .is the free electron mass)

R_y, meV	M_{hh} m_0	M_{lh} m_0	$\hbar\omega_{P,hh}$, meV	$\hbar\omega_{P,lh}$, meV	$\hbar\Gamma_{hh}$, meV	$\hbar\Gamma_{lh}$, meV	$\hbar\Gamma_g$, meV	ε_S
17.0	2.2	0.33	202.1	116.7	0.75	0.75	1.0	22+i10

Table 2 Barrier parameter values of "dark" (non-illuminated) and "light" (illuminated) sample regions.

Sample region	ε_{00}	L_B, μm	$\hbar\omega_{T,hh}^B$, eV	$\hbar\omega_{T,lh}^B$, eV	$\hbar\omega_g^B$, eV
"dark"	8.1+i0.06	3.510	2.8040	2.8015	2.8121
"light"	8.1+i0.03	2.009	2.8224	2.8199	2.8394

Table 3   Near-surface dead layer and potential well parameter values.

ε_D	$L_D,$ nm	ε_{00}	$L_W,$ nm	$\hbar\omega_{T,hh}^{W},$ eV	$\hbar\omega_{T,lh}^{W},$ eV	$\hbar\omega_{g}^{W},$ eV
8.1	6.0	8.1+i0.03	40.0	2.8054	2.8054	2.8224

For the near-surface potential well we chose the same value for both *hh*- and *lh*-exciton resonance frequencies since the film (barrier) between the quantum well and the GaAs substrate is thick enough to neglect the strain-induced splitting of the heavy and light excitons.

As is seen from Fig. 1, the theoretical spectrum reproduces the main features of the experimental reflection spectrum for the non-illuminated ("dark") sample region. The Fabry-Perot oscillations are well identified at the low frequency region (below the exciton resonance region). On the other side of the exciton resonance, such oscillations are considerably damped, disappearing above the gap frequency $\omega = \omega_g$. The line-shape of the reflectivity shows clearly the resonances of the *hh*- and *lh*-excitons. The resonance of the *hh*-exciton characterized by larger oscillator strength is more prominent.

A good agreement between the theoretical reflectivity spectrum of Fig. 2 and the experimental one for the illuminated ("light") sample region allows one to explain the complicated fine structure observed. Outside the well-exciton resonance region (below $\omega_{T,hh}^{W}$ and above $\omega_{T,hh}^{B}$) the spectra exhibit Fabry-Perot resonances associated with the exciton propagation within the thick film. In the range $\omega_{T,hh}^{W} < \omega < \omega_{T,hh}^{B}$, a rich resonant structure appears due to the quantization of the translational motion of both *hh*- and *lh*-excitons inside the well.

5 Summary   We have experimentally and theoretically investigated the reflection spectra of the ZnS_xSe_{1-x} ternary solution films deposited on the GaAs substrates in the photo-induced MOCVD process. Two kinds of the heterostructures, with and without near-surface excitonic quantum wells, were studied. We have developed here a theoretical model which explains the principal features of the experimental spectra. It is shown that the resonant structure of the reflectivity spectra is due to quantization of both light- and heavy-hole free excitons inside the photo-induced near-surface quantum well and their confinement in the thick ZnS_xSe_{1-x} film. It was found that the depth of the near-surface quantum well changes with the lateral variation in the composition x and determines the frequency range where the resonant fine structure of reflectivity spectra is observed. The approach applied here to analyzing the reflectance in the excitonic spectral region seems to be a powerful tool for characterization of many-layer planar semiconductor heterostructures.

Acknowledgements   Work partially supported by CONACYT (Grant 36047-E) and RFBR (Grant 02-02-17601).

References

[1] A. E. Cherednichenko and V. A. Kisilev, Prog. Surf. Sci. 36, 179 (1991).
[2] P. Halevi, in Spatial Dispersion in Solids and Plasmas, edited by P. Halevi, Electromagnetic Waves – Recent Developments in Research Vol. 1 (Elsevier, Amsterdam, 1992), chap. 6.
[3] F. Pérez-Rodríguez and P. Halevi, Phys. Rev. B **45**, 11854 (1992); *ibid* **48**, 2016 (1993); *ibid* **53**, 10086 (1996).
[4] B. Flores-Desirena, F. Pérez-Rodríguez, and P. Halevi, Phys. Rev. B **50**, 5404 (1994).
[5] A. S. Batyrev, N. V. Karasenko, and A. V. Sel'kin, Phys. Solid State **35**, 1525 (1993).
[6] A. S. Batyrev, B. V. Novikov, and A. V. Sel'kin, JETP Lett. **61**, 809 (1995).
[7] A. S. Batyrev, B. V. Novikov, A. V. Sel'kin, and L. N. Tenishev, JETP Lett. **62**, 408 (1995).
[8] J. Madrigal-Melchor, et al., Appl. Phys. Lett. **71**, 69 (1997).
[9] H. Azucena-Coyotécatl, et al., Thin Solid Films **373**, 227 (2000).
[10] J. Madrigal-Melchor, et al., Phys. Rev. B **61**, 15993 (2000); Phys. Solid State **40**, 796 (1998).
[11] B. Flores-Desirena, A. Silva-Castillo, and F. Pérez-Rodríguez, J. Appl. Phys. **93**, 3308 (2003).
[12] A. Stahl and I. Balslev, Electrodynamics of the semiconductor band edge (Springer-Verlag, Berlin, 1987).

phys. stat. sol. (c) **0**, No. 8, 2931–2937 (2003) / **DOI** 10.1002/pssc.200303828

High-resolution measurements
of the bulk dielectric constants of single crystal gold
with application to reflection anisotropy spectroscopy

N. P. Blanchard[1], **C. Smith**[1], **D. S. Martin**[1], **D. J. Hayton**[2], **T. E. Jenkins**[2],
and **P. Weightman**[*, 1]

[1] Department of Physics, University of Liverpool, Oxford Street, Liverpool L69 7ZE, England, UK
[2] Department of Physics, University of Aberystwyth, Ceredigion SY23 3BZ, Wales, UK

Received 30 May 2003, revised 4 August 2003, accepted 11 August 2003
Published online 10 November 2003

PACS 07.60.Fs, 78.20.Ci, 78.40.Kc

We report measurements of the optical constants of single crystal Au(110) in the range of 1.5 to 5.0 eV using Spectroscopic Ellipsometry. The results are compared with previously published data. In addition we have used the data to carry out a fit to RAS data obtained from the same crystal.

1 Introduction

The growing interest [1, 2, 3] in optical studies of single crystal metal surfaces together with the development of empirical methods of interpreting experimental results [4, 5] has lead to renewed interest in the bulk dielectric function of metals. The bulk dielectric constant can be calculated from the refraction coefficient, n, and the extinction coefficient, k, using the relation $\varepsilon = N^2$, where N, the refractive index is defined as $N = n - ik$. The n and k optical constants can be obtained empirically as a function of photon energy using the technique of spectroscopic ellipsometry (SE) [6, 7].

SE is a non-destructive optical technique with a long history of application, in which polarised light is reflected from the surface of a material and the change in the polarization of the light on reflection is measured. Unlike the reflectometry and optical absorption techniques, SE measures both the amplitude and phase change of the reflected light and this makes it possible to determine the complex dielectric constant, or, equivalently the complex refractive index, over the range of optical wavelengths employed without the need to employ a Kramers-Krönig analysis. Typically, for bulk materials, ellipsometry can be used to determine the bulk refractive index and optical absorption coefficient of the material.

Numerous studies of the optical properties of the noble metals have been reported [8, 9, 10, 11, 12 13] and Palik has reviewed a large body of work on the optical constants of Au(110). While Paliks' [14] data sets have proved very useful in interpreting results obtained in reflection anisotropy spectroscopy (RAS) studies of Au [15, 16], they lack resolution in the 1.5 to 5.0 eV range of importance to RAS. Another concern is that the data sets reviewed by Palik [14] were obtained by measurements in air on thin films evaporated on to glass substrates whereas RA spectra are obtained for clean single crystal surfaces, often in UHV or in electrochemical cells.

Stahrenberg et al [17] have recognised the need for improved results for the bulk dielectric constants of metals in order to facilitate the interpretation of RAS measurements and they recently published spectra of the optical constants of Cu and Ag determined from SE measurements on (110) faces of single

* Corresponding author: e-mail: peterw@liv.ac.uk, Phone: 44 (0)151 794 3871, Fax: 44 (0)151 794 3348

crystals in the energy range 2.5 to 9.0 eV. Both air exposure and surface roughness are expected to reduce the measured value of the imaginary part of the dielectric function, especially in the visible region where reflectivity is high and skin depth is small. Subsequently they carried out measurements in UHV and were able to monitor sample cleanliness using RAS and surface roughness using atomic force microscopy.

In this work we determine the optical constants of Au from SE measurements on the (110) crystal face in the energy range 1.41eV to 4.96eV. We use these values to simulate the RA spectra of a clean Au(110) surface in terms of a Fresnel model [15] and compare the results with measurements of the RA spectrum of the clean (110) surface of the same crystal in UHV.

It is not in the remit of this paper to give a detailed description of the origin of the features in the dielectric function of Au as this has been done in numerous publications [8, 9, 10, 11, 12, 13].

2 Experimental

The Au(110) single crystals where mechanically polished and aligned to ~1° accuracy using Laue x-ray diffraction. Prior to carrying out the SE the Au crystal was flame annealed in air, then rapidly quenched in ultra pure water. The ellipsometry measurements were carried out in ambient atmosphere immediately following the evaporation of the water covering the Au(110) surface using a rotating polariser spectroscopic ellipsometer (SOPRA model GESP5) over the wavelengths range 250 nm to 850 nm.

Following the SE measurements the Au(110) crystal was placed in a UHV environment and the clean surface was prepared by cycles of Ar ion bombardment (7 mA, 0.5 kV, 300 K) and subsequent annealing to 580K. Surface order was confirmed by a sharp (1×2) low-energy electron diffraction pattern. The RA spectrometer of the Aspnes design [18] was coupled to the UHV chamber through a low-strain window. RAS, measures as a function of photon energy, the anisotropy in the normal incidence Fresnel reflection coefficient for light linearly polarised in two orthogonal directions (Δr) normalised to the mean reflection (r):

$$\frac{\Delta r}{r} = \frac{2(r_{[001]} - r_{[1\bar{1}0]})}{r_{[1\bar{1}0]} + r_{[001]}} . \tag{1}$$

We measured the real part of the complex reflection anisotropy defined in Equation 1. The results are shown in figure 1. The narrow peak at ~2.5 eV is better resolved than in our previous study [15].

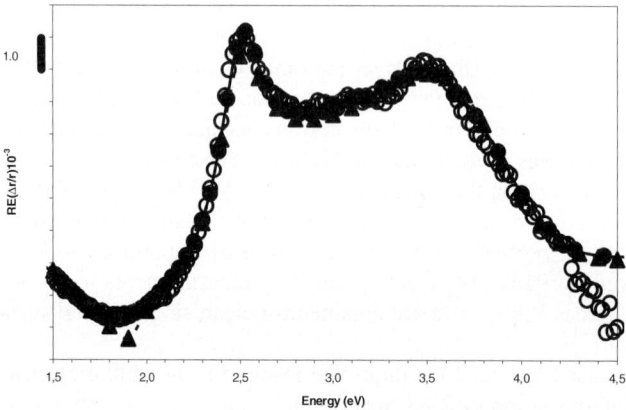

Fig. 1 Comparison between experimental and simulated RA spectra for a clean well ordered Au(110) surface. Open Circles: RA data, Closed Circles: Simulated spectra using data from this work, Closed Triangles: Simulated spectra using data from Palik [14].

3 Results

The results obtained from the SE measurement for n and k for the Au(110) crystals in the range of 1.41eV to 4.96eV are shown in table 1, this is data obtained from multiple angle of incidence readings [7]. These values have been used to calculate the real, ε', and complex, ε'', parts of the bulk dielectric function, $\varepsilon = \varepsilon' + i\varepsilon''$.

Table 1   Table of n & k measured for a Au(110) single crystal.

Energy (eV)	n	%Δn	k	%Δk	Energy (eV)	n	%Δn	k	%Δk
4.96	1.4531	0.6673	1.6538	0.4625	2.18	0.2995	3.0745	2.7776	0.3228
4.77	1.4926	0.0003	1.7618	0.0002	2.14	0.2837	4.1909	2.9003	0.4043
4.59	1.5442	0.8948	1.8425	0.6084	2.10	0.2630	0.0013	3.0024	0.0001
4.43	1.6239	0.7420	1.8715	0.5331	2.07	0.2575	4.4321	3.0756	0.3731
4.28	1.7084	0.6524	1.8845	0.4986	2.03	0.2367	5.7295	3.1772	0.4348
4.13	1.7586	0.7494	1.8912	0.5984	2.00	0.2237	0.0017	3.2817	0.0001
4.00	1.7888	0.6699	1.8880	0.5525	1.97	0.2136	5.9620	3.3800	0.3918
3.88	1.8300	0.6451	1.8769	0.5563	1.94	0.1922	0.0023	3.4837	0.0001
3.76	1.8577	0.5754	1.8467	0.5191	1.91	0.1866	0.0024	3.5767	0.0001
3.65	1.8549	0.6319	1.8058	0.5840	1.88	0.1782	9.2687	3.6797	0.4817
3.54	1.7967	0.6168	1.7813	0.5508	1.85	0.1672	8.8602	3.7803	0.4250
3.44	1.7402	0.6060	1.8073	0.5063	1.82	0.1529	12.4200	3.8948	0.5351
3.35	1.7167	0.6042	1.8353	0.4859	1.80	0.1524	0.0032	3.9843	0.0001
3.26	1.7059	0.6385	1.8614	0.5011	1.77	0.1451	12.9398	4.0863	0.5128
3.18	1.6994	0.5535	1.8790	0.4266	1.75	0.1533	0.0037	4.1926	0.0002
3.10	1.6807	0.5972	1.8896	0.4511	1.72	0.1546	0.0043	4.2828	0.0002
3.02	1.6831	0.5281	1.8834	0.4007	1.70	0.1642	0.0031	4.3872	0.0001
2.95	1.6612	0.6993	1.8799	0.5206	1.68	0.1721	12.9642	4.4769	0.5713
2.88	1.6338	0.7360	1.8664	0.5400	1.65	0.1749	15.0479	4.5824	0.6640
2.82	1.5920	0.7458	1.8469	0.5350	1.63	0.1847	10.8299	4.6804	0.4976
2.76	1.5449	0.6705	1.8178	0.4680	1.61	0.1957	12.5300	4.7511	0.6032
2.70	1.4883	0.5745	1.7769	0.3886	1.59	0.2114	0.0038	4.8290	0.0002
2.64	1.3770	0.5349	1.7292	0.3354	1.57	0.1939	0.0052	4.8983	0.0002
2.58	1.2298	0.4736	1.6871	0.2646	1.55	0.2121	13.9555	5.0046	0.7051
2.53	1.0365	0.7824	1.6876	0.3652	1.53	0.2285	15.6596	5.0977	0.8416
2.48	0.8195	0.0003	1.7715	0.0001	1.51	0.2002	0.0046	5.1671	0.0002
2.43	0.6484	0.0005	1.9152	0.0001	1.49	0.2349	12.9367	5.2660	0.7022
2.38	0.5327	0.0007	2.0765	0.0002	1.48	0.2335	12.5838	5.3058	0.6767
2.34	0.4368	0.0007	2.2308	0.0001	1.46	0.2806	11.1112	5.4018	0.7061
2.30	0.3966	2.4718	2.3852	0.3793	1.44	0.2835	0.0042	5.4718	0.0003
2.25	0.3517	2.9442	2.5219	0.3874	1.43	0.3065	0.0037	5.5501	0.0003
2.21	0.3326	3.2679	2.6557	0.3915	1.41	0.2865	0.0034	5.6208	0.0002

Our results for the real and imaginary parts of the dielectric constants of Au are shown in figure 2. In general our results for the real part of the dielectric constant of Au are in very good agreement with previously published data [9, 10, 13, 14]. However except for the work of Rosei [11], who presents results for the imaginary part only and which is limited to the range 1.0 to 3.0 eV, we have obtained results at a finer mesh of energies. There is an important difference between our results for the real part of the dielectric function of Au and that derived from the tables in Palik's review [14] in the range 1.8 eV to 2.3 eV (see inset in figure 2a). We have not been able to establish the origin of this difference. The results obtained from tables in Palik's review [14] do not lie on a smooth curve and this creates problems of a statistical nature when applying the Fresnel model [15] to the analysis of high resolution RAS data. The need to obtain higher resolution data on the bulk dielectric constant of Au in this spectral range in order to facilitate the interpretation of RAS was one of the reasons for undertaking this study. Our results for the imaginary part of the bulk dielectric function of Au(110) are in good agreement to that of previously published work [9, 14]. While it would be useful to check the accuracy of our results for the bulk dielectric constant by SE measurements on other low index faces of single crystal Au we do not anticipate significant differences to arise from such experiments since our results are in general in good agreement with results obtained on polycrystalline Au.

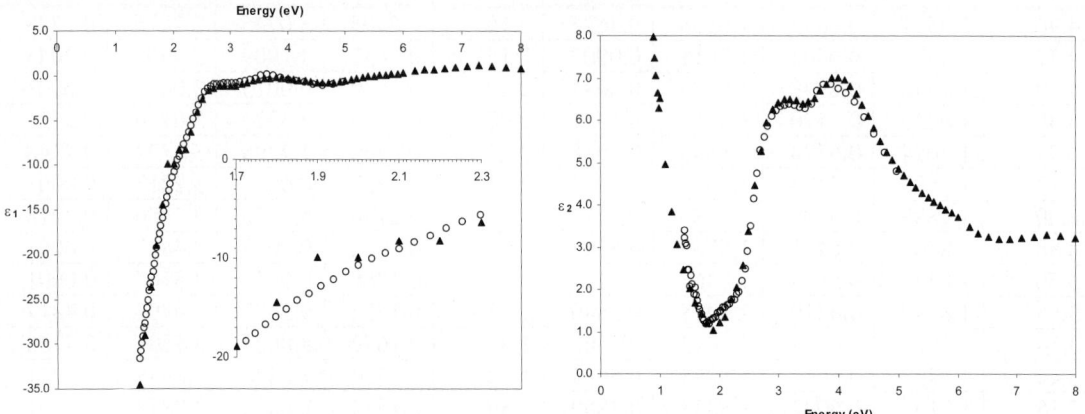

Fig. 2 (a) Real and (b) Imaginary parts of the dielectric function calculated from the measured n & k of single crystal Au(110). Triangles: Data From Palik [14], Circles: This work.

4 Discussion

RA spectra can be simulated in terms of a three-phase model [5], consisting of vacuum, $\varepsilon_v(\omega) = 1$, a biaxial anisotropic surface layer, $\varepsilon_x(\omega) \neq \varepsilon_y(\omega)$, and an isotropic substrate, $\varepsilon_b(\omega)$. Fresnel analysis [19] shows that the reflectance amplitudes of this system are related to the surface, $\varepsilon_s(\omega)$, and bulk dielectric, $\varepsilon_b(\omega)$, functions by:

$$\frac{\Delta r}{r} = \frac{2i\omega d}{c} \left(\frac{\Delta \varepsilon_s(\omega)}{\varepsilon_b(\omega) - 1} \right) \tag{2}$$

where d is the surface layer thickness acting as a scaling factor, $\Delta \varepsilon_s(\omega) = \varepsilon_y(\omega) - \varepsilon_x(\omega)$ is the surface dielectric anisotropy (SDA), and $\hbar\omega$ is the photon energy. This approximation is valid for small anisotropy and within the thin film approximation $(d \ll \lambda)$. It is convenient to introduce the bulk dielectric function in terms of the functions $A(\omega)$ and $B(\omega)$ defined by:

$$A(\omega) - iB(\omega) = \frac{1}{1 - \varepsilon_b(\omega)}. \tag{3}$$

Figure 3 shows the $A(\omega)$ and $B(\omega)$ functions for Au calculated from the results obtained for the bulk dielectric function in our work and from the results in Palik's review [14]. Clearly our data has a finer energy resolution and consequently a smoother profile. The smooth profile obtained for the $A(\omega)$ function from our results is particularly useful in the interpretation of RAS results.

The real part of equation 2 can be written in terms of $A(\omega)$ and $B(\omega)$:

$$\mathrm{Re}\left(\frac{\Delta r}{r}\right) = -\frac{2\omega d}{c}\left[A(\omega)\,\Delta\varepsilon_s''(\omega) + B(\omega)\,\Delta\varepsilon_s'(\omega)\right] \tag{4}$$

where $\Delta\varepsilon_s'$ and $\Delta\varepsilon_s''$ indicate the real and imaginary parts of the complex SDA respectively. This expression generates simulated RA spectra in terms of the bulk optical properties of Au and a parameterised representation of surface electronic transitions. The SDA is simulated by selecting transitions within the surface layer for x and y directions of appropriate energy (ω_t), relative strength (S), and line width (Γ). Each transition has a Lorentzian form given by:

$$\varepsilon_{x,y} = 1 + \sum_{n=1}^{m} \frac{S_n/\pi}{\omega_{tn} - \omega + i\Gamma\big/2}. \tag{5}$$

In Figure 1 we compare the experimental RA spectra obtained for a clean, well-ordered Au sample in UHV with a simulated spectra calculated from the $A(\omega)$ and $B(\omega)$ functions deduced from our SE measurements. While in this work we are seeking to simulate the RA spectrum of Au(110) in terms of empirical parameters we note that an earlier study [15] provides a good interpretation of the origin of the transitions necessary to simulate the Au(110) RA spectrum in terms of the electronic structure of the Au(110)-(1×2) surface. We also show the results of a similar simulation deduced from the $A(\omega)$ and $B(\omega)$ functions derived from the data tabulated by Palik [14]. The individual components of each fit are shown in figure 4 and the parameters used in the fits are shown in table 2. The simulation obtained using our results for the $A(\omega)$ and $B(\omega)$ functions produces a better fit to the experimental RAS data, especially in the range 1.5 to 2.5eV and in the region of the narrow peak at ~2.5eV. Figure 4 shows that the coarser data obtained for $A(\omega)$ and $B(\omega)$ from Paliks [14] tabulation makes it particularly difficult to model the contribution to the simulation from the component at ~1.7eV (Table 2) which has a strong influence on the shape of the peak observed at ~2.5eV.

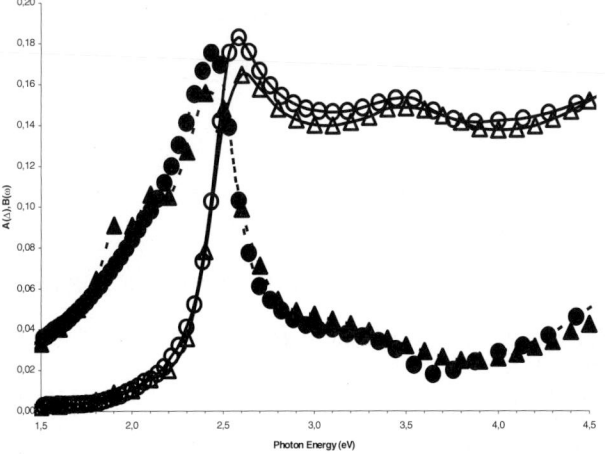

Fig. 3 A(ω) (*solid marker, dotted line*) and B(ω) (*hollow marker, solid line*) for Au calculated from this work (*circles*) and Palik [14] (*triangles*).

© 2003 WILEY-VCH Verlag GmbH & Co. KGaA, Weinheim

Table 2 Parameters of fits to clean Au(110).

Transition	1	2	3
Direction	$[001]$	$[1\bar{1}0]$	$[1\bar{1}0]$
ω_t (eV)	1.70	2.51	3.87
Γ (eV)	0.7	0.4	0.9
S	1110	140	500

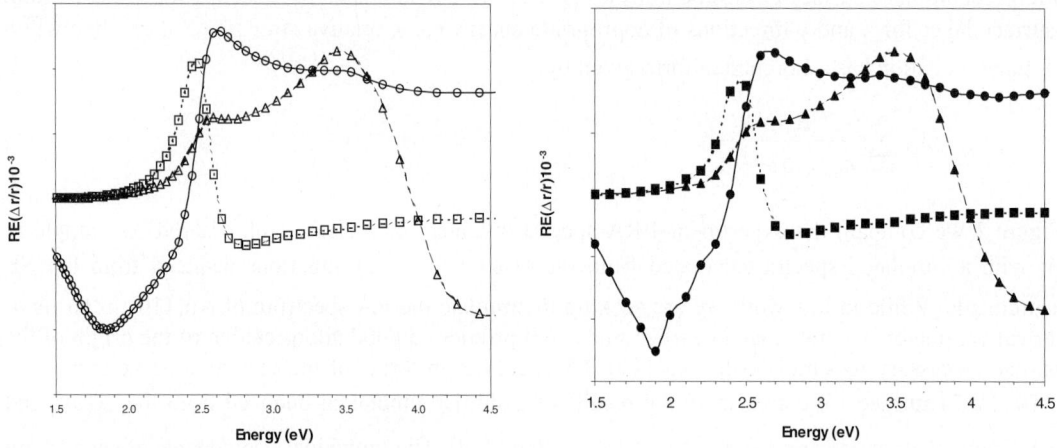

Fig. 4 Components for simulated RA spectra: - Transition 1 (Circles), Transition 2 (Squares), Transition 3 (Triangles); (a) Data from this work; (b) Data from Palik [14].

5　Conclusion

We have provided high-resolution results for the $A(\omega)$ and $B(\omega)$ functions used to represent the dielectric constants of bulk Au. We anticipate these results will be useful in the interpretation of the RA spectra of Au(110) [16] and of molecules absorbed on the Au(110) surface [20].

Acknowledgement　The authors thank the UK EPSRC for support of this work.

References

[1] D.S. Martin and P. Weightman, Surf. Rev. Lett. **7** (2000) 389-97
[2] D.S. Martin and P. Weightman, Surf. Interface Anal. **31** (2001) 915-26
[3] P. Weightman, phys. stat. sol. (a) **188** (2001) 1443-53
[4] B.G. Frederick, J.R. Power, R.J. Cole, C.C. Perry, Q. Chen, S.Haq, Th. Bertrams, N.V. Richardson, and P. Weightman, Phys. Rev. Lett. **80** (1998) 4490-3
[5] R.J. Cole, B.G. Frederick, and P. Weightman, J. Vac. Sci. Technol. A **16** (1998) 3088-95
[6] K. Vedam, Thin Solid Films **313** (1998) 1-9
[7] T E Jenkins, J. Phys. D: Appl. Phys. **32** (1999) R45-R56
[8] G.P. Pells and M. Shiga, J. Phys. C **2** (1969) 1835
[9] Marie-Luce Théye, Phys. Rev. B **2** (1970) 3060 - 3078
[10] P.B. Johnson and R.W. Christy, Phys. Rev. B **6** (1972) 4370-4379
[11] M. Guerrisi, R. Rosei, and P. Winsemius, Phys. Rev. B **12** (1975) 557

[12] P. Winsemius, H.P. Lengkeek, and F.F. Van Kamper, Physica (the Hague) **79B** (1975) 529
[13] Yu Wang, Liang-Yao Chen, Bo Xu, Wei-Ming Zheng, Rong_Jun Zhang, Dong-liang Qian, Shi-Ming Zhou, Yu_Xiang Zheng, Ning Dai, Yu-Mei Yang, Kou-Bao Ding, and Xui-Miao Zhang, Thin Solid Films **313** –**314** (1998) 232-236
[14] *Handbook of Optical Constants of Solids*, edited by E. D. Palik (Academic, New York, 1985 and 1991), Vol. 1.
[15] B. Sheridan, D. S. Martin, J. R. Power, S. D. Barrett, C. I. Smith, C. A. Lucas, R. J. Nichols, and P. Weightman, Phys. Rev. Lett. **85** (2000) 4618 – 4621
[16] D. S. Martin, N. P. Blanchard, and P. Weightman, Surf. Sci., *in press*
[17] K. Stahrenberg, Th. Herrmann, K. Wilmers, N. Esser, W. Richter, and M. J. G. Lee, Phys. Rev. B **64** (2001) 115111
[18] D.E. Aspnes, J.P. Harbison, A.A. Studna, and L.T. Florez, J. Vac. Sci. Technol. A **6** (1988) 1327
[19] J.D.E McIntyre and D.E. Aspnes, Surf. Sci. **24** (1971) 417
[20] C. I. Smith, A. J. Maunder, C. A. Lucas, R. J. Nichols, and P. Weightman, J. Electrochem. Soc. **150** (2003) E233

phys. stat. sol. (c) **0**, No. 8, 2938–2943 (2003) / **DOI** 10.1002/pssc.200303833

In-situ ellipsometry:
Identification of surface terminations during GaN growth

C. Cobet[*,1], **T. Schmidtling**[1], **M. Drago**[1], **N. Wollschläger**[1], **N. Esser**[1], **W. Richter**[1], and **R. M. Feenstra**[2]

[1] Institut für Festkörperphysik, Technische Universität Berlin, Hardenbergstr. 36, 10623 Berlin, Germany
[2] Dept. Physics, Carnegie Mellon University, Pittsburgh, PA 15213, USA

Received 30 May 2003, revised 4 August 2003, accepted 11 August 2003
Published online 10 November 2003

PACS 78.20.Ci, 78.55.Cr, 78.66.Fd, 81.15.Gh, 81.15.Hi

Spectroscopic ellipsometry (SE) is used to determine GaN surface termination during growth with metal-organic vapor phase epitaxy (MOVPE) by a correlation to well known results of plasma-assisted molecular beam epitaxy (PAMBE). The results manifest that in MOVPE under typical growth conditions the surface is not terminated by a Ga-bilayer as suggested for MBE. Moreover, it turns out that ellipsometry can be used to characterize the surface reconstruction in wurtzite GaN similar as reflectance anisotropy does for cubic III–V-compounds. The optical spectra for the PAMBE reveal clear differences between growth under Ga-rich and N-rich conditions, which are attributed to the presence of a Ga-bilayer and various N-rich reconstructions on the surface [1].

1 Introduction

The GaN growth by metal-organic vapor phase epitaxy (MOVPE) as well as by plasma-assisted molecular beam epitaxy (PAMBE) is actually widely used and investigated. For the latter case much is already known concerning the surface structures during growth by varying the surface stoichiometry from Ga-rich to N-rich conditions [1, 2]. Under Ga-rich conditions it has been established that slightly more than 2 ML (ML = monolayer = 1.14×10^{15} atoms/cm^2) of Ga resides on the surface in a metallic fluid-like state [2]. This Ga bilayer has been shown to have a dramatic influence on the kinetics of the growing surface [1]. Finally, it is known, that the presence of this Ga bilayer is necessary for a smooth layer by layer growth of GaN in PAMBE. However, the common device production of nitride semiconductors is based on MOVPE. In contrast to the PAMBE the surface structure and chemistry during GaN growth by MOVPE is still unknown. The associated high pressures preclude in-situ observation by electron based techniques and therefore disable a direct investigation of the surface structure with these standard methods. It would be therefore of considerable interest to determine with a optical method whether such a metallic bilayer found in PAMBE is also present during GaN growth by MOVPE.

Surface optical probes have been developed as a very sensitive tool in means of monitoring, and thereby controlling the growth of semiconductor films [3–8]. In particular, spectroscopic ellipsometry has proven its ability to measure sub-nm thicknesses of roughness or overlayers on surfaces [4, 5], while the widely applied reflectance anisotropy spectroscopy is specifically selective to surface signals

---

[*] Corresponding author: e-mail: chrcobet@physik.tu-berlin.de, Phone: +49 30 314 23262, Fax: +49 30 314 21769

in case of cubic materials [7]. Such a measurement is not applicable to the threefold-symmetric (0001) surface of GaN. Using SE, one is not limited to any special bulk or surface symmetry for optical characterisation. In PAMBE ellipsometry reveals clear differences between growth under Ga-rich and N-rich conditions, which can be attributed to the presence of the Ga-bilayer on the surface. Results for MOVPE surfaces obtained during growth are found to be practically the same as those obtained from NH_3 stabilized surfaces, which are very similar to the N-rich PAMBE results. We thus will show that, under normal MOVPE growth conditions, the GaN surface is *not* terminated by a Ga-bilayer.

2 Experiment

The MOVPE growth and surface preparation was performed in a standard horizontal rf heated reactor equipped with a rotating polarizer spectral ellipsometer [9] for in-situ measurements of the optical properties. At first a GaN nucleation layer was deposited on the (0001) sapphire substrate. Afterwards a GaN layer of about 1.5 µm thickness was grown at an substrate temperature of 1040 °C with tri-methylgallium (TMGa) and ammonia precursors at a V/III ratio of 2,000 [9]. GaN grows in MOVPE on sapphire in the (0001) direction with Ga terminated basal plans (Ga-face) [10]. On the base of these substrates we prepare different surface stoichiometries by a variation of the precursor amount in nitrogen carrier gas without deterioration of the surface. In case of an increased Ga supply this can be achieved through short TMGa pulses to avoid Ga droplet formation. For these experiments we choose a substrate temperatur of 900 °C to achieve similar energetic positions of the band critical points in the GaN dielectric function for easy comparison to the PAMBE results. Also surface deterioration due to hydrogen etching (from the dissociated ammonia) occurs more slowly than at usual growth tem-perature of 1040 °C while the general behavior of the surface to precursor switching is nearly the same.

The PAMBE preparation of GaN films where performed in an ultra high vacuum (UHV) chamber providing low energy electron diffraction (LEED) and Auger electron spectroscopy (AES). A spectral ellipsometer as described before was used for monitoring of optical properties during growth. The GaN films were deposited at 850 °C on the MOVPE grown GaN substrates. Thus their polarity was Ga-face, like on the substrate. Reactive nitrogen was supplied by an rf cracker cell using a flow of approx. 5 sccm/min molecular nitrogen at pressure in the range of 8×10^{-5} mbar during epitaxy. Ga was supplied via a standard Knudsen effusion cell. For a variation of the GaN surface stoichiometry the Ga cell temperature was varied between 910 °C (low Ga-flux) and 950 °C (high Ga-flux) which corresponds to a gallium flux ratio of 1:3 estimated from the gallium vapor pressure at 910 °C and 950 °C, respectively.

3 Results and discussion

Figure 1 shows ellipsometry transients acquired at 4.8 eV during PAMBE growth at approx. 850 °C sample temperature with various Ga and N supply. We plot the imaginary part of the effective dielec-tric function $\langle \varepsilon_2 \rangle$, which describes the dissipative part of the reflected light including overlayer effects like roughness and layer – substrate interferences. Changes in the surface properties generated by the variations in beam fluxes and growth temperature produce large changes of 5–40% in the value of $\langle \varepsilon_2 \rangle$. At a given temperature these changes are immediately reversible when the fluxes are set to their original states, indicating that they do not arise simply from roughening of the surface. This is also confirmed by the fact that an increase in $\langle \varepsilon_2 \rangle$ here occurs for higher gallium fluxes where AFM has proven a decrease in surface roughness. We attribute this difference purely to the difference in the surface properties since the changes occur in less than one second upon varying the growth param-eters.

During growth periods with a stable optical response we record spectra for different beam fluxes as shown in Fig. 2. Below the band gap energy (about 3 eV at 850 °C) all spectra display Fabry–Perot

Transient at 4.8 eV

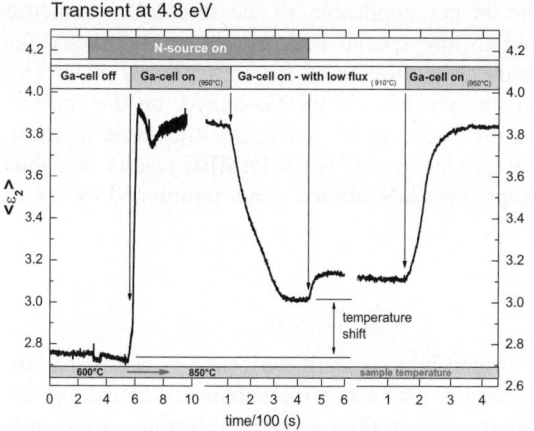

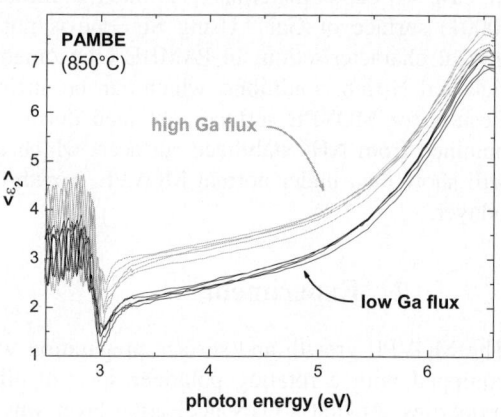

Fig. 1 Transient measurement of the imaginary part of the effective dielectric function, $\langle\varepsilon_2\rangle$, at 4.8 eV during variations in PAMBE growth parameters. First the N-source is turned on. Then the sample temperature is increased from 600 to 850 °C and simultaneously the Ga shutter is opened with the Ga cell being at 950 °C. The Ga-cell is then ramped down to 910 °C and then the N-course is turned off. Finally the Ga-cell is ramped back up to 950 °C.

Fig. 2 Imaginary part of the dielectric function $\langle\varepsilon_2\rangle$, for GaN(0001) films during PAMBE growth under Ga-rich and N-rich conditions.

resonances within the GaN film. Above the band gap these $\langle\varepsilon_2\rangle$ spectra are clearly arranged in two distinct bunches. Thereby, the spectra with the higher overall amplitude are correlated to higher Ga beam fluxes using a Ga effusion cell temperature of 950 °C. Low-energy electron diffraction (LEED) images of surfaces stabilized under these conditions reveal in all cases the presence of a "1 × 1" 1 + 1/6 LEED pattern with the six characteristic spots around the hexagonal bulk spots as known for Ga-bilayer on the surface [1]. In contrast, for N-rich conditions the $\langle\varepsilon_2\rangle$ spectra show a lower overall amplitude which refer to Ga effusion cell temperature of 910 °C. In the associated LEED images the characteristic Ga-bilayer structure is vanished. We obtain a $\sqrt{3} \times \sqrt{3}$-R30° LEED pattern as reported in prior studies from N-rich GaN surfaces [11]. However, the surface roughening after continuous growth under N-rich conditions disable the stabilization of smooth N-rich surfaces and LEED measurements were difficult in many cases. The surface morphologies after growth under Ga- and N-rich conditions, however, agree well with prior observations [1, 2, 11, 12]. The Ga-rich case shows smooth step-flow growth together with Ga droplets on a larger length scale while the N-rich case displays rough, 3-dimensional growth. Between the most Ga- and N-rich spectra we find distinct changes in the dielectric function connected to the various Ga effusion cell temperatures between 950 °C and 910 °C with continuous nitrogen supply. But a further increase of the Ga cell temperature above 950 °C or a further decrease below 910 °C leaves $\langle\varepsilon\rangle$ almost unaffected and finally ends up in irreversible changes due to the formation of Ga droplets or surface roughening, respectively.

Using optical simulations for the effective dielectric function of layered samples we find that the observed differences in $\langle\varepsilon_2\rangle$ and $\langle\varepsilon_1\rangle$ between the most Ga- and N-rich spectra can be explained with a thin metallic overlayer having a Drude like behavior of free carriers [13] (Fig. 3). Therefore we use optical spectra of the most N-rich surface as a substrate and add a metallic overlayer to fit the most Ga-rich spectra in the spectral range above the bandgap. In that calculation we obtain a layer thickness of about 0.4 nm, close to the known Ga bilayer thickness of 0.48 nm [1]. Further on, the calculated electron density of 1.2×10^{22} cm^{-1} ($m_{eff}/m_e = 1$) corresponds very well to the bulk electron density of metals like Ga. We believe that the remaining differences in the $\langle\varepsilon\rangle$ spectrum below the bandgap and in the vicinity of the E_1 transition around 6.5 eV may also include effects of interface

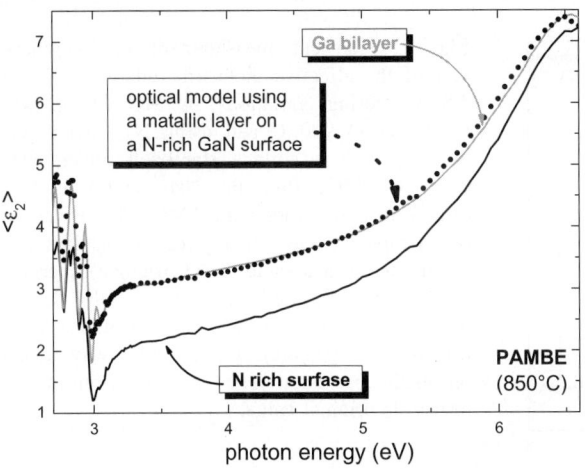

Fig. 3 Imaginary part of the dielectric function $\langle \varepsilon_2 \rangle$, for GaN(0001) films during PAMBE growth under most Ga- and N-rich conditions in comparison with a optical model calculation using the optical spectra of the most N-rich surface plus a metallic overlayer with a Drude like behavior of free carriers.

electronic states. But by assuming a metallic layer, significant differences remain only below the fundamental bandgap of GaN where a quantitative modelling is hardly possible. From the previous results we conclude that the Ga-bilayer corresponds to the thickest Ga layer which forms on the GaN surface and a further deposition of Ga just leads to the formation of Ga droplets.

Finally, the high sensitivity of ellipsometry to different GaN surface conditions found in the PAMBE experiments (Figs. 1 and 4) clearly demonstrates its usefulness as a real-time monitor of the GaN surface conditions during MOVPE growth. Figure 4b shows ellipsometry results for a MOVPE prepared GaN surface in comparison with the PAMBE results shown in Fig. 4a. The measurements were made at 900 °C close to regular MOVPE growth conditions (V/III ratio of 2,000) and during an interruption of that growth when the TMGa source is turned off. Apart from a slight differences due to sample variations, the respective results for $\langle \varepsilon_2 \rangle$ in Fig. 4b are nearly identical, indicating that during MOVPE growth the GaN surface is not terminated by a Ga bilayer. Rather, the surface is presumably terminated by a structure involving N and Ga, and possibly H. This conclusion is supported by noting that the magnitude of $\langle \varepsilon_2 \rangle$ in MOVPE (Fig. 4b) is quite close to that of the N rich PAMBE results in Fig. 4a, consistent with our interpretation that both surfaces are composed of a surface structure which does not include a metallic Ga-bilayer. Beyond this conclusion, we cannot deduce more about the detailed structure of the NH$_3$ stabilized surface from our results. Grazing incidence X-ray diffraction measurements (GIXS) have indicated a $\sqrt{3} \times 2\sqrt{3}$-R30° surface structure

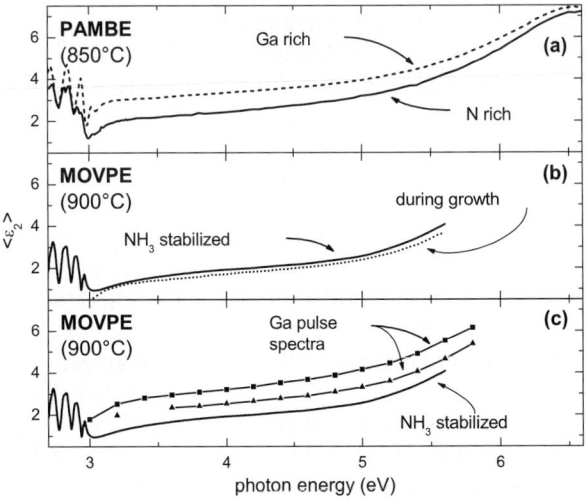

Fig. 4 Imaginary part of the dielectric function $\langle \varepsilon_2 \rangle$, for GaN(0001) films comparing results a) during PAMBE under Ga-rich or N-rich conditions, b) during MOVPE growth or stabilized under an NH$_3$-flux, and c) during MOVPE on NH$_3$-stabilized surfaces or on surfaces exposed to a pulse of TMGa after ≈ 1 s and ≈ 8 s in the absence of NH$_3$.

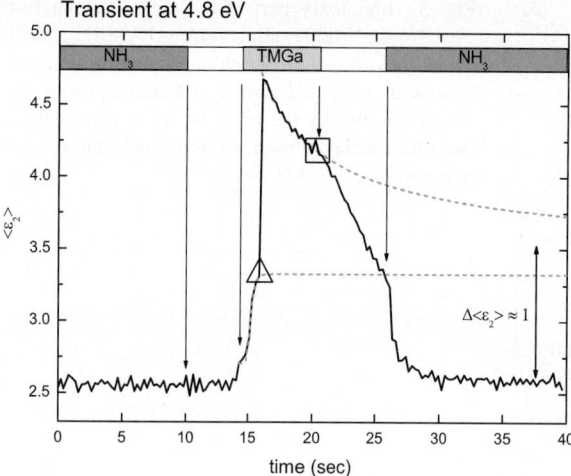

Fig. 5 Transient measurement of imaginary part of the effective dielectric function, $\langle \varepsilon_2 \rangle$, at 4.8 eV, during variations in MOVPE growth parameter at 900 °C and using N_2 carrier gas. Starting with a surface stabilized under NH_3 with no TMGa flux, the NH_3 supply is first turned off and 5 s later the TMGa flux is turned on. A rapid increase in $\langle \varepsilon_2 \rangle$ results, displaying a small knee near a value of 3.3 (triangle), increasing to about 4.6, and then exponentially decaying. The TMGa supply was shut off after a duration of ≈ 8 s (square), and the NH_3 was turned on again 5 s later. The $\langle \varepsilon_2 \rangle$ then returns to nearly its original value.

during MOVPE growth [14], and recent theoretical work provides insight into possible surface structures [15], but to prove the existence of these structures additional experimental work is needed. This can be done by examining ellipsometric transients during MOVPE growth and/or growth interruptions, respectively. Thereby gallium rich as well as gallium poor surfaces should be obtained.

As noted above, turning off the TMGa source during regular MOVPE growth produces very little change in the ellipsometric signal. However, starting from an NH_3 terminated surface and then attempting to produce a Ga terminated surface by shutting off the NH_3 supply and turning on the TMGa supply produces rather complicated results with various changes in the optical response, as shown in Fig. 5. The value of $\langle \varepsilon_2 \rangle$ increases rapidly when the TMGa is turned on, reproducibly displaying a small knee in the first second after the turn-on and then continuing to rapidly increase for another 1–2 seconds. The signal then reaches a maximum and begins to decrease exponentially. In the example shown in Fig. 5 the TMGa supply is then turned off and the NH_3 supply is turned on after 5 s. The $\langle \varepsilon_2 \rangle$ signal finally reach a level nearly equal to that at the start of the sequence. If, alternatively, the TMGa supply had been left on, the $\langle \varepsilon_2 \rangle$ signal seems to exponential approach a constant value below 4 followed by the formation of Ga droplets. The final difference between the $\langle \varepsilon_2 \rangle$ signal after a continuous TMGa supply and that of the NH_3 terminated surface is very similar to the difference observed in the PAMBE. However, this procedure leads to a non reversible roughening of the surface after switching on again ammonia for stabilization through growths of crystallites around the Ga droplets. This result is in accordance to the previous finding that smooth surfaces in MOVPE can be grown only under N-rich conditions. We tentatively interpret these results as follows: the initial increase in $\langle \varepsilon_2 \rangle$ up to the above-mentioned knee is likely due to the formation of a Ga rich surface. This is proven by an estimation of the Ga supply which is in the range of $\approx 5 \times 10^{15}$ Ga-atoms/s for the used 2 μmol min^{-1} TMGa flow. Additional changes then may arise from a transition of this bilayer into clusters and Ga droplets involving the formation of adsorbate assisted surface reconstructions [15, 16]. Thereby, the complex behavior possibly could be ascribed to the presence of hydrogen and the formation of hydrogen (H, NH, ...) induced reconstructions [15]. This is also supported by the fact, that the described behavior occurs more distinct but on a larger time scale by using hydrogen as carrier gas. In contrast hydrogen can be neglected in the PAMBE due to the sputter effect of the nitrogen plasma.

Finally, in Fig. 4c we show spectral results obtained during the MOVPE growth interrupts. The NH_3-stabilized surface results are very similar to those for the surface during growth, as noted above. The Ga-pulse results were obtained during transient measurements at different photon energies of the type shown in Fig. 5, and then constructing a spectrum from those values. We choose data points taken at the small knee position as shown in the transient measurements after ≈ 1 s and after ≈ 8 s where the TMGa supply is switched off (encircled positions in Fig. 5). The results of $\langle \varepsilon_2 \rangle$ are shown in Fig. 4c. The values of $\langle \varepsilon_2 \rangle$ are somewhat higher than those for the Ga-rich spectrum of Fig. 4a, presumably arising from the different surface structures and morphology as discussed above. From these data we conclude that Ga terminated surface in MOVPE should be found somewhere in between, close to these spectra found in the MBE prepared Ga terminated surface.

4 Summary

In summary, we have used spectroscopic ellipsometry to probe the (0001) surface of GaN during growth by MOVPE and PAMBE. Large differences in $\langle \varepsilon_2 \rangle$ are found between Ga-rich and N-rich surfaces and are attributed to the presence of the surface Ga-bilayer in the former and its absence in the latter case. Results of surfaces during MOVPE growth or with NH_3-stabilisation only are very similar to the N-rich PAMBE spectrum, indicating that the MOVPE surface is not terminated by a Ga bilayer under normal growth conditions. Gallium layers can, however, be deposited on the MOVPE surface by exposure to TMGa in the absence of NH_3, although these layers display a rather complex behavior involving adsorption, roughening and reformation at the growth temperature.

Acknowledgements We gratefully acknowledge the technical support from the Institut für Physik, Technische Universität Chemnitz and especially the help of T. U. Kampen by performing the PAMBE preparation of high quality GaN films in UHV. We also thank the BMBF for the financial support under 05 KS1KTB/2.

References

[1] R. M. Feenstra, J. Neugebauer, and J. Northrup, MRS Internet J. Nitride Semicond. Res. **7**, 3 (2002).
[2] A. R. Smith, R. M. Feenstra, D. W. Greve, M.-S. Shin, M. Skowronski, J. Neugebauer, and J. Northrup, J. Vac. Sci. Technol. B **16**, 2242 (1999).
[3] K. Vedam, Thin Solid Films **313–314**, 1 (1998).
[4] D. E. Aspnes, J. B. Theeten, and F. Hottier, Phys. Rev. B **20**, 3292 (1979).
[5] D. E. Aspnes, In B. O. Seraphin, Optical properties of solids: new developments (North-Holland Publishing Company, Amsterdam, 1975).
[6] W. Richter, Appl. Phys. A **75**, 129 (2002).
[7] K. Hingerl, D. E. Aspnes, I. Kamiya, and L. T. Florez, Appl. Phys. Lett. **63**, 885 (1993).
[8] T. Wethkamp, K. Wilmers, N. Esser, W. Richter, O. Ambacher, H. Angerer, G. Jungk, R. L. Johnson, and M. Cardona, Thin Solid Films **313–314**, 745 (1998).
[9] S. Peters, T. Schmidtling, T. Trepk, U. W. Pohl, J.-T. Zettler, and W. Richter, J. Appl. Phys. **88**, 4085 (2000).
[10] F. A. Ponce, D. P. Bour, W. T. Young, M. Saunders, and J. W. Steeds, Appl. Phys. Lett. **69**, 337 (1996).
[11] A. Pavlovska and E. Bauer, Surf. Sci. **480**, 128 (2001).
[12] B. Heying, R. Averbeck, L. F. Chen, E. Haus, H. Riechert, and J. S. Speck, J. Appl. Phys. **88**, 1855 (2000).
[13] Robert W. Collins and K. Vedam, Appl. Phys. **12**, 285 (1995).
[14] A. Munkholm, G. B. Stephenson, J. A. Eastman, C. Thompson, P. Fini, J. S. Speck, O. Auciello, P. H. Fuoss, and S. P. DenBaars, Phys. Rev. Lett. **83**, 741 (1999).
[15] C. G. Van de Walle and J. Neugebauer, Phys. Rev. B **61**, 9932 (2000).
[16] W. G. Schmidt, P. H. Hahn, F. Bechstedt, N. Esser, P. Vogt, A.Wange, and W. Richter, Phys. Rev. Lett. **90**, 126101 (2003).

phys. stat. sol. (c) **0**, No. 8, 2944–2948 (2003) / **DOI** 10.1002/pssc.200303847

Drastic improvement of electrical properties of Nafion® 112 membrane on impregnation of bimetallic Au/Pd nanoclusters

U. Pal[*, 1, 3], **J. F. Sánchez-Ramírez**[2], **S. A. Gamboa**[3], **P. J. Sebastian**[3, 4], and **R. Pérez**[3]

[1] Instituto de Física, BUAP, apdo. postal J-48, Puebla 72570, Mexico
[2] Instituto de Ciencias Químicas, BUAP, Puebla 72570, Mexico
[3] Instituto Mexicano del Petroleo, Programa de Investigación y Desarrollo de Ductos,
   Eje Central Lazaro cardenas 152, Col. San Bartolo Atepehuacan, 07730, Deleg. GAM, Mexico, D.F.,
   Mexico
[4] Centro de Investigación en Energía, UNAM, 62580 Temixco, Morelos, Mexico

Received 30 May 2003, revised 4 August 2003, accepted 11 August 2003
Published online 10 November 2003

PACS 61.46.+w, 68.47.Mn, 72.80.Le, 72.80.Tm

We report on the improvement in electrical conductivity of Nafion® 112 membrane by the impregnation of bimetallic Au/Pd nanometric clusters. Colloidal Au/Pd nanoclusters of different molar ratios were prepared by simultaneous reduction of corresponding metal salts. The colloidal bimetallic clusters were characterized by TEM and optical spectroscopy techniques. The Au/Pd nanoclusters of different molar ratios were impregnated into the well hydrated Nafion® membranes for 3 hrs. Optical absorption spectroscopy was used to verify the impregnation process. The electrical characteristics of the impregnated membranes were investigated by electrochemical impedance spectroscopy (EIS). The experiments were performed in a solid state electrochemical cell consisted of the impregnated Nafion® in electrical contact with Au electrodes. The EIS response of the impregnated membranes showed a drastic improvement in their electrical conductivity, which promises a better performance for the proton exchange membrane fuel cell (PEMFC) when Au/Pd nanocluster impregnated Nafion® is used as solid electrolyte.

1 Introduction   Perfluorosulfonate ion-exchange membranes have received a considerable attention in recent years because of their application in various electrochemical systems such as electrolysis, proton exchange membrane fuel cells and gas sensors. This kind of membrane has good electrochemical properties that permit it to be used as an adequate material for conducting electricity in ionic form [1, 2]. Nafion® is a commercially available perfluorinated proton-exchange membrane developed by Dupont de Nemours Co., USA. This membrane is commonly used in electrochemical systems where the electrical transport is through electronic-ionic phenomena, those systems are for example the PEMFCs and gas sensors. Nafion® membrane provides the electric contact between anode and cathode (electrodes) taking charge of proton conduction from the oxidation to reduction reaction couple.

The electrical mechanism occurring at Nafion® and electrode surfaces is not well understood in electrochemical systems. Nafion® consists of a microscopically phase-separated structure formed by hydrophilic ionic clusters, hydrophobic perfluorocarbon backbones and an intermediate region [3-4]. The three regions interact with the electrode surface and the total charge transfer process depends on the membrane characteristics and the physico-chemical properties of the electrode [5].

---

[*] Corresponding author: e-mail: upal@imp.mx, Phone + 52-222295500 extn 2047

In this study we present a novel process to obtain modified Nafion® 112 using nanostructured materials based on Au/Pd and Pd clusters for improving its electrical characteristics.

2 Experimental Bimetallic Au/Pd and monometallic Pd nanostructures were used to obtain modified Nafion® 112 membranes with improved electrical conducting properties. Au/Pd bimetallic nanostructures were prepared by simultaneous reduction of two metal ions in presence of poly(N-vinyl-2-pyrrolidone) (PVP). Ethanol solution of palladium (II) chloride (0.033 mmol in 25 ml de ethanol) was prepared by stirring dispersion of $PdCl_2$ powder in ethanol for 48 hrs. Solution of tetrachloroauric acid was prepared by dissolving crystals in water (0.033 mmol in 25 ml of water). For preparing bimetallic clusters, solutions containing two metal ions were mixed in 50 ml of 1/1 volume ratio of pure ethanol/water, 151 mg of PVP (K-30, average molecular weight 10,000) was added to the total metal ion content of 6.66×10^{-5} mol. The mixture solution was stirred and refluxed at about 100 °C for 2 hrs. For the preparation of bimetallic clusters with different Au/Pd content ratios (1/1 and 5/1), metal ion solutions of corresponding ratios were mixed in ethanol/water-PVP, maintaining the total metal ion concentration (6.66×10^{-5}) fixed, and then refluxed at 100°C for 2 hrs. For the preparation of monometallic palladium clusters, 75.5 mg of PVP was added to the ethanol solution of $PdCl_2$ and refluxed for 2 hrs at 100 °C.

For Impregnation of monometallic Pd clusters in Nafion® 112 (modified membranes), Nafion® 112 membrane (by Du-Pont) was cut into 1 cm^2 pieces and hydrated as a pre-treatment procedure mentioned elsewhere [6]. The hydrated Nafion® sheets were then dipped into 10 ml of monometallic colloidal solution for 3 hrs. For preservation of the water content of the hydrated membranes, the modified Nafion® membranes were kept in plastic ampoules immersed in deionized water at room temperature. The same procedure was carried out for bimetallic Au/Pd cluster impregnation at different content ratios (1/1 and 5/1).

A JEOL-JEM200 microscope was used for the analysis by transmission electron microscopy (TEM) of the colloidal mono- and bimetallic clusters. A Shimadzu UV-VIS 3191PC double beam spectrophotometer was used for measuring the absorption spectra of the colloids. For the study of absorption spectra of the modified Nafion®, the membranes were dehydrated in a vacuum chamber. A non-modified Nafion® 112 membrane was used as reference for absorption measurements. Electrochemical studies based on cyclic voltammetry and electrochemical impedance spectroscopy were carried out with a Voltamaster4-EG&G potenciostat/galvanostat coupled to a computer interface.

3 Results and discussion Figure 1 shows the TEM micrographs of colloid clusters obtained by reduction. We can see the formation of clusters of nanometer size with average particle size varying with metal content ratios. Figure 1(a) corresponds to Pd monometallic clusters with 2.26nm average particle size. 1(b) Au/Pd bimetallic clusters with 3.5 nm average size. 1(c) Au/Pd bimetallic clusters with 5.2 nm average size. The metallic clusters obtained by reduction were used for impregnation of Nafion® 112 membrane for modifying its electrical characteristic.

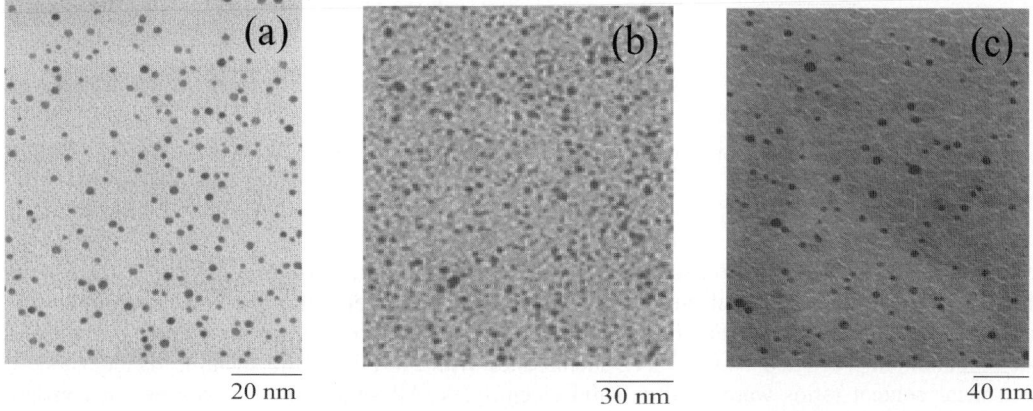

20 nm	30 nm	40 nm

Fig. 1 TEM micrographs of colloids, (a) Pd, (b) Au/Pd(1/1), (c) Au/Pd(5/1) clusters.

In Fig. 2(a), the absorption spectra of the bimetallic colloids with different metal content ratios are presented. The surface plasmon resonance (SPR) peak for the bimetallic nanoparticles appeared at about 530nm. There was no evidence of SPR peak below the Au/Pd ratio 5/1. In Fig. 2(b), the absorption spectra for the bimetallic cluster Au/Pd (5/1) impregnated Nafion® 112 membrane are shown. The appearance of a SPR band in impregnated Nafion® at about 560 nm for the impregnation time of 1 hr, establishes the formation of modified membranes by colloidal nanoclusters based on Au/Pd. For an impregnation time of 3 hrs the SPR peak position shifted to 570 nm and the band became broader than that for 1 hr. The same procedure was carried out for analyzing the impregnation of clusters on Nafion® 112 with other Au/Pd and Pd content ratios. 3 hrs impregnation time was considered as optimum for this study.

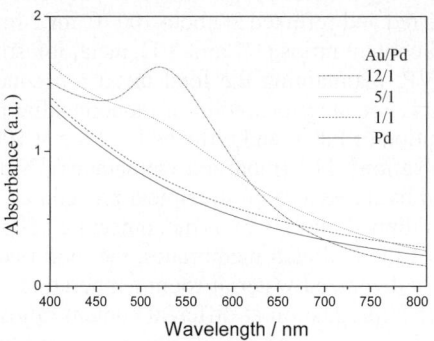

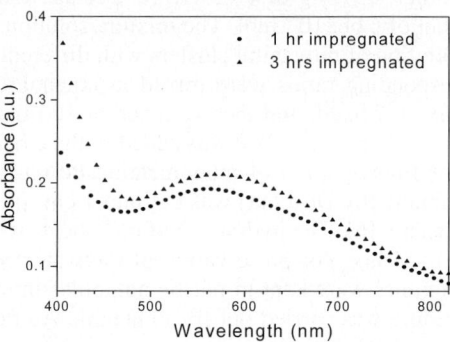

Fig. 2   a) Absorption spectra of bimetallic colloids with different metal content ratios.

Fig. 2   b) Absorption spectra of Au/Pd (5/1) cluster impregnated Nafion membrane.

The electrocatalytic properties of metal cluster modified Nafion® 112 were investigated electrochemically by cyclic voltammetry and EIS using a related electrochemical cell in solid state configuration described by Ch. Yu et al. [7] and shown in Fig. 3.

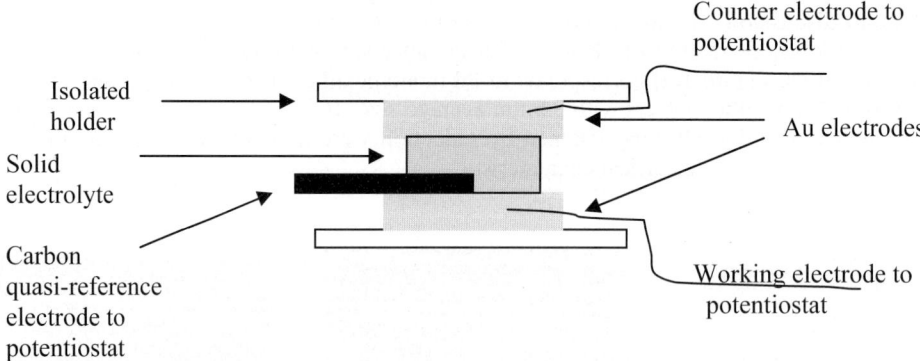

Fig. 3 Schematic representation of the solid state electrochemical cell used for characterizing modified Nafion.

Sputtered Au electrodes deposited on stainless steel sheets (0.15 mm thickness and 2 cm² surface area) were used as counter and working electrodes in the electrochemical cell. High purity carbon ribbon was used as quasi-reference potential. For further studies the plots can be referred to normal hydrogen electrode potential. Impregnated Nafion® 112 membranes with Au/Pd and Pd metallic nanoclusters at different metal content ratios were used as solid electrolytes. All studies were performed at constant temperature (30 °C).

A non-modified Nafion® 112 membrane was used as reference in electrochemical investigations. The stability of the modified membranes was made sure by monitoring the open circuit potential during the experiments.

The voltammogram plot of the non-modified Nafion® 112 for a scan rate of 10 mV/sec did not show the presence of redox reactions along the electrochemical window in the range of -0.91 ≤ Potential (V/ref) ≤ 0.84. For lower and upper potentials out of the range, the hydrogen evolution reaction (HER) and oxygen evolution reaction (OER) took place respectively. For the case of impregnated Nafion® 112 with Au/Pd bimetallic clusters at ratio 5/1 (Fig. 4), a defined hydrogen reduction peak appeared at around 0.489 V/ref. A broad oxidation zone could be observed in anodic direction at 0.8 V/ref. The threshold potential for the HER decreased in presence of Au/Pd=5/1. It suggests that Au/Pd is acting as electro-catalyst, and we can obtain some qualitative information about the improvement in electrical characteristics of the modified Nafion® membrane. In Table 1 is presented the general results obtained for the Pd and both Au/Pd metal content ratios using characterization by cyclic voltammetry.

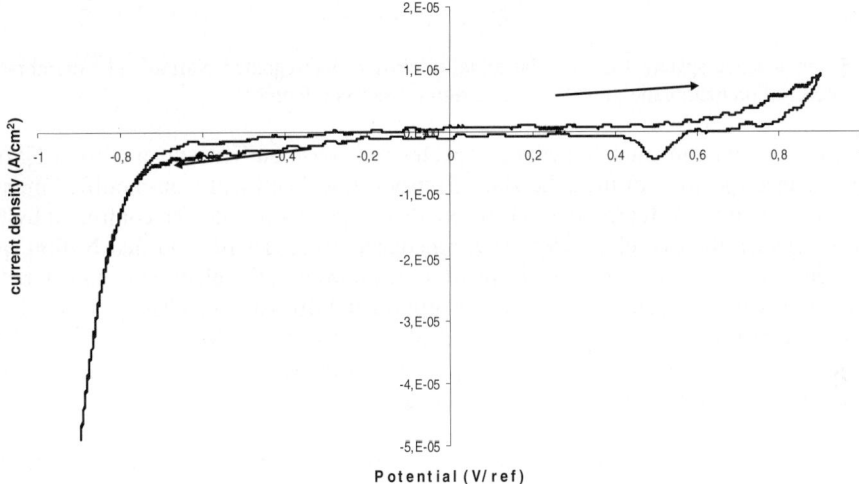

Fig. 4   Voltammogram plot for modified Nafion using Au/Pd (5/1) metallic clusters. 10 mV/sec scan rate, 30 °C.

Table 1   Resume of electrochemical parameters obtained by cyclic voltammetry of modified Nafion® 112 membranes. All potentials are referred to quasi-reference potential.

Structure (Ratio)	$P_{OC}(V)$	$P_e(V)$	$P_{ads}(V)$	HER threshold (V)
Nafion (non)	−0.25	−0.95	–	−0.91
Pd(1)	−0.13	−0.80	0.34	−0.77
Au/Pd(1/1)	−0.22	−0.50	0.37	−0.47
Au/Pd(5/1)	−0.02	−0.70	0.48	−0.62

Where P_{OC} is the open circuit potential, P_e is the equilibrium potential at activated H^+ scatter potential (for very low HER overpotential), P_{ads} is the adsorption potential of hydrogen in modified membranes.

From Table 1, all modified membranes showed better performance than non-modified Nafion® respect to HER threshold potential. These results mean that the electrical properties of the Nafion® 112 can be improved by using metallic clusters based on Au/Pd nanoparticles.

Electrochemical impedance spectroscopy technique was used to evaluate the improvement in electrical properties of modified Nafion® 112 membranes. The EIS spectra were recorded in the frequency range of $0.1 \leq f\,(Hz) \leq 1000$. The equilibrium potentials shown in Table 1 were used as initial potentials in EIS studies for each system.

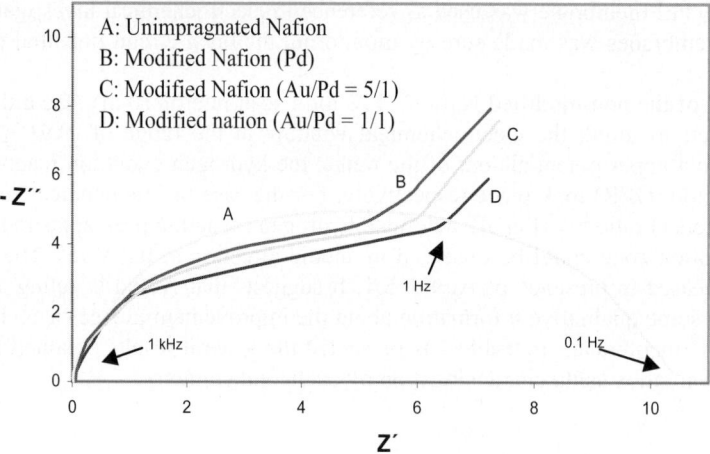

Fig. 5 Impedance spectra for colloidal metallic clusters impregnated Nafion® 112 membranes. Unimpregnated Nafion membrane impedance spectrum is used as reference.

Figure 5 shows the experimental impedance spectra for non-modified and modified Nafion membranes. The impedance spectrum of the solid state electrochemical cell with non-modified membrane (as electrolyte) is shown as fig. 5A. Its response corresponds to a polarizable device controlled by the charge transfer process between the two electrodes. The impedance responses of modified Nafion (plots B, C and D) show a decrease in the charge transfer resistance, improving the electrical characteristics of the Nafion due to the presence of metallic clusters. Adsorption and diffusion mechanisms can also be interacting with the charge transfer process in this kind of systems at low frequencies (around 1 Hz and lower). The influence of adsorption states in impregnated Nafion needs to be investigated before considering the use of these membranes in PEM fuel cell and gas sensor.

4 Conclusions   Colloidal monometallic and bimetallic clusters based on Au/Pd and prepared by chemical reduction were used for obtaining modified membranes with better electrical characteristics. The use of Au/Pd based nanoclusters is capable of improving the conduction properties of membranes like Nafion®. Membranes with better electrical properties are necessary for fuel cell applications. In this work we also observed the relation between the change in HER threshold and the decrease in charge transfer resistance, as a function of the metal content ratio of the clusters. The presence of Pd and the existence of adsorption states in modified membranes need to be investigated.

Acknowledgements   We would like to thank Mr. Arturo Rivera and Mr. Alejandro del Valle for their help to perform the electrochemical studies. The authors also thank to CONACyT (42146-Y) for partial financial support.

References

[1] K. Kanamura, H. Morikawa, and T. Umegaki, J. Electrochem. Soc. **150**, A193(2003).
[2] J. Maruyama and I. Abe, Electrochim. Acta, **46**, 3381(2001).
[3] P.J. James, T.J. McMaster, J.M. Newton, and M.J. Miles, Polymer **41**, 4223(2000).
[4] Z. Ogumi, T. Kuroe, and Z-I- Takehara, J. Electrochem. Soc. **132**, 2601(1985).
[5] J. Maruyama, M. Inaba, and Z. Ogumi, J. Electroanal. Chem. **458**, 175(1998).
[6] V.A. Paganin, E.A. Ticianelli, and E.R. Gonzalez, J. Appl. Electrochem. **26**, 297(1996).
[7] Ch. Yu, Y. Wang, K. Hua, W. Xing, and T. Lu, Sensors and Actuators B **86**, 259(2002).

phys. stat. sol. (c) **0**, No. 8, 2949–2955 (2003) / **DOI** 10.1002/pssc.200303851

In-situ Raman Spectroscopy on III–V semiconductors at high temperature in MOVPE

E. Speiser*, **T. Schmidtling**, **K. Fleischer**, **N. Esser**, and **W. Richter**

[1] Institut für Festkörperphysik, Sekr. PN 6-1, Hardenbergstr. 36, D-10623 Berlin, Germany

Received 30 May 2003, revised 4 August 2003, accepted 11 August 2003
Published online 10 November 2003

PACS 63.20.Dj, 78.30.Fs, 81.15.Gh

Through the implementation of a Raman spectroscopic equipment into a metalorganic vapor phase epitaxy setup (MOVPE) via optical fibers we determine surface and bulk related properties of III–V semiconductors in a temperature range up to 1200 K. Surface damages due to high temperature are avoided by a stabilization with gaseous group V elements which allows for reproducible measurements. The temperature dependent changes are monitored and analyzed through the change in the vibrational properties which are sensitive indicators of the sample and surface status.

The results can be grouped into two categories: (i) irreversible changes with temperature (crystal quality, doping) due to sample annealing during the measuring process at high temperatures, and (ii) anharmonic effects (reversible) on phonon frequencies and line widths. We show that the TO and LO-phonon shifts at high temperature can be described well including fourth order phonon decay. The maximum temperature which can be reached is at present only limited by the subscale heating system.

1 Introduction

Raman spectroscopy (RS) is a well known method for the analysis of vibrational bulk properties. It has also been established in the last decade for vibrational analysis of surface processes. Particularly Raman spectroscopy, which permits the detection of surface phonons [1] and local vibrations of adsorbates, may be useful to distinguish surface reconstructions and to analyze the different atomic arrangements of the surface. RS, as an optical technique, has the advantage to be usable in all environments with little requirements on window materials. With the high resolution (0.1 meV) compared to the competing technique HREELS (1 meV) it has the potential to monitor anharmonic effects of the bulk properties and surface modifications with high sensitivity. Many of technologically relevant processes (like deoxidation or epitaxial growth) require high temperatures. Raman experiments at high temperature are therefore important for an deeper understanding of these processes. However, not much has been done in the field of high temperature Raman analysis on III–V compound semiconductors. This is because their surfaces deteriorate easily at high temperature due to the high vapor pressure of the group V elements. Therefore measurements at elevated temperatures can only be performed under active stabilization of the volatile components. For III–V semiconductors like GaAs this can be accomplished by offering additional arsenic (MBE: As_2, As_4; MOVPE: AsH_3) to the sample surface. We use here Metal Organic Vapor Phase Epitaxy (MOVPE).

---

* Corresponding author: e-mail: espeiser@gift.physik.tu-berlin.de, Phone: +49 30 314 22079, Fax: +49 30 314 21769

2 Experimental

Combining a MOVPE growth reactor and a Raman spectrometer is not easy because of their size and complexity. In order to solve this problem we used optical fibers to guide the excitation laser light to the MOVPE reactor, as well as the scattered Raman light from the reactor to the spectrometer (Fig.2). This approach in general creates a number of problems: (i) the loss of intensity in the fibers requires a proper choice of fiber diameter and numerical aperture, (ii) the laser light normally generates Raman scattering signals in the fiber and (iii) thermal radiation from the heated sample/susceptor will generate additional background signals.

The excitation laser light reaches the sample by a single mode (polarization retentive) fiber. For the scattered light we use a multi mode fiber. The choice of numerical aperture and the correct diameter of the scattered light guiding quartz fiber is a compromise between best focus into the fiber (larger numerical aperture (NA) and core dimension) and best transmission to the spectrometer (smaller NA and core dimension). The quartz fibers themselves produce a specially structured background signal due to Raman scattering in the fiber. This signal can obscure weak signals from the sample (eg. surface phonons). To eliminate the fiber related background we use a narrow line filter at the excitation side and a notch filter at the scattered light side (fig.2). Both filters have to be positioned in a parallel optical light path in order to minimize the signal loss and the spectral bandwidth of the filter. To collect the scattered light from the sample with maximum intensity outside the reactor chamber this had to be equipped with a large aperture window suitable for Raman spectroscopy. To protect this window against depositions and the high temperature inside the reactor, it is flushed with carrier gas (H_2 or N_2) during measurements.

The graphite susceptor is electrically heated. Currently the maximum temperature (1200 K) is limited by the maximum power of the heating element but is not a principle limit for the Raman spectroscopic measurement. The thermal radiation produced by the sample at high temperature generates an intense background. This background, however, has its maximum in the infrared and appears as a constant offset in the Raman spectra. This offset decreases using shorter laser excitation wavelengths (514.5, 488.0 nm).

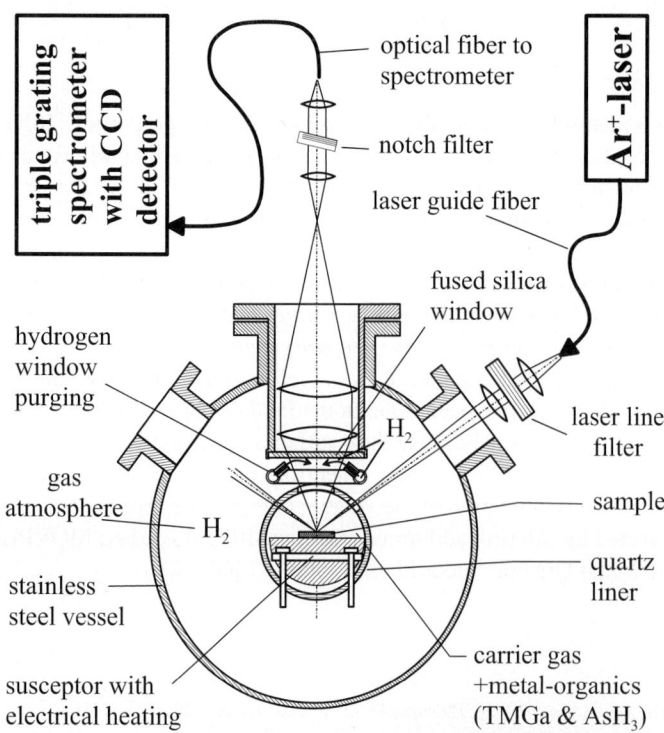

Fig. 1 Cross section of the MOVPE reactor with integrated Raman setup. The necessary modifications for Raman measurements are depicted. The Ar^+-laser for excitation is fiber guided and focused through a side view port on the sample. The scattered light passes the large aperture top view port made from fused silica and is led through a multi mode fiber to the triple grating DILOR spectrometer with a multichannel CCD detector.

The special MOVPE reactor used in our study is embedded in a stainless steel vessel and also connected to an UHV analysis chamber (equipped with LEED and Auger spectroscopy) allowing for additional characterization of surface reconstructions and possible contaminations at room temperature. Reflection Anisotropy Spectroscopy (RAS) measurements are also possible there.

The necessity of group V stabilization and the principle working ability of the setup is demonstrated in fig.2. Two samples of the same wafer and with identical room temperature Raman spectra were heated to 870 K with and without providing AsH_3 for stabilization. Without AsH_3 stabilization we observe a breakdown of the selection rules (forbidden TO mode) due to a large roughness of the surface. AFM measurements afterwards show a 130 nm rms roughness for the unstabilized case while 0.8 nm rms was obtained for the stabilized sample. As a consequence, the laser beam will have different directions (k-vectors) inside the sample which leads to different scattering geometries and polarizations with respect to the crystal structure. This results in a strong reduction of the LO/TO phonon intensity ratio (lower curve in fig.2) [2]. Such a surface damage by strong roughening can not be reversed anymore by subsequent AsH_3 stabilization.

3 Results

During the RS measurements the samples are hold at high temperatures for an extended time (30-60 min), a procedure which is equivalent to annealing. Therefore, irreversible changes are observed in some samples during the measurements depending on the crystal quality and doping of the samples used. In the next section we point out two of such changes occurring in our measurements.

3.1 Influence of crystal quality and doping

High temperature Raman experiments where performed using highly Te-doped ($2*10^{18}cm^{-3}$) GaAs (100) samples (fig.3). It was found that at room temperature the LO and TO frequencies appear already at values close to those in the literature [3] but shifted for approx. 2 cm^{-1} to lower energies. A high defect density (etch pit density, $EPD < 4*10^4 cm^{-2}$) was found to be the cause. Below 570 K we observe temperature induced shifts of the LO and TO frequencies, similar to those observed in unstrained GaAs (same slope as in fig.5). Above 620 K the shift of LO and TO deviates from the expected behavior which was not observed for GaAs of better crystalline quality. It seems obvious that this is caused by an improvement in crystal quality during the annealing process. Hence with this experiment direct observation of the defect healing during annealing by MOVPE is demonstrated through RS.

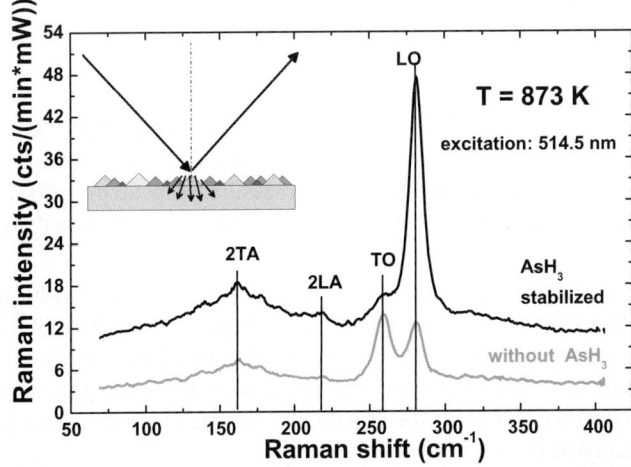

Fig. 2 Raman spectra of GaAs(100) under stabilized and non stabilized conditions. The rough surface induces changes of the k-vector of the incident light (inset sketch), causing a breakdown in the selection rules with decrease in LO and TO intensity and change in LO/TO ratio.

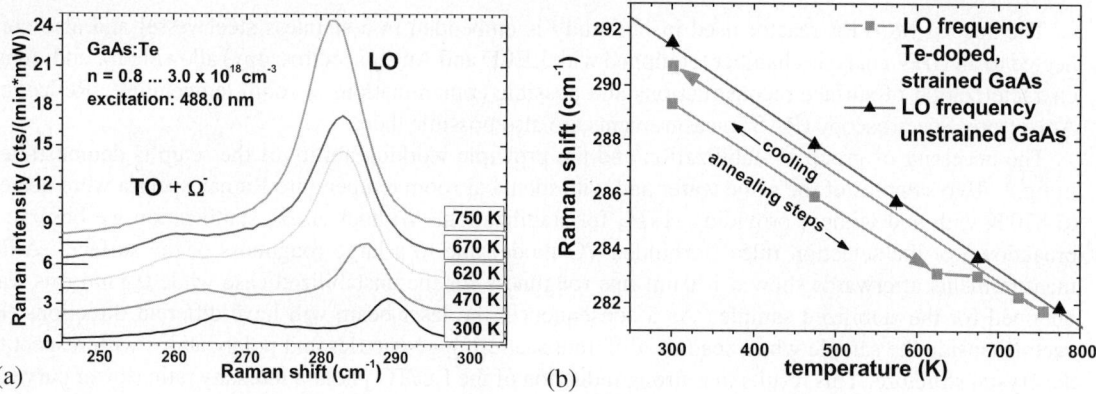

Fig. 3 Raman spectra of GaAs:Te with high defect density during extended time at high temperature (annealing time for each step: 30 min). (a) spectra, (b) LO frequencies measured with temperature. Deviation from the expected shift of the LO frequencies above 620 K is obvious. In (b) the peak position of the LO phonon, derived from the spectra shown in (a), are compared to the expected shift of unstrained GaAs.

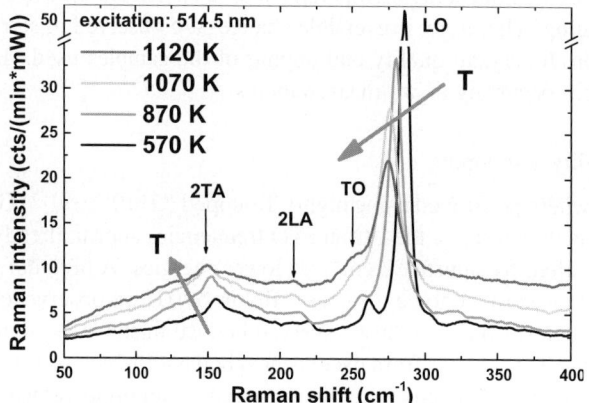

Fig. 4 Raman spectra of GaAs at elevated temperatures. One observes a shift in the energetic position of the LO and TO mode together with a thermal broadening of the mode. The scattering intensity of the 2 phonon modes (e.g 2TA) is increased, the intensity of single phonon scattering processes is decreased

Apart from the changes in the Raman shift of the bulk phonon modes we also observes changes in the free carrier concentrations, indicated by the disappearance of the plasmon–LO-phonon modes and increase in LO-phonon intensity.

3.2   Anharmonicity of bulk phonons

By selecting GaAs samples of high crystalline quality and low doping levels, or by taking freshly grown undoped epitaxial layers one is in the position to study the intrinsic temperature effects on the phonons. Within this study we were able to extend the temperature range for Raman measurements up to 1200 K as shown in fig.4. The decrease in LO-phonon frequency and intensity as well as the line width broadening due to the high temperatures are clearly seen.

The corresponding values for LO- and TO-frequencies as well as for the LO-phonon line width are plotted in fig.5 versus temperature. These temperature induced changes are caused by the anharmonicity of the lattice potential. In the terminology of phonons higher than harmonic terms in the lattice potential of third (fourth) order describe the decay of phonons into two (three) other phonons by conserving of course energy and momentum. After the first proposal by Cowley [4] for Si with two acoustical phonons of equal energy these calculations where later substantiated [5, 6, 7].

The phonon dispersion curves and the 1-phonon density of states suggest [8] that a decay of the LO-phonon into two phonons with

$$\omega_{ac}^1 = 0.26\omega_{LO} \quad \text{and} \quad \omega_{ac}^2 = 0.74\omega_{LO}$$

is the most important third order channel. For the fourth order contributions we assumed decay into three acoustical phonons with equal energy

$$\omega_{ac}^3 = \frac{1}{3}\omega_{LO}.$$

The occupation numbers (n_i) for the phonons which are involved in the decay process (ω_{ac}^i, i=1,2,3) are given by:

$$n_i = \frac{1}{\exp\left(\dfrac{\hbar\omega_{ac}^i}{k_B T}\right) - 1}. \tag{1}$$

With these decay processes one can now describe the broadening and shift of the phonon modes. Following the notation in [9] and [6] one can write for the line width $2\Gamma_{LO}$:

$$2\Gamma_{LO}(T) = A\left[1 + n_1 + n_2\right] + B\left[1 + 3n_3 + 3n_3^2\right] \tag{2}$$

where $A + B$ is the LO line width at T = 0 K. A is treating the influence of a decay into two phonons while B treats the case of decay into three phonons.

Fig.5(a) shows fits of (2) to the measured line widths (solid line). The dashed line shows the contribution of the third order term to the broadening ($B = 0$). Nevertheless, above 900 K the line width increases stronger than predicted and additional phonon decay channels possibly caused also by even higher terms in the lattice potential must be responsible for the larger line width.

Similarly the frequency shifts can be described with [9, 10, 11]

$$
\begin{aligned}
\omega(T) &= \omega_0 - \Delta^{(1)}(T) - \Delta^{(2)}(T), \\
\Delta^{(1)}(T) &= \omega_0^{TO,LO}\left(\exp\left(-3\gamma^{TO,LO}(\Gamma)\int_0^T \alpha(T')dT'\right) - 1\right), \\
\Delta^{(2)}(T) &= C\left[1 + n_1 + n_2\right] + D\left[1 + 3n_3 + 3n_3^2\right].
\end{aligned}
\tag{3}
$$

Where $\Delta^{(1)}(T)$ is the term describing the line shift through thermal expansion [6, 10], with $\alpha(T)$ as the linear thermal expansion coefficient from [3] used for the numerical calculation of $\int_0^T \alpha(T')dT'$. The Grüneisen parameters $\gamma^{LO}(\Gamma) = 1.23$ and $\gamma^{TO}(\Gamma) = 1.39$ where obtained from [12]. $\Delta^{(2)}(T)$ is the energy shift resulting from multiple phonon decays, taking in account up to the fourth order contribution of anharmonicity. ω_0 was chosen with the assumption that $\omega(T = 0) = \omega_0 - C - D$ has to be fulfilled and $\omega(T = 0)$ was taken from [3].

In fig.5(b) the experimental values of the phonon energy is given with a fit of the parameters C, D according to eqn. (3). In order to deconvolute the different contributions to the overall line shift plots for $D = 0$ (dotted lines) and $D = 0, C = 0$ (dashed lines) are shown as well. In the latter case only the contribution of the thermal extension ($\Delta^{(1)}(T)$) is considered. Even if the coefficient ratio between the third ($C = 0.43$) and fourth ($D = 0.009$) order anharmonic terms indicates a very low fraction of four phonon processes they can not be neglected for a proper description in case of the temperature induced line shift of GaAs.

Within the error of the fitting procedure the values for C^{LO} and C^{TO} as well as D^{LO} and D^{TO} are very similar. Consequently the LO-TO splitting is unchanged in the temperature range investigated. Earlier

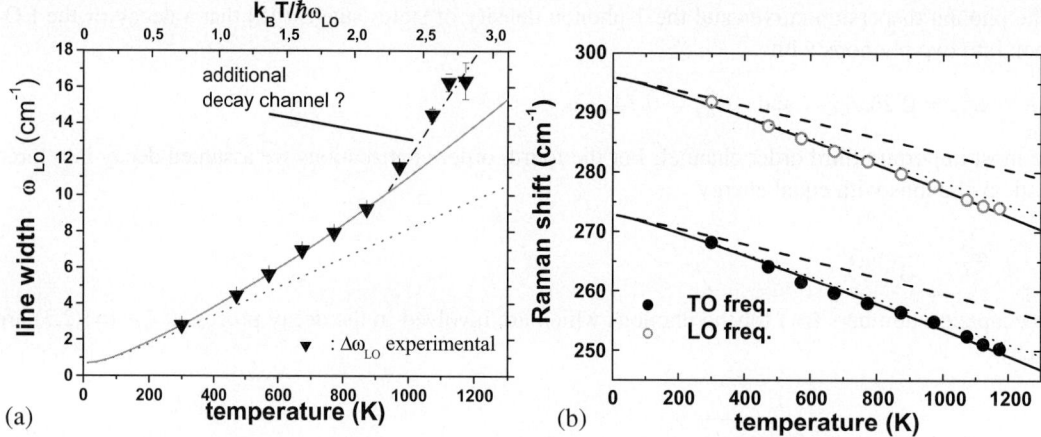

Fig. 5 Temperature dependence of the LO and TO phonon modes. In (a) the dependence of the LO line width in comparison to the theoretical model described above is shown. Triangles show experimental values, dotted line the contribution of the decay into two phonons, solid line the complete description of eq.2. Fitting parameters are: $A = 0.662 \pm 0.026 cm^{-1}$, $B = 0.019 \pm 0.003 cm^{-1}$. Above 900 K strong increase in line width occurs. (b) shows the shift in energetic position of the LO and TO modes together with a fit of C and D according to equ.3 (solid lines). fitting parameter for the LO-shift: $\omega_0^{LO} = 296.4 cm^{-1}$, $C^{LO} = 0.381 \pm 0.018 cm^{-1}$, $D^{LO} = 0,008 \pm 0.006 cm^{-1}$ and for TO-shift: $\omega_0^{TO} = 273.2 cm^{-1}$, $C^{TO} = 0.429 \pm 0.046 cm^{-1}$, $D^{TO} = 0,009 \pm 0.004 cm^{-1}$. The dotted line shows the line shift if the three phonon decay is neglected ($D^{LO,TO} = 0$), and the dashed curve shows the contribution of the thermal expansion ($C^{LO,TO}, D^{LO,TO} = 0$)

reports for InSb [13] showed a decrease in the LO-TO splitting almost to zero. This was interpreted as a screening of the macroscopic electric phonon field by thermally generated free carriers. However, the much larger fundamental gap of GaAs ($E_0 = 0.98$ eV at 1200 K [14]) leads to thermal carrier concentrations which are several orders of magnitude lower than those in InSb ($E_0 = 0.044$ eV at 800 K [15]). Hence the screening effect and reduction in LO-TO gap does not occur.

4 Conclusion

We have shown that the current *in-situ* MOVPE-Raman scattering setup is capable to measure vibrational properties of GaAs at high temperatures, used in semiconductor growth environments. We are able to observe changes due to annealing which were related to improvement of the crystal quality as well as redistribution of dopants. Details of the GaAs spectra show that the usual models of phonon anharmonic effects have to be modified by additional channels at high temperatures (>900 K) in order to explain the observed line widths in GaAs. We attribute this to different decay channels at such high temperatures. This is confirmed by the fact that the LO-phonon line shift can be well described including a fourth order anharmonicity term.

Our main goal of this work is the measurement of the surface phonons in different surface reconstructions and possibly the observations of phase transitions. Since by now the properties of the bulk related features are known in detail, we are optimistic concerning the outcome of such measurements on surface vibrational properties.

References

[1] N. Esser and W. Richter, *Raman Scattering from Surface Phonons*, Eds.: M. Cardona, G. Güntherodt, Springer-Verlag Berlin Heidelberg New York (2000).

[2] J. Shealy and G. Wicks, Appl. Phys. Lett. **50**, 1173 (1987).

[3] Landolt-Börnstein data collection, http://www.springer.de/phys/laboe/.
[4] R. Cowley, J. Phys. (Paris) **26**, 659 (1965).
[5] P. G. Klemens, Phys. Rev. **148**, 854 (1966).
[6] M. Balkanski, R. Wallis and E. Haro, Phys. Rev. B **28**, 1928 (1983).
[7] T. Hart, R. Aggarwal and B. Lax, Phys. Rev. B **2**, 638 (1970).
[8] H. Bilz and W. Kress, *Phonon Dispersion Relations in Insulators*, Eds.: M. Cardona, P. Fulde, Springer-Verlag Berlin Heidelberg New York (1979).
[9] H. Tang and I. Herman, Phys. Rev. B **43**, 2299 (1991).
[10] J. Menendez and M. Cardona, Phys. Rev. B **29**, 2051 (1984).
[11] A. Debernardi, Phys. Rev. B **57**, 12847 (1998).
[12] P. Wickboldt, E. Annastassakis, R. Sauer and M. Cardona, Phys. Rev. B **35**, 1362 (1987).
[13] E. Liarokapis and E. Anastassakis, Phys. Rev. B **30**, 2270 (1984).
[14] H. Shen, S. Pan, Z. Hang, J. Leng, F. Pollak, J. Woodall, R. Sacks, Appl. Phys. Lett **53**, 1080 (1988).
[15] P. Liu and J. Maan, Phys. Rev. B **47**, 16274 (1993).

phys. stat. sol. (c) **0**, No. 8, 2956–2960 (2003) / **DOI** 10.1002/pssc.200303848

Formation of Cu$_x$ clusters in Cu/ZnO nanocomposites studied by IR spectroscopy

U. Pal[1,2*], **O. Vázquez-Cuchillo**[3], **A. Bautista-Hernández**[2], **and J. F. Rivas-Silva**[2]

[1] Instituto Mexicano del Petrıleo, Programa de Investigacıın y Desarrollo de Ductos, Eje Central Lázaro Cárdenas 152, Col. San Bartolo Atepehuacan, 07730 Mexico D.F, Mexico
[2] Instituto de Física, Universidad Autınom a de Puebla, Apdo. Postal J-48, Puebla, Pue. 72570, Mexico
[3] Laboratorio de Catálisis Ambiental, IC-BUAP, 14 Sur 6301, Puebla, Pue., 72550 Mexico

Received 30 May 2003, revised 4 August 2003, accepted 11 August 2003
Published online 10 November 2003

PACS: 61.46.+w, 63.22.+m, 78.30.Er, 78.67.Bf

Nanocomposites of Cu/ZnO with different Cu contents were prepared by radio frequency co-sputtering technique. The composite films were annealed at different temperatures in argon atmosphere for 2 hrs. Transmission electron microscopy and x-ray diffraction studies revealed the formation of partially oxidized Cu nanoparticles in ZnO matrix. The size of the Cu nanoparticles depended strongly on the annealing temperature. Infrared spectroscopy study revealed that the nanoparticles consist of an elemental Cu core surrounded by the oxidized copper shell. The elemental core consists of Cu$_x$ clusters. IR spectroscopy and DFT calculation were used to identify the Cu clusters.

1 Introduction

Formation of metal nanoparticles such as Cu, Ag and Au in glass matrix has attracted much attention recently due to their potential applications in non-linear optical devices [1-6]. Several techniques, for example, ion-implantation [2,6,7,8], chemical [9], sol-gel [10] and sputtering [11-13] have been employed by several workers to prepare metal nanocomposites. For the production of nonlinear optical devices, it is important to ensure a certain concentration of such colloidal particles in the matrix. Though a large amount of colloidal particles can be incorporated in the matrix by ion-implantation, the defects created by high energy ions modify their optical properties drastically. On the other hand, by alternate or co-sputtering technique, we can control the amount of dopant easily without modifying the optical properties of the host material. Though, silica or quartz glass have been used vastly as matrix material to prepare colloidal metal particle composites, there are only a few reports on the use of functional matrix material like ZnO for this purpose [11,12,14].

In the present work, an evidence of the formation of Cu nano-clusters in ZnO matrix is presented. Cu/ZnO nanocomposites were grown on quartz glass substrates by r.f. co-sputtering technique and annealed in argon at different temperatures. Transmission Electron Microscopy (TEM), X-ray diffraction and Infra-red (IR) absorption studies revealed that the nanoparticles consist of a metallic Cu core surrounded by a partially oxidized cap layer. Density Function Theory (DFT) in combination with the effective core potentials was used to calculate the optimum geometry of the Cu isomers and their corresponding vibrational frequencies. A comparison of the experimental results with the theoretical calculation revealed that the elemental Cu in the core of the nanoparticles remains mainly in dimer (Cu$_2$) trimer (Cu$_3$) and tetramer (Cu$_4$) cluster forms.

---

* Corresponding author: e-mail: upal@sirio.ifuap.buap.mx, Fax: +52-22 2295611

2   Experimental

The Cu/ZnO composite films were prepared on well cleaned quartz glass substrates by sputtering of ZnO and Cu wires simultaneously. Different numbers of Cu wires (2 mm length and 0.5 mm of diameter) were placed symmetrically on a ZnO target (50 cm diameter) and sputtered with 200W r.f. power at 20 mTorr Ar pressure. The content of Cu in the films was varied by changing the number of Cu pieces on the ZnO target, keeping the time of sputtering fixed (2 hrs). Depending on the number of Cu pieces, the thickness of the composite films varied from 0.19 μm to 0.22 μm. The as-grown composite films were annealed at different temperatures for 2 hrs in argon atmosphere. A Siemens D5000 X-ray diffractometer with Cuka source was used to record the XRD pattern of the samples. A JEOL 2010 electron microscope was used for obtaining TEM micrographs. A FT-IR Vector 22 spectrometer was used to record the IR absorption spectra of the samples in diffuse mode.

3   Computational details

For the calculation of geometry optimization and vibrational frequencies of Cu clusters, we have used DFT technique in combination with the effective core potentials. This combination has been used currently for the study of metallic clusters with few atoms [15]. The basis set used for the copper atom is the Los Alamos laboratory (LANL) set for effective core potential (ECP) of double-ζ type [16] with relativistic corrections. This pseudopotential consists of small core ECP with 3s and 3p orbitals in the valence space. The functional that we used is the B3PW91 [17], a hybrid functional which define the exchange as a linear combination of Hartree-Fock and gradient-corrected exchange terms [18]. All the calculations were performed using the Gaussian-98 program [19].

Full geometry optimizations were performed via the Berny algorithm in redundant internal coordinates. The thresholds for the convergence were 0.00045 au, 0.0003 au, 0.0018 au and 0.0012 au for the maximum force, root-mean-square (RMS) force, maximum displacement, and RMS displacement, respectively. Once the optimization of cluster geometry is done, the vibrational frequencies were calculated as the second derivative of the energy with respect to the nuclear positions.

4   Results and discussion

Figure 1 shows the typical TEM micrographs of the Cu/ZnO composite films. Formation of nano-clusters in the matrix is clear from the contrast of the micrographs. Most of the nanoparticles were in the range of 3-14 nm in diameter and dispersed homogeneously in the matrix. The average diameter of the nano-clusters increased with the increase of Cu content and also with the annealing temperature [12].

The XRD patterns of the composite films revaled the presence of Cu, both in elemental and oxide form in the composites [12].

To extract the effect of Cu incorporation in the composite films precisely, for the recording of the IR spectra, a quartz substrate with ZnO film on it was used as the reference. For the measurement of IR spectra of the composite films annealed at different temperatures, the ZnO films on quartz substrate annealed at corresponding temperatures were used as the reference. Therefore, the features appeared in the IR spectra are only due to the incorporation of Cu in ZnO.

As the absorption peaks related to the oxides of copper appears in the frequency range 1000-400 cm^{-1} and of elemental Cu in 400-200 cm^{-1}, we divided the full range of measured spectra in two parts: 1000-400 cm^{-1} and 400-200 cm^{-1}.

In figures 2a and 2b, the IR spectra of the as-grown Cu/ZnO composites prepared with different Cu contents are presented for the 1000-400 cm^{-1} and 400-200 cm^{-1} spectral range respectively. For all the samples, we can observe the appearance of two absorption bands at around 530 and 800 cm^{-1} for the 1000-400 cm^{-1} spectral range. There appeared three absorption bands in the low frequency spectral range at around 252, 240 and 226 cm^{-1}. The intensity of the bands increased with the increase of Cu content in the films. Figures 3a and 3b show the IR spectra for the composite films prepared with different Cu contents and annealed at 400^{0}C for the two spectral ranges. There appeared three absorption bands at around

530, 655 and 800 cm^{-1} for the first spectral range. The absorption band observed at around 655 cm^{-1} was assigned to the symmetric stretching mode of Cu(O2), 2A" with C_{2n} symmetry by several workers [20-23]. The absorption detected at around 530 cm^{-1} was generally assigned to the symmetric stretching vibration of Cu-O bond of the CuOO molecule [24-26]. The last absorption band appeared close to 800 cm^{-1} was assigned to the vibration of Cu-O bond in the CuO_3 molecules [24, 26].

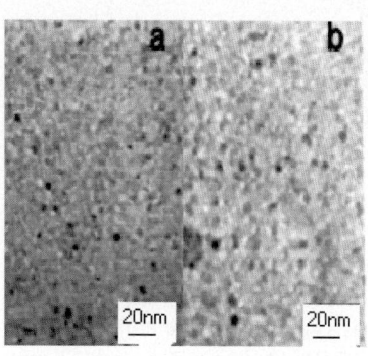

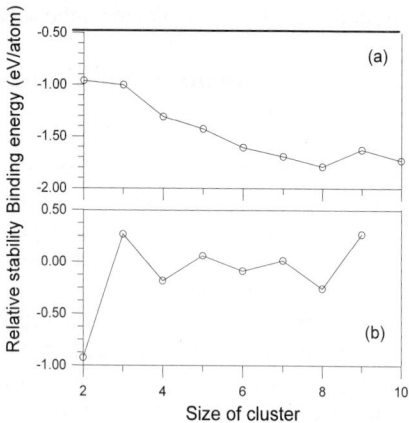

Figure 1. TEM micrographs of the Cu/ZnO composites prepared with 16 pieces of Cu wires a) as-grown and b) annealed at 400^{0}C.

Figure 2. IR spectra of the as-grown composite films prepared with different Cu contents a) for the 1000-400 cm^{-1} and b) 400-200 cm^{-1} spectral ranges.

The peak position of the absorption bands in the 400-200 cm^{-1} spectral range did not change on annealing the sample at higher temperatures. However, on annealing the samples at 400^{0}C, the relative inten-

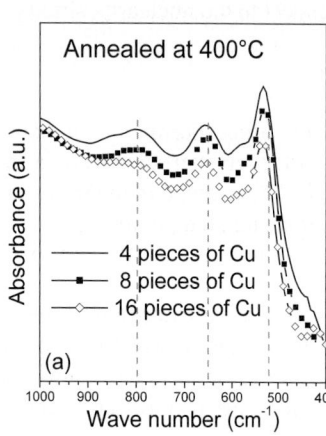

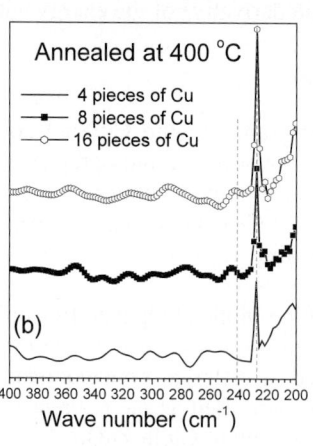

Figure 3. IR spectra of Cu/ZnO nanocomposites prepared with different Cu contents and annealed at 400^{0}C a) for 1000-400 cm^{-1} and b) for 400-200 cm^{-1} spectral ranges.

Figure 4. Binding energy (a) and (b) relative stabilities (eV) of Cu clusters calculated by DFT

sity of the 226 cm^{-1} band increased drastically. The first band in this spectral range at about 252 cm^{-1} is assigned to the vibration of Cu_2 linear clusters [27]. The second one appeared at around 240 cm^{-1} is generally assigned to the stretching asymmetric vibration of triangular Cu_3 clusters [28]. However, for the last one appeared at about 226 cm^{-1}, there had been no experimental or theoretical report in the literature. We assigned this band to the vibration of Cu_4 clusters from our theoretical study of vibrational frequencies of Cu_x clusters.

5 Theoretical results

Figure 4 shows the a plot of binding energy versus cluster size (number of Cu atoms). The values of the binding energy for the 2 and 3 atomic clusters are in good agreement with the experimental results [29], and for the other higher size clusters, the values are also in agreement with the reported theoretical results [29]. In the same figure (4b), the relative stability of the Cu clusters is ploted against the number of copper atoms. The relative stability of the Cu clusters was calculated by the relation E=2E(N)-E(N+1)-E(N-1); where, N is the number of Cu atoms in the cluster. We can observe that the clusters formed with N=2,4,6 and 8 are the most stable. In table 1, the most stable spatial geometry and corresponding frequencies of vibration for the Cu clusters for N=2-10 are presented. The highest intensity (more probable) frequencies for each cluster are given in the parentheses.

By comparing our experimental results with the theoretical calculations, we can associate the absorption band appeared at around 252 cm^{-1} to the formation of Cu_2 clusters in the composites. Though the frequency is very close to the calculated frequencies of Cu_5 and Cu_6 clusters, the relative stability of the Cu_2 clusters is higher than the others.

Table 1. Point group and frequencies for the Cu clusters calculated with DFT. The most intense frequencies are given in the parentheses.

Cluster	Point group	Frequencies (cm^{-1})
Cu_2	D_h	(259)
Cu_3	C_{2v}	80, (153), 239
Cu_4	D_2	(220), 58, 112, 149, 255
Cu_5	C_2	(251), 38, 40, 99, 111, 135, 160, 197, 212
Cu_6	D_{3h}	(252), 26, 33, 43, 96, 97, 118, 119, 164, 187, 193, 254
Cu_7	D_{5h}	(208), 60, 105, 106, 119, 123, 127, 146, 147, 153, 155, 209, 223
Cu_8	T_d	(220), 57, 71, 73, 100, 101, 107, 110, 116, 159, 167, 175, 203, 211
Cu_9	C_{3v}	(234), 9, 12, 33, 38, 40, 51, 58, 70, 88, 93, 100, 131, 145, 149, 164, 178, 182, 210, 253, 270
Cu_{10}	D_{4d}	(242), 8, 16, 31, 36, 39, 45, 48, 62, 81, 87, 97, 106, 130, 136, 141, 157, 161, 184, 190, 196, 208, 253, 266

Though our calculation revealed the most intensity frequency for Cu_3 clusters at around 153 cm^{-1}, which is in agreement with the other theoretical works [30]. J. Szeyrbowski et al. [28] have attributed the 240 cm^{-1} frequency to the Cu_3 clusters from their experimental results. It is to be noted that, our theoretical calculation also revealed a frequency at around 239 cm^{-1} but with lower frequency.

The band appeared at around 226 cm^{-1} is very close to the most intensity vibrational frequency for Cu_4 clusters (Table 1). Though the frequency is very close to the most probable frequency for Cu_8 clusters, from the binding energy value in the figure 8a, we can see the formation of Cu_8 cluster is less probable than Cu_4 clusters. Therefore, we assigned our experimental 226 cm^{-1} band to the formation of Cu_4 clusters in the composite films.

6 Conclusions

Incorporation of Cu in ZnO matrix by sputtering resulted the formation of partially oxidized Cu nanoparticles in ZnO matrix. Infra-red absorption and our theoretical calculation using DFT revealed that the nanoparticles consist of an elemental Cu core surrounded by an oxide cap layer. The elemental cores consist of Cu dimer, trimer and tetramer clusters. However, the annealing of the films at 400^{0}C favored the formation of Cu_4 clusters. A futher theoretical investigation is needed to study the mecanism of this transformation.

Acknowledgements   Partial support from VIEP, BUAP (No. II013I02) and CONACyT are gratefully acknowledged.

References

[1] G.I. Stegeman, R.H. Stolen, J. Opt. Soc. B **6**, 652 (1989).
[2] R.F. Haglung Jr., R.H. Magruder III, S.H. Morgan, D.O. Henderson, R.A. Zuhr,L. Yang, R.A. Weller, L. Yang, R.A. Zhur, Nucl. Instrum. Methods B **65**, 405 (1992).
[3] R.H. Magruder III, R.A. Zuhr, R.A. Weeks, Nucl. Instum. and Meth. B **59/60**, 1306 (1991).
[4] R.A. Wood, P.D. Townsend, N.D. Skelland, D.E. Hole, J. Barton, C.N. Afonso, J. Appl. Phys. **74**, 5754 (1993).
[5] N.D. Skelland, J. Sharp, P.D. Townsend, Nucl. Instrum. and Meth. B **90**, 446 (1994).
[6] R.H. Magruder III, R.F. Haglund Jr., L. Yang, J.E. Witting, R. A. Zuhr, J. Appl. Phys. **76**, 708 (1994).
[7] G.W. Arnold, J.A. Border, J. Appl. Phys. **48**, 1488 (1977).
[8] U. Pal, A. Bautista-Hernández, L. Rodríguez-Fernández, J.C. Cheang-Wong, J. Non-Cryst. Sol. **275**, 65 (2000).
[9] M.J. Bolemer, J.W. Hans, P.R. Ashley, J. Opt. Soc. Am. B **7**, 790 (1990).
[10] A. Chatterjee, D. Chakravorty, J. Phys. D: Appl. Phys. **22**, 1386 (1989).
[11] Y. Yoshino, S. Takanezawa, T. Ohmori, Hideki Masuda, Jpn. J. Appl. Phys. **35**, L1512 (1996).
[12] O. Vazquez-Cuchillo, U. Pal, C. Vazquez-Lopez, Solar Energy Mat. & Solar Cells **70**, 369 (2001).
[13] J. Garcia-Serrano, U. Pal, International J. Hydrogen Ener. **28**, 637 (2003).
[14] U. Pal, E. Aguila Almanza, O. Vazquez Cuchillo, N. Koshizaki, T. Sasaki, S. Terauchi, Solar Ener. Mat. & Solar Cells **70**, 363 (2001).
[15] P. B. Balbuena, P.A. Derosa, and J.M. Seminario, J. Phys. Chem. **103**, 2830 (1999).
[16] P. J. Hay, W.R. Wadt, J. Chem. Phys. **82**, 270 (1985); W.R. Wadt, P.J. Hay, J. Chem. Phys. **82**, 284 (1985).
[17] A. D. Becke, J. Chem. Phys. **98**, 5648 (1993); J.P. Perdew, J.A. Chevary, S.H. Vosko, K.A. Jackson, M.R. Pederson, D.J. Singh, C. Fiolhais, Phys. Rev. B **46**, 6671 (1992).
[18] J. P. Perdew, Y. Wang, Phys. Rev. B **45**, 13244 (1992).
[19] Gaussian 98, Revision A.9, M.J. Frisch et al. Gaussian, Inc., Pittsburgh PA, 1998.
[20] V. E. Bondybey, J.H. English, J. Phys. Chem. **88**, 2247 (1984).
[21] W. Hongbin, R. Sunil, and W. Lai-Sheng, J. Chem. Phys. **101**, 3898 (1994).
[22] W. Hongbin, S.R. Desai, and W. Lai-Sheng, J. Phys. Chem.A **101**, 2103 (1997).
[23] D.E. Tevault, J. Chem. Phys. **76**, 2859 (1982).
[24] G.V. Chertihin, L. Andrews, C.W. Bauschlicher, Jr., J. Phys. Chem. A. **101**, 4026 (1997).
[25] K.P. Muthe, J.C. Vyas, N.S. Narag, D.K. Aswal, S.K. Gupta, D. Bhattacharya, R. Pinto, S.C. Sabharawal, Thin Solid Films **324**, 37 (1998).
[26] D.E. Tevault, R.C. Mowery, R.A. Demarco, R.R. Smardzewsky, J. Chem. Phys. **34**, 244 (1981).
[27] H. Hongbin, R. Sunil and W. Lai-Sheng, J. Chem. Phys. **101**, 3898 (1994).
[28] J. Szeyrbowski and A. Czapla, Thin Solid Flims **46**, 127 (1977).
[29] A. J. Bhattacharyya, A. Mookerjee and A. K. Bhattacharyya, arXiv:cond-mat/0001285 v1 20 Jan 2000.
[30] P. Calaminici, A. M. Koster, N. Russo, D. R. Salahub, J. Chem. Phys. **105**, 9546 (1996).

phys. stat. sol. (c) **0**, No. 8, 2961–2965 (2003) / **DOI** 10.1002/pssc.200303858

Infrared spectroscopy of high k thin layer
by multiple internal reflection and attenuated total reflection

N. Rochat[*, 1], **K. Dabertrand**[2], **V. Cosnier**[2], **S. Zoll**[2], **P. Besson**[2], and **U. Weber**[3]

[1] CEA Grenoble, DRT/DTS-LETI/SCPC, 17 rue des martyrs, 38054 Grenoble, France
[2] STmicroelectronics, 850 rue Jean Monnet, 38926 Crolles Cedex, France
[3] AIXTRON AG, Kackerstr. 15-17, 52072 Aachen, Germany

Received 30 May 2003, revised 5 September 2003, accepted 8 September 2003
Published online 10 November 2003

PACS 78.30.Hv, 78.66.Nk, 81.15.Gh, 85.40.Sz

A complete analysis of HfO_2 high k thin layers deposited by metal organic chemical vapour deposition (MOCVD) is made by infrared spectroscopy using different optical configurations. Multiple internal reflection (MIR) and attenuated total reflection (ATR) are used, giving information about morphology and composition of the layers. Samples under investigation are HfO_2 layer of 3 to 12 nm thick deposited at 400 and 550°C on thin silicon oxide. MIR technique elucidates carbon content in the layer and ATR informs about the interface evolution, and the crystalline character of the HfO_2 layer. The high sensitivity of MIR coupled with the large spectral range available of ATR allow the full characterisation of these layers.

1 Introduction

Aggressive scaling down of silicon based microelectronic devices for sub-100 nm CMOS technology requires ultra thin gate insulators. To overcome leakage current due to direct tunnelling limitation of SiO_2, HfO_2 is considered as a serious alternative dielectric gate due to its higher dielectric constant (high k) and good chemical compatibility with silicon. Introduction of such material in microelectronic devices stimulates interest for non destructive surface characterisation providing structural and chemical information. The lack of sensitivity of standard infrared spectroscopy (IRS) compared to other surface techniques such as x-ray photoelectron spectroscopy is motivating the development of optical configurations able to enhance its surface sensitivity enough to detect one monolayer. Multiple internal reflection (MIR) [1, 2, 3] and attenuated total reflection (ATR)[4, 5, 6] are the most useful configurations. The most sensitive is the MIR arrangement [7], however ATR allows oxide bond spectral range investigations [8, 9]. So the use of both of them allows the sensitive study of thin films deposited on silicon in a wide range of wave number.

Metal-organic chemical vapour deposition (MOCVD) is a potential technology to deposit thin HfO_2 layers. Moreover the electrical characteristics are very sensitive to the layer morphology and composition. So the effect of thickness and deposition temperature of HfO_2 from 3 to 12 nm thick, deposited at 400 and 550°C, on thin silicon oxide is studied using MIR and ATR arrangement. The layer morphology and the evolution of the interfacial layer between the silicon substrate and the hafnium oxide layer are under investigation.

---

[*] Corresponding author: e mail: nevine.rochat@cea.fr, Phone: +33 4 38 78 93 31, Fax : +33 4 38 78 94 85

2 Experimental

The ATR set up [4] consist of a germanium prism pressed against the silicon sample as shown Figure 1. A pressure tip mounted on a micrometric screw allows a tight contact between the prism and the wafer. This enable a reproducibility better than +/- 10%. A P polarized IR beam coming from a Bruker IFS55 FTIR spectrometer is directed onto the prism base with an angle of 65° which ensures a total reflection on the germanium prism dioptre. After one reflection on the prism base, the IR light is focused onto a liquid N_2 cooled HgCdTe detector. The sensitivity of ATR measurement is due to the enhancement of the E_{pz} electric field component, orthogonal to the sample surface. The E_{pz} component can be increased by a factor 50 under total reflection in the air gap which is present between the germanium prism and the sample. For this reason ATR measurements are always performed using P polarized IR light, instead of S polarization, that is not sensitive enough to detect such thin layers. Moreover theoretical development shows that P polarization ATR spectra gives an image of the energy loss function $\varepsilon_f'' / |\varepsilon_f|^2$ of the absorbing thin film on silicon substrate [4]. Sample spectrum is referenced to the spectrum obtain when no sample is coupled to the prism. Infrared data are acquired between 600 and 4500 cm^{-1}, but ATR lost its sensitivity with increasing wave number due to the penetration depth decrease. So ATR spectra are exploited between 600 and 2000 cm^{-1}.

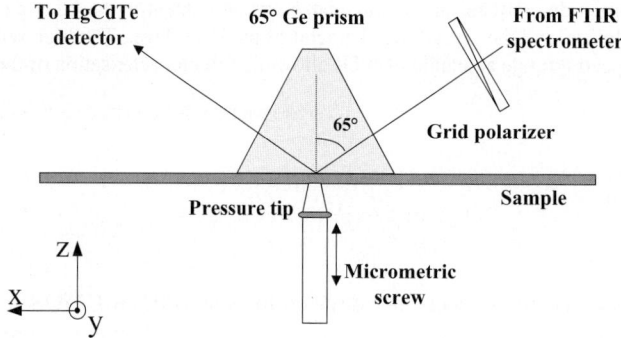

Fig. 1 Schematic representation of ATR set up

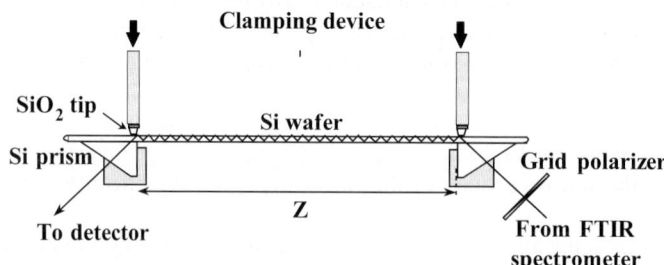

Fig. 2 Schematic representation of MIR set up

The MIR set up [1] consist of a work platform on which two silicon prisms couplers are mounted. The distance Z between the two prisms can vary from 5 mm to the full diameter of the sample. As shown in Figure 2 the silicon wafer is putdown horizontally on the prism facets. Two fused silica pressure tips supported by a clamping device allow a tight contact between the prism and the wafer. An S polarized IR beam coming from a Bruker IFS 55 FTIR spectrometer is directed on the coupling area of the input prism which ensure optical tunnelling inside the wafer. After being internally reflected W times (W=280 for θ=34°, substrate thickness e = 750 μm and propagation distance Z = 150 mm) the IR beam is coupled out of the wafer by the second prism and focused onto the liquid N_2 cooled HgCdTe detector. The sensitivity of MIR measurement is due to the multiple interaction of light with the sample surface (280 times in our case). The number of internal reflection is controlled through the propagation distance Z between the prisms. As described in reference [1] the MIR absorbance of a chemical bond X is expressed by $ABS_X = K \cdot [X] + cte$, where [X] is the surface density of the specie and $K = \dfrac{\gamma \cdot \sigma_X \cdot Z}{2.3 \cdot e \cdot \tan \theta}$, with γ the internal reflection field sensitivity factor [10], e the substrate thickness, θ the internal reflection angle, Z the propagation distance and σ_X the relevant absorption cross-section. Because the IR beam propagates inside the silicon wafer, samples have to be double side polished, and low doped. The S polarization is used to avoid optical loses at the in and out coming of the light by optical tunneling. A reference sample is needed to eliminate the optical bench contribution to the MIR spectrum. In our case the sample with the thinnest layer of each deposition temperature is used as a reference. Spectra are acquired using 150 mm propagation distance between 1800 and 4500 cm^{-1} due to the silicon absorption which enable measurement below 1800 cm^{-1}.

To sum up ATR is sensitive to silicon surface from 600 to 2000 cm^{-1} and MIR from 1800 to 4500cm^{-1}. As a result, the combine use of MIR and ATR allows characterisation of thin layers deposited on silicon. In this paper samples under investigation are HfO$_2$ layers deposited in an Aixtron Tricent MOCVD tool using double side polished high-resistivity silicon as substrate (with 1 nm chemical silicon oxide) and tetra diethyl amino hafnium (TDEAH) as Hf precursor. The high k films are grown in a thickness range between 3 to 12 nm at 400°C and 550°C. Layers thicknesses are determined by ellipsometry.

3   Results and discussion

3.1   ATR results

HfO$_2$ ATR spectra show a large absorption band between 650 and 800 cm^{-1} attributed to Hf-O stretching vibration (figure 3). Two different shapes are observed. 400°C deposited HfO$_2$ (figure 3.a) shows one maximum of absorption at 680 cm^{-1}, 550°C deposited HfO$_2$ (figure 4.b) exhibit another maximum near 770 cm^{-1}. The presence of two maxima at 550°C is related to the monoclinic phase of the hafnium oxide, previously mentioned by N. Mc Devitt *et al.* [11]. The disappearance of the second maxima at 400°C can be related to the amorphous character of the film [12]. These results are confirmed by XRD measurement which are performed on other samples with a Bruker D500 diffractometer (Figure 4). The 550°C XRD spectrum shows diffraction peaks of monoclinic HfO$_2$ [13], and the 400°C spectrum shows no diffraction peaks, representative of an amorphous layer. For low temperature deposition the crystallisation takes place beyond 8 nm. For higher temperature the crystallisation occurs earlier, at 550°C the 4.9 nm thick layer is already crystallised. So the crystalline character of the HfO$_2$ film is sensitive to temperature deposition and thickness.

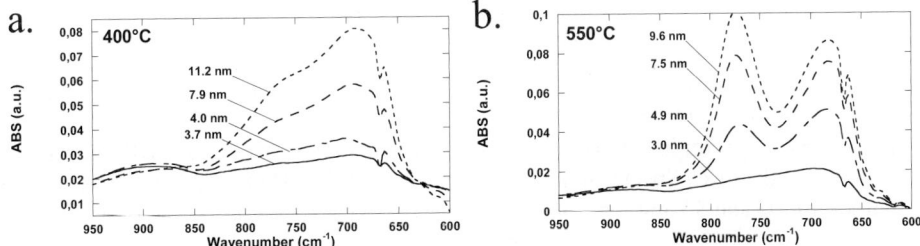

Fig. 3   ATR spectra in Hf-O stretching region of various HfO$_2$ layers deposited at a. 400°C and b. 550°C

SiO$_2$ interface layer is also analysed in the 1050 - 1250 cm^{-1} spectral range. It has been observed (figure 5) that for 400°C deposition SiO$_2$ absorption peak amplitude clearly decreases while HfO$_2$ thickness increases above 4 nm. This trend is weaker for the 550°c deposition but is still present for thicknesses above 8 nm (see 3.2).

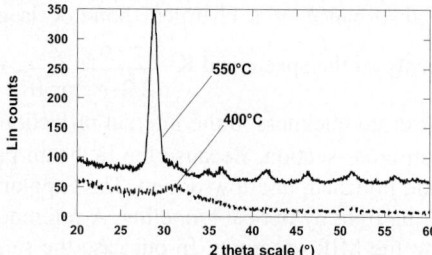

Fig. 4   X Ray diffraction spectra of 9 and 11 nm thick samples respectively deposited at 400 and 550°C.

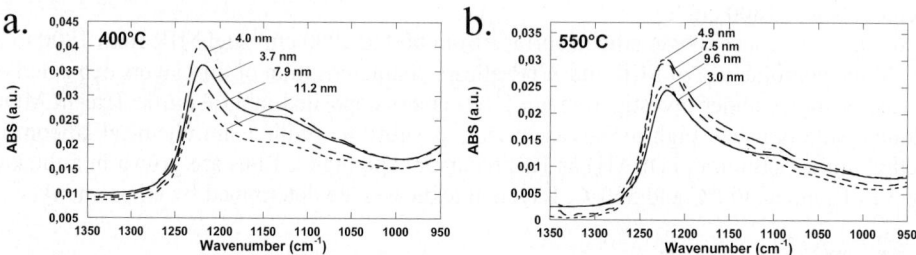

Fig. 5   ATR spectra in Si-O stretching region of various HfO₂ layers deposited at a. 400°C and b. 550°C

3.2   MIR results

A well known drawback of MOCVD deposition is the carbon content inside the layer [14]. As shown Figure 6.a CH$_x$ bonds are detected between 2800 and 3000 cm^{-1} only for the thicker layer deposited at 400°C. No carbonated groups are detected for higher temperature deposition. So the carbon is directly related to the deposition temperature. The surface concentration of CH$_2$ groups located at 2920 cm^{-1} in the thicker layer deposited at 400°C is evaluated to be around 2 10^{-13} cm^{-2}, using 2.9 10^{-19} cm² as absorption cross section values of CH$_2$ groups [15]. To draw conclusion out of these observation special care has to be taken. Since the thinnest sample of each temperature is used as a reference, as mentioned before, the correction of the CHx amplitude introduces an error of +/- 30% for the 8 nm thick sample.

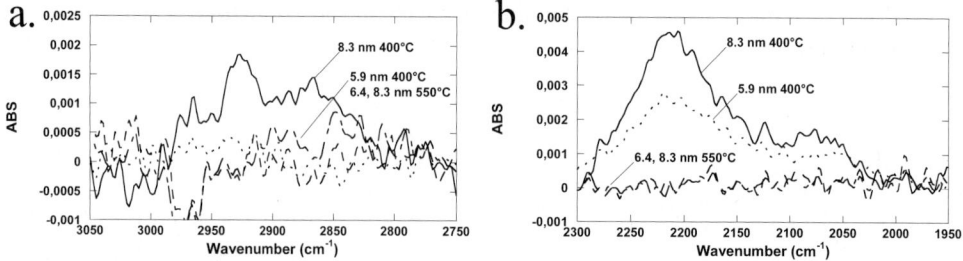

Fig. 6   MIR spectra in a. C-H and b. Si-H stretching region of various HfO₂ layers deposited at 400°C and 550°C. Spectrum are referenced to the thinnest sample for each temperature (2.9 nm for 400°C and 3.9 nm for 550°C)

Figure 6.b shows, located between 2000 and 2300 cm^{-1}, that Si-H peak amplitude increases with increasing HfO₂ thickness at 400°C only. The Si-H peak is located around 2210 cm^{-1} meaning that the silicon atom is back bonded to 2 oxygen atoms (Si-H$_2$(O$_2$)) [16]. So Si-H bonds are created during the HfO₂ growth at 400°C only. This observation has to be correlate with the decrease of the SiO₂ interfacial layer during HfO₂ growth seen by ATR. The thickness range of SiO₂ consumption at 400°C (above

4 nm) and at 550°C (above 8 nm) correspond to the thickness range of Si-H creation at 400°C (seen at 6 and 8 nm) and at 550°C (not seen before 8 nm). The SiO_2 interfacial layer decrease coupled with the Si-$H_2(O_2)$ increase might be due to the oxidation of metallic hafnium by the reduction of the SiO_2 interfacial layer. Such reaction is thermodynamically favourable, oxidation potential of HfO_2/Hf and SiO_2/Si are respectively equal to -1.56 V and –0.84 V [17]. The reduction of SiO_2 may lead to Si dangling bonds, saturated by hydrogen [18], creating Si-H bonds in the silicon oxide layer so Si-$H_x(O_y)$ bonds. This hypothesis have to be confirmed by others characterisation results to determine if the HfO_2 is substochiometric or not, that may confirmed a favourable kinetic of silica reduction. Moreover results about the localisation of the Si-H bonds, located near Si/SiO_2 or HfO_2/SiO_2 interface should informed if the metallic hafnium reduced silica or if it is a non radiative recombination of Si/SiO_2 interfacial defect [19].

4 Conclusion

HfO_2 layers of various thicknesses and structures were investigated using ATR and MIR measurements. These techniques are able to detect silicon oxide, hafnium oxide, and carbon or hydrogen defects in layers as thin as one nanometer. Study of the HfO_2 growth and its interfacial layer evolution shows that the HfO_2 crystallization depends on the growth temperature and the layer thickness. For thicker films and higher deposition temperature the layers are more crystalline. Moreover the carbon content, dependent on the growth temperature of the layer, has been quantified. Furthermore, the MIR results indicate that a reaction takes place in the interfacial layer, and might be attributed to a reaction between metallic hafnium and SiO_2 interfacial layer during deposition.

This highlights the extreme sensitivity and the large spectral range available which is achieved because of the combined use of ATR and MIR homemade set up. So the couple use of MIR and ATR allows the full characterization of ultra thin layers deposited on silicon substrate. However, other characterization are needed to confirmed these results.

Acknowledgements The authors thank G. Rolland from CEA-LETI/SCPC for the XRD measurement. This work has been carried out in the frame of CCMC consortium between CEA-LETI, ST Microelectronics and FRANCE TELECOM-R&D.

References

[1] N. Rochat, M. Olivier, F. Conne, G. Lefeuvre, and C. Boll-Burdet, Appl. Phys. Lett. **77(14)**, 2249 (2000).

[2] M. Olivier, N. Rochat, A. Chabli, G. Lefeuvre, and F. Conne, Mat. Sc. in Semicon. Process. **4**, 15 (2001).

[3] N. Rochat, A. Troussier, A. Hoang, and F. Vinet, Mat. Sc. Eng. **C23**, 99 (2003).

[4] N. Rochat, A. Chabli, F. Bertin, M. Olivier, C. Vergnaud, and P. Mur, J. Appl. Phys. **91/8**, 5029 (2002).

[5] N. Rochat, A. Chabli, F. Bertin, C. Vergnaud, P. Mur, S. Petitdidier, and P. Besson, to be published in Mater.Sci.Eng C.

[6] V. Cosnier, H. Bender, M. Caymax, J. Chen, T. Conard, H. Nohira, O. Richard, W. Tsai, W. Vandervorst, E. Yuoung, C. Zhao, S. De Gendt, M. Heyns, J. W. H. Maes, M. Tuominen, N. Rochat, M. Olivier, and A. Chabli, IWGI 2001, IEEE N°01EX537, 226 (2001).

[7] N. Rochat, M. Olivier, A. Chabli, F. Conne, and G. Lefeuvre Appl. Phys. Lett. **77**, 2249 (2000).

[8] J. E. Olsen and F. Shimura, J. Appl. Phys. **66**, 1353 (1989).

[9] S. Petitdidier, N. Rochat, D. Rouchon, P. Besson, Solid state phenomena, **92**, 187 (2003).

[10] N. J. Harrick, internal reflection spectroscopy, Wiley, New York 1967.

[11] N. T. Mc Devitt, W. L. Baun, Spectrochem. Acta, **20** 799 (1964).

[12] Y. Morisaki, Y. Sugita, K. Irino, T. Aoyama, IWGI 2001, 184].

[13] JCPDS n°34-0104 of the American Society for Testing and Materials.

[14] M. Morstein, I. Pozsgai, N. D. Spencer, Chem. Vap. Dep. **5(4)**, 151, 1999.

[15] K. Nakanishi and P. H. Solomon, Infrared absorption spectroscopy, Holden Day, Oakland, CA, 1977.

[16] G. Lucovsky, J. Vac. Sci. Technol., **16** 1225 (1979).

[17] Handbook of chemistry and physics, 63rd ed., CRC press, Boca Raton, 1982-1983.

[18] R. R. Razouk and B. E. Deal, J. Electrochem. Soc. **129(4)**, 806 (1982).

[19] T. Dittrich, T. Burke, and F. Koch, J. Appl. Phys. **89(8)**, 4636 (2001).

phys. stat. sol. (c) **0**, No. 8, 2966–2970 (2003) / **DOI** 10.1002/pssc.200303830

Isotropic and anisotropic optical reflectances of clean and hydrogen-covered Si(001)2x1 surfaces

Y. Borensztein*, N. Witkowski, and S. Royer

Laboratoire d'Optique des Solides, UMR CNRS 7601, University Pierre et Marie Curie, 4 place Jussieu, F-75252 Paris cedex05, France

Received 30 May 2003, revised 4 August 2003, accepted 11 August 2003
Published online 10 November 2003

PACS 68.35.Bs, 68.43.-h, 68.47.Fg, 73.20.–r, 78.68.+m

A combination of two different surface optical spectroscopies, Surface Differential Reflectance Spectroscopies and Reflectance Anisotropy Spectroscopies, is used to investigate the optical response of the clean and hydrogenated Si(001)2x1 surface. It is shown that the isotropic contribution is about 4 times more intense than the anisotropic one, and is composed of two main peaks. The measured spectra are in excellent agreement with previous ab-initio calculations The components of the surface dielectric tensor for polarizations of light parallel and perpendicular to the dimers of silicon are determined from the experimental reflectances.

1 Introduction

Surface optical spectroscopies have been proved, in the last decade, to be efficient techniques for in-situ investigations of structural, electronic and optical properties of clean single crystals [1-3], and can be used as well in vacuum as in transparent media like gas or liquids [4]. The following linear optical techniques have been widely used: Reflectance Anisotropy Spectroscopy (RAS) (also called by some authors Reflectance Difference Spectroscopy) and Surface Differential Reflectance Spectroscopy (SDRS). However, the difficulty of interpreting the data requires a strong interaction between experiments and theories [3]. Important progress has been obtained by different groups in the understanding of the optical response of semiconductor surfaces, either by use of the semi-empirical tight-binding approach (TB) [5,6], or more recently by use of first-principle approaches like the density-functional theory in the local-density approximation (DFT-LDA) [7]. Recent improvements have been made by introducing corrections due to excitonic and local-field effects [8]. Although TB approach gave excellent agreement with experiment, for example in the case of the complicated Si(111)7x7 surface [6,9], which can be understood by the fact that the surface electronic states are well localized on the given atomic features of this surface, the situation is more difficult with the apparently simpler Si(001)2x1 surface, which is formed by rows of parallel Si dimers, for which TB approach failed [5] and first-principle models are required. This can be explained by the presence of surface states delocalized along the dimer rows, which cannot be described within the TB model.

However, although RAS is a very sensitive technique, which gives well-reproducible spectra (it is an easy method for controlling the quality of a Si(001)2x1 surface, alternative to the usual Low-Energy Electron Diffraction), it provides only the *anisotropy* of the surface reflectance, and not the intrinsic optical response of the surface. In particular, features in the surface optical response which would have no or small anisotropy would not be determined by RAS, and there is no certainty that a theoretical approach, which would reproduce correctly the RA spectra, would also reproduce such features. The moti-

---

* Corresponding author: e-mail: borens@ccr.jussieu.fr

vation of this work is to gain both the isotropic and anisotropic optical responses of the Si(001) surface and to compare them to recent theoretical results. Accordingly, we combine RAS measurements on clean and H-adsorbed surfaces, with SDRS measurements that compare clean and H-covered surfaces.

2 Experimental

The Si(001) samples were vicinal surfaces, 4°-miscut towards the $[110]$ direction. They were prepared by direct-current heating in ultra-high vacuum of base pressure 5.10^{-11} mbar, with flashes up to 1050 °C with pressure maintained in the 10^{-10} mbar range, leading to surfaces with 3.9 nm-wide 2x1 terraces separated by double-steps in the $[\bar{1}10]$ direction, and covering about 80 % of the surface, as checked by LEED and scanning tunneling microscopy. The terraces are covered by Si-dimers parallel to the steps, forming rows in the $[110]$ direction. The RAS measurements were performed by use of an homemade apparatus similar to the one developed by Aspnes [10]. The SDRS were performed by use of a spectrometer, based on an optical multichannel analyzer consisting in a Si photodiode array, described in detail previously [11]. It delivers the relative change of reflectance of the substrate upon adsorption of hydrogen : $\dfrac{\Delta R_s}{R_s} = \dfrac{R_{Si} - R_{Si:H}}{R_{Si}}$, where R_{Si} and $R_{Si:H}$ are the reflectances (in intensity) of the clean Si surface and of the H-covered Si surface. The measurements were performed in s-polarization (electric field parallel to the surface, either aligned along or perpendicular to the dimer rows).

3 Results and discussion

When atomic hydrogen, obtained by H molecule dissociation from a hot W wire located in front of the sample, is adsorbed on the Si surface maintained at room temperature, it results in a poorly ordered surface with flat di-hydride Si(001)1x1:H regions, where the Si dimers are broken and the surface Si atoms are bounded to two H atoms, equivalent to an hydrogenated ideal Si(001) surface. When H is adsorbed at higher temperatures (300 °C), it results in a perfect mono-hydride Si(001)2x1:H surface, where the dimers are preserved [12]. Figure 1 present the RA spectra obtained for these different surfaces, defined by $\dfrac{\Delta R}{R} = \dfrac{R_{[\bar{1}10]} - R_{[110]}}{R}$,

where $R_{[\bar{1}10]}$ and $R_{[110]}$ are the reflectances of the surface for the electric field parallel to the $[\bar{1}10]$ and $[110]$ directions respectively [13]. The spectrum of the clean surface is similar, both in shape and in intensity, to spectra obtained previously by different authors [14-16], and has been already discussed in details. To our knowledge, two different reports have been made previously of the RA spectra for the Si(001)2x1:H surface [17,18], and ours is similar to the one obtained by Shioda and van der Weide [18]. Our RA spectrum for the Si(001)1x1:H surface is also close to the one

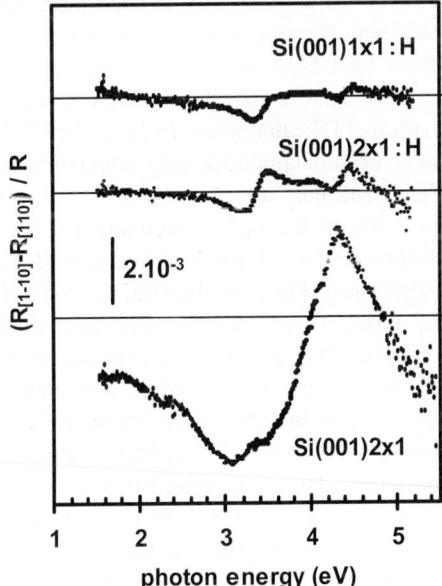

Fig. 1 Reflectance anisotropy spectra, defined by $\dfrac{R_{[\bar{1}10]} - R_{[110]}}{R}$ for clean Si(001)2x1 and hydrogenated Si(001)2x1:H and Si(001)1x1:H surfaces.

published in [19], although in our case an additional small feature appears close to the E_2 critical point. The shape of the hydrogenated surface spectra are dominated by features located close to the bulk critical points of silicon E_1 at 3.4 eV and E_2 at 4.3 eV. As proposed, they originate probably from modulation due to the change of the surface strain induced by the adsorption of H [19,20].

In order to get the intrinsic optical response of the clean surface, we measured the SDR spectra corresponding to the difference of reflectances of the clean surface and of the di-hydride one which allows direct comparison with previous calculations [7,21]. Figure 2.b displays the SDR spectra obtained in s-polarization for two orientations of the electric field: parallel and perpendicular to the dimers [22]. Both spectra are very similar, being dominated by broad features around 3 and 4 eV, with a minimum around 3.4 eV. The intensity almost reaches 20.10^{-3}, while the intensity of the RA spectrum for the clean Si(001)2x1 surface, reproduced in this figure, is smaller than 5.10^{-3}. This means that the observed anisotropy is a rather small effect, which is superimposed to an overall 4 times larger isotropic optical response. These spectra are to be compared with calculations obtained recently in Del Sole's group [7,21]. In [7] the comparison has been made between DFT-LDA calculation and TB one for difference in reflectances of the clean 2x1 surface and the (1x1):H surface, for unpolarized light. Both the intensity and the shape of the calculated signal obtained from DFT-LDA are in excellent agreement with our experimental results, whereas there is poor agreement with the TB calculation. In [21], the DFT-LDA results have been obtained for both orientations of the electric field, and they are reproduced in Fig. 2.a. The larger intensity of the optical response for the electric field perpendicular to the dimers agrees well with the present experiment (Fig 2.b), although there is a shift in energy of about 0.6 eV between the main double structure observed at 3 and 4 eV in experiment and the one at 2.4 and 3.4 eV in calculation. This underestimation of the gaps in the DFT-LDA methods is a well-known effect [7]. In order to compare better theory and experiments, the same group has been applying in [7] an upward rigid shift $\Delta=0.5$ eV to the conduction band, in agreement with GW calculations, which yields an

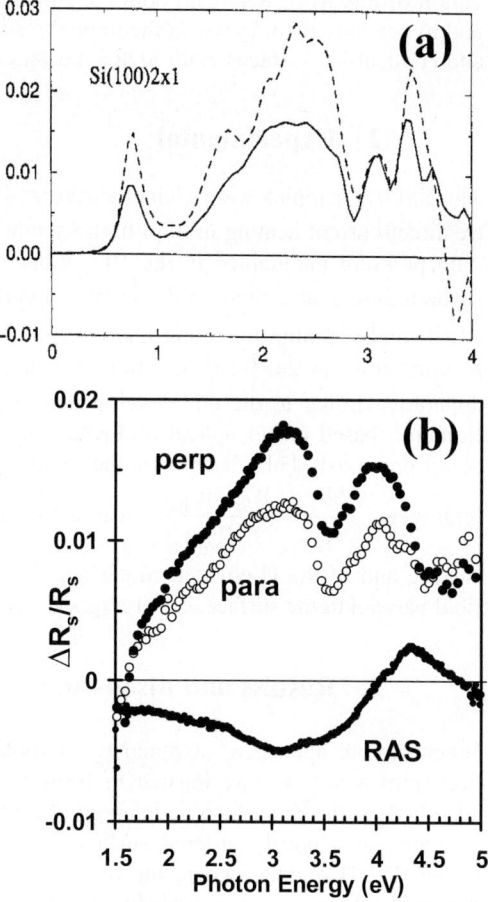

Fig. 2 Surface differential reflectance spectra, defined by $\Delta R_s/R_s = (R_{Si} - R_{Si:H})/R_{Si}$, for light polarization perpendicular (perp) and parallel (para) to the dimers. (a): DFT-LDA calculation, reproduced from [21], with authorization. (b): experiments; the RA spectrum for clean Si(001) surface is reproduced for comparison from Fig. 1 (dashed line: perp; continuous line: para).

excellent agreement in energy position with the present experiments (but for unpolarized light). Accordingly, we have shifted Fig. 2.a with respect to Fig. 2.b, in order to reproduce an energy shift of 0.5 eV. This permits to demonstrate the good agreement between experiment and theory, both in shape, intensity and energy position, after having taken into account the gap underestimation. The main difference between experiment and theory is the presence, in the calculated curves, of an optical absorption peak around 0.9 eV (1.4 eV with the shift), which is interpreted as due to optical transitions between π and π^* surface states, strongly delocalized along the dimer rows, which is not obtained in our experiments, neither in RA or SDR spectra. Actually, this state could be experimentally obtained in RAS only for well-ordered 2x1 surfaces with broad terraces [15]. It is likely that the small width of the 2x1 terraces in the vicinal surfaces leads to a strong modification or even to a disappearance of these states. It must be noticed that the present experiments have been performed on vicinal surfaces, for which the presence of steps have been shown to produce some anisotropic response [16,23]. Although it is difficult to clearly separate the contribution to the RAS due to the steps from the change of the signal of the terraces due to their reduced size, experiments and theory have shown that the intrinsic anisotropy of the steps is not larger than about $2 \cdot 10^{-3}$ [16,23]. Accordingly, the presence of steps is likely not to have a large influence on the intrinsic optical response of the surface, which justifies the comparison between our experimental results on vicinal surfaces and the calculations for ideal 2x1 surfaces.

The differential reflectance can be expressed by : $\dfrac{\Delta R_s}{R_s} = 4 \dfrac{\omega}{c} \cos\theta \operatorname{Im}\left\{ \dfrac{\Delta\varepsilon_{surf}}{\varepsilon_{bulk} - 1} \right\}$ where θ is the angle

of incidence, ε_{bulk} the dielectric function of bulk silicon and $\Delta\varepsilon_{surf}$ the difference between the dielectric responses of the clean surface and the hydrogenated one [3]. Assuming that optical response of the ideal 1x1 hydrogenated surface is small, because the surface states are removed by the presence of H, $\Delta\varepsilon_{surf}$ gives essentially the dielectric response of the clean 2x1 surface. $\Delta\varepsilon_{surf}$, which has the dimension of length, is a complex quantity, which fulfills Kramers-Kroenig relation. Consequently, it can be ex-

pressed, for example, as a sum of lorentzian functions: $\Delta\varepsilon_{surf} = \displaystyle\sum_j \dfrac{f_j}{\omega_j^2 - \omega^2 - i\Gamma_j\omega}$. The fitting of the

lorentzian function parameters, in order to reproduce correctly the DR spectra, allows us to get the $\Delta\varepsilon_{surf}$ quantities along and perpendicular to the dimers [24]. Their imaginary parts, multiplied by the frequency of light ω, which gives the surface optical absorptions, are drawn in Fig. 3 in comparison with the bulk dielectric function. Both surface dielectric responses display a main peak lying below 3.5 eV, and a secondary peak centered around 4 eV. It can be shown that the 3 eV peak originates essentially from the dangling bond states of the dimers, while the 4 eV peak is related to the internal bonding of the dimers [25]. These peaks correspond clearly to the DR peaks observed in Fig.2. It is worthwhile noticing that,

although the DR spectrum for light polarization along the dimer ($[\bar{1}10]$ direction) has a smaller overall intensity than the spectrum for light perpendicular to the dimers ($[110]$ direction), this is not the case for $\omega.\Delta\varepsilon_{surf}$, the 4 eV peak being more intense for the parallel case, while the 3 eV one is more intense in the perpendicular case. It is known that $\omega.\Delta\varepsilon''_{surf}$ fulfills the following sum rule, adapted for the surface response:

$$n_{eff} = \frac{m}{2\pi^2 e^2 N_s} \int_{\omega_1}^{\omega_2} \omega' \left[\varepsilon''_{surf}(\omega') \right] d\omega' \text{ , where } N_s \text{ is}$$

the number of surface atoms and n_{eff} the effective number of electrons per atom involved in the optical transitions occurring between light frequencies ω_1 and ω_2 [26,27]. The use of this sum rule, integrated between 1.7 and 5 eV, for the previously determined $\Delta\varepsilon_{surf}^{perp}$ and $\Delta\varepsilon_{surf}^{para}$, where perp and para refers to the orientation with respect to the dimers, gives $n_{eff}^{perp} = n_{eff}^{para} \approx 2.5$. This number is close to the effective number of electrons per atom, which is obtained from the sum rule for bulk Si between 1.7 and 5 eV : $n_{eff}^{bulk} \approx 2$.

Moreover, the fact that $n_{eff}^{perp} = n_{eff}^{para}$ indicates

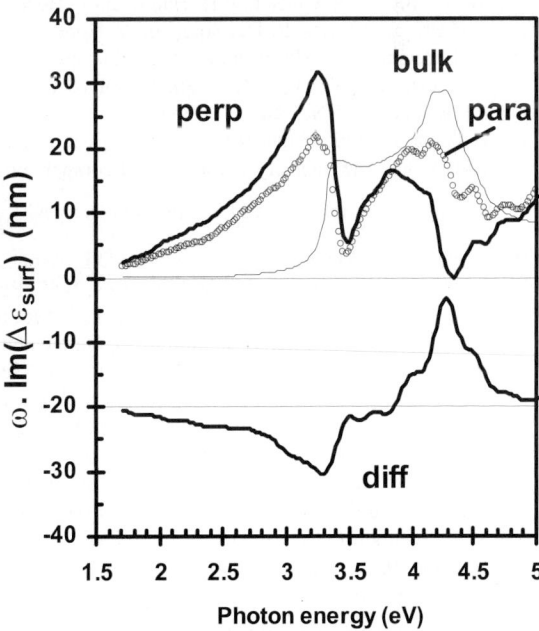

Fig. 3 $\omega\operatorname{Im}(\Delta\varepsilon_{surf})$ obtained from the fitting of the experimental SDR spectra, either perpendicular (thick line) or parallel (empty circles) to the dimers, given in nanometers. The difference is also drawn, with a vertical shift of 20. $\omega\operatorname{Im}(\varepsilon_{bulk})$ (divided by 5) is shown in thin line for comparison.

© 2003 WILEY-VCH Verlag GmbH & Co. KGaA, Weinheim

that, as it is expected, the same number of electrons per atom is involved in the interband transitions, whatever the polarization of light is. This can be seen, actually, by the difference between $\Delta\varepsilon_{surf}^{perp}$ and $\Delta\varepsilon_{surf}^{para}$ drawn in Fig.3, which displays a negative peak around 3.2 eV balanced by a positive one around 4.2 eV. From the previous discussion, it can be concluded that the effective number of electrons per Si atom involved in the surface optical response is approximately the same, in the energy range under investigation, as the one for the bulk optical response. Finally, it can be seen that the difference curve: $\omega\,\mathrm{Im}\left\{\Delta\varepsilon_{surf}^{para} - \Delta\varepsilon_{surf}^{perp}\right\}$ (Fig.3) has a similar shape as the RAS spectrum for the clean surface (Fig.1).

This was expected, as the RAS is given by the formula: $\dfrac{\Delta R}{R} = 4\,\dfrac{\omega}{c}\,\mathrm{Im}\left\{\dfrac{\varepsilon_{surf}^{para} - \varepsilon_{surf}^{perp}}{\varepsilon_{bulk} - 1}\right\}$, showing the consistency of these two different optical approaches [28].

References

[1] P. Chiaradia and G. Chiarotti, in: Photonic Probes of Surfaces, edited by P. Halevi (Elsevier Science, 1995), chap. 3.
[2] Y. Borensztein, Surf. Rev. Lett. 7, 399-410 (2000).
[3] R. Del Sole, see [1], chap. 4.
[4] V. Mazine and Y. Borensztein, Phys. Rev. Lett. **88**, 147403 (2002).
[5] A.I. Shkrebtii and R. Del Sole, Phys. Rev. Lett. **70**, 2645 (1993).
[6] C. Noguez, C. Beitia, W. Preyss, A. I. Shkrebtii, M. Roy, Y. Borensztein, and R. Del Sole, Phys. Rev. Lett. **76**, (1996) 49
[7] M. Palummo, G. Onida, R. Del Sole, and B. S. Mendoza, Phys. Rev. B **60**, 2522 (1999).
[8] W.G. Schmidt, S. Glutsch, P.H. Hahn, and F. Bechstedt, Phys. Rev. B **67**, 085307 (2003).
[9] C. Beitia, W. Preyss, R. Del Sole, and Y. Borensztein, Phys. Rev. B **56**, R4371 (1997).
[10] D. E. Aspnes, J. P. Harbison, A. A. Studna, and L. T. Florez, J. Vac. Sci. Technol. A **6**, 1327 (1988).
[11] Y. Borensztein, T. Lopez Rios, and G. Vuye, Appl. Surf. Sci. **41-42**, 439 (1989).
[12] J.J. Boland, Surf. Sci. **261**, 17 (1992).
[13] It should be noticed that RA is usually given in terms of the amplitude r of the reflectance, whereas we have chosen here the intensity R for a direct comparison with the SDRS data. The relation between the reflectance anisotropies in intensity and in amplitude is : $\left[R_{[\bar{1}10]} - R_{[110]}\right]\big/R = 2\,\mathrm{Re}\left\{\left[r_{[\bar{1}10]} - r_{[110]}\right]\big/r\right\}$
[14] T. Yasuda, L. Mantese, U. Rossow, and D.E. Aspnes, Phys. Rev. Lett. **74**, 3431 (1995).
[15] R. Shioda and J. van der Weide, Phys. Rev. B **57**, R6823 (1998).
[16] S.G. Jaloviar, J.-L. Lin, F. Liu, V. Zielastek, L. McCaughan, and M.G. Lagally, Phys. Rev. Lett. **82**, 791 (1999)
[17] J.R. Power, W. Richter, M. Palummo, G. Onida, and R. Del Sole, phys. stat. sol. (a) **175**, 63 (1999).
[18] R. Shioda and J. van der Weide, Appl. Surf. Sci. **130-132**, 266 (1998).
[19] U. Rossow, L. Mantese, T. Yasuda and D.E. Aspnes, Appl. Surf. Sci. **104/105**, 137 (1996).
[20] V.I. Gavrilenko and F.H.Pollak, Phys. Rev. B **58**, 12964 (1998).
[21] C. Kress, A. Shkrebtii, and R. Del Sole, Surf. Sci. **377**, 398 (1997).
[22] The spectra have been renormalized to the normal incidence case, by dividing the spectra by cosθ following the expression of the differential reflectance, in order to compare with the RA spectra and the theoretical results.
[23] W.G. Schmidt, F. Bechstedt, and J. Bernholc, Phys. Rev. B **63**, 045322 (2001).
[24] Actually, there is some uncertainty on the intensity of $\Delta\varepsilon_{surf}$: it can be seen from the equation that, if $\Delta\varepsilon_{surf}$ is solution of the fitting, any expression $\varepsilon_{surf} + \alpha(\varepsilon_{bulk} - 1)$ is also solution. We have chosen the constant α so that $\mathrm{Im}(\Delta\varepsilon_{surf})$ does not display negative values.
[25] N. Witkowski, O. Pluchery, S. Royer, and Y. Borensztein, to be published.
[26] F. Wooten, Optical Properties of Solids, Academic Press, New York (1972).
[27] G. Chiarotti, P. Chiaradia, E. Faiella, and C. Goletti, Surf. Sci. **453**, 112 (2000).
[28] Actually, the right comparison should be made between the dielectric function difference and the difference between RAS measured on the clean and on the 1x1:H surfaces.

phys. stat. sol. (c) **0**, No. 8, 2971–2975 (2003) / **DOI** 10.1002/pssc.200303859

Reflectance anisotropy for porphyrin octaester Langmuir-Schaefer films

C. Castillo, **R. A. Vázquez-Nava**, and **Bernardo S. Mendoza**[*]

Centro de Investigaciones en Optica, A.C. León, México

Received 30 May 2003, revised 4 August 2003, accepted 11 August 2003
Published online 12 November 2003

PACS 73.40.Me, 78.55.Qr, 78.66.Qn, 78.68.+m

We present a theoretical study for the reflectance anisotropy of porphyrin layers deposited onto gold substrates by Langmuir-Schaefer technique. We have obtained reflectance anisotropy spectra as function of the number of porphyrin layers, and show that the line shape goes from peak-like to derivative-like as the geometrical arrangement of the porphyrin molecules is changed. Our results coincide qualitatively with those of the experiments.

1 Introduction Optical spectroscopic techniques are increasingly used nowadays to investigate surfaces and interfaces. Both linear and non-linear optical probes are employed to investigate very different physical aspects of surfaces with great success [1]. In particular reflectance anisotropy spectroscopy (RAS) has received a growing attention from the experimental and theoretical sides, since it is one of the few optical techniques that probes directly the surface and interface structure of cubic materials. It measures the difference between the normal-incidence optical reflectance of light polarized along the two principal axes in the surface plane as a function of the photon energy. RAS data are typically obtained in the visible-ultraviolet spectral range, thus providing information about electronic structure modifications due to the creation of the surface, reconstructions, adsorbates, surface electric fields etc. [1].

Molecular materials have aroused considerable interest in recent years due to their large potential impact on nanotechnology. The characterization of electronic states in the fabricated molecular structures is essential. Recently, RAS has been applied to organic layers, showing that the spectra are reliably connected to the electronic properties of the molecule and to the morphological characteristics of the layer [2, 3, 4]. Optical techniques have demonstrated to be particularly useful to characterize the arrangements of porphyrins in Langmuir-Schaefer (LS) films. In particular, the use of polarized light offers the possibility of studying system that exhibit anisotropies due to electronic or morphological characteristics of the organic layer grown onto an isotropic substrate, in analogy to what has been successfully done in the case of semiconductor growth by means of RAS [1].

In this paper, we present a theoretical study of the reflectance anisotropy of porphyrin layers deposited onto a gold isotropic substrates by the Langmuir-Schaefer technique, which ensures a highly ordered deposition. Goletti et al. [3], measured RAS as a function of the number of porphyrin layers, and found an abrupt change in the RAS line shape at 8-10 monolayer coverage: the line shape, which at lower coverage is essentially proportional to the Soret band absorption and thus peak-like, becomes derivative-like. In their paper it is argued that this change may be due to a structural change in the porphyrin orientation. Indeed, with a polarizable dipole model, in which the effect of the electric local field is incorporated, we find that this change could be readily related to morphological changes in the layers of the films.

---

[*] Corresponding author: e-mail: bms@cio.mx, Fax: + 52 477 717 5000

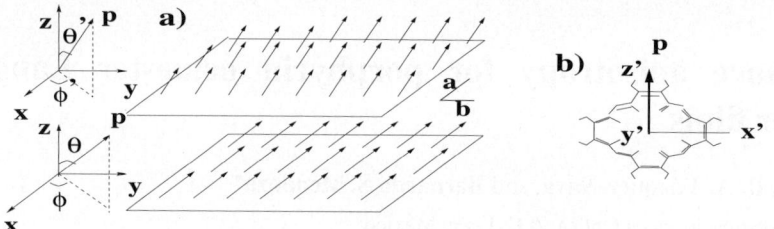

Fig. 1 a) The arrows represent the polarizable entities with a tilt angle θ and a twist angle ϕ. The planes with arrows represent a stack of several layers, with equal θ and ϕ, thus the upper plane shows a change in tilt and twist with respect to the lower plane. x, y, z are the system's coordinates. b) Structure of the porphyrin and its total polarizable dipole moment along the principal axis, z', of the porphyrin.

2 Theory

The optical activity of the porphyrin molecule is represented by a polarizable dipole that responds to the incident light like a harmonic oscillator. The induced dipole moment, $\vec{p}$, is given by

$$p_i(\omega) = \alpha_{ij}(\omega)E_j(\omega), \tag{1}$$

where $\vec{E}$ is the incident electric field of frequency ω, and $\overset{\leftrightarrow}{\alpha}$ is the polarizability of the porphyrin. Using the porphyrin's coordinate system, we assume that $\overset{\leftrightarrow}{\alpha}$ is diagonal and has only one resonant frequency, implying that the molecule only polarizes along its principal axis (see Fig. 1). Thus we can write

$$\alpha(\omega) = \alpha_0 / \left[1 - (\omega/\omega_0)^2 - i(\omega/\omega_0)(1/(\omega_0\tau)) \right], \tag{2}$$

where ω_0 is the unique resonant frequency, τ is a damping parameter related to the width of the resonance, and α_0 is the value of the static dipole moment of the porphyrin.

To reproduce the layers of porphyrins, we consider a system consisting of N identical non-overlapping polarizable entities regularly distributed in a plane with the same dipole orientation characterized by angles θ and ϕ, with respect to the plane's coordinate system (see Fig. 1). The polarizable entities, or layers, characterized with a dielectric function $\epsilon_1(\omega)$, sit on top of an isotropic substrate, also characterized by a dielectric function $\epsilon_2(\omega)$. The Langmuir-Schaefer technique ensures a highly ordered deposition of the porphyrin molecules, which we assume sitting in a rectangular lattice, with lattice parameter a and b along the two mutually perpendicular x and y directions along the plane, and even interlayer spacing d perpendicular to the substrate, i.e. along z. For simplicity we assume that the first layer sits a distance d on top of the substrate. Since all the dipoles in a given layer, ℓ, are identical, the induced dipole moment, $p_i(\ell, \omega)$, would be identical too, and is given through

$$
\begin{aligned}
p_i(\ell, \omega) = {} & \alpha_{ij}(\ell, \omega) \left[E_j + \sum_{\ell'} \left(T_{jk}(a,b)\delta_{kl}\delta_{\ell\ell'} + \mathcal{T}_{jk}(a,b,|d_\ell - d_{\ell'}|)\delta_{kl}(1 - \delta_{\ell\ell'}) \right. \right. \\
& \left. \left. + \frac{\epsilon_2 - \epsilon_1}{\epsilon_2 + \epsilon_1} \mathcal{T}_{jk}(a,b,|d_\ell + d_{\ell'}|)S_{kl} \right) \right] p_l(\ell', \omega),
\end{aligned}
\tag{3}
$$

where ℓ, ℓ' go from one till the total number of layers, L. Sum over repeated Cartesian indexes is assumed. Note that this equation is the same as Eq. 1, but we have replaced the incident field by the local field, which is given in the square bracket, and is composed by the same incident field and the field coming from the system's dipoles. Then, the term $T_{ij}(a,b)$ is recognized as the dipolar interaction tensor that gives the local field contribution of the entities on the same plane, and $\mathcal{T}_{ij}(a,b,z)$, as that of the entities in plane ℓ with those of plane ℓ', with z the inter-plane separation and d_ℓ the vertical position of the ℓ-th plane. However the third term gives the contribution of the planes on top of the substrate, whereas the fourth term gives the contribution coming from the images located inside the substrate, as can be recognized by the screening

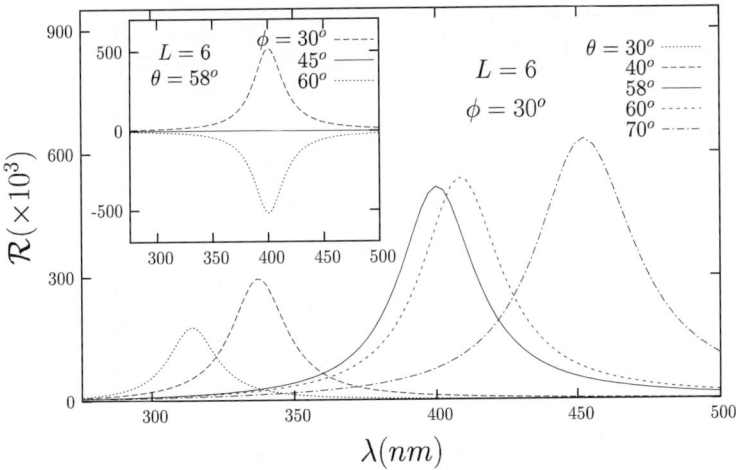

Fig. 2 $\mathcal{R}$ vs. λ for $L = 6$ for several values of θ and $\phi = 30°$, whereas the inset is for for several values of ϕ and $\theta = 58°$.

factor given by the dielectric functions, and $S_{ij} = \text{diag}(-1, -1, 1)$ to give the correct orientation of the image dipoles. Both $\overset{\leftrightarrow}{T}$ and $\overset{\leftrightarrow}{\mathcal{T}}$ have been summed over all the entities in the corresponding planes using the plane-wise summation technique of [5].

To solve Eq. 3, we need to go from $\alpha(\omega)$ of Eq. 2, into $\alpha_{ij}(\ell, \omega)$ which is done by a simple transformation from the porphyrin axis to the coordinate system of the sample (see Fig. 1), then

$$\alpha_{ij}(\ell, \omega) = R_{ik}(\ell)R_{jl}(\ell)\alpha'_{kl}(\omega), \tag{4}$$

where R_{ij} is the rotation matrix that takes us from one system to the other, and $\alpha'_{kl}(\omega) = \text{diag}(0, 0, \alpha(\omega))$. The orientation of the molecule given through the angles θ (polar-tilt) and α (azimuthal-twist) would be a function of the plane ℓ, and thus the dependence on $\mathcal{R}_{ij}$. Then, once we have $\vec{p}(\ell, \omega)$, we follow [6], to calculate the RAS signal of the system, $\mathcal{R}$, as the normalized change in reflectance through

$$\mathcal{R} \equiv 4\pi \left(\frac{d}{\lambda}\right) \frac{1}{p_o} \Im m \sum_{\ell=1}^{L} [p_x(\ell, \omega) - p_y(\ell, \omega)], \tag{5}$$

with λ the wavelength of the incident light, and p_o a normalization factor with the units of dipole moment and proportional to α_o, which is used to set the scale of $\mathcal{R}$.

3 Results

For simplicity we assume $a = b = d$, i.e. a cubic lattice, and $d_\ell = \ell d$, $\epsilon_1(\omega) = 1$ and $\epsilon_2(\omega)$ given by the experimental values for gold [7]. Then, besides the number of layers L, we leave θ and ϕ as the only variables in the model. Since $a = b$ one would expect $\mathcal{R} = 0$ since the system is isotropic, but if we allow for the porphyrins to be tilted and twisted with respect to the substrate, i.e. $\theta \neq 0$ and $\phi \neq 0$, the RAS signal would be different from zero. Since the substrate is isotropic the x and y axis are chosen arbitrarily. To chose ω_0 and τ we use the experimental results of the UV-visible absorption spectrum of [3], where a dominant Soret band is seen at $\omega_0 = 387$ nm^{-1}, with $\omega_0\tau = 0.25$.

In Fig. 2 we show $\mathcal{R}$ for $L = 6$ as a function of θ. We see that as θ grows, the peak-like structure shifts to long wavelengths and its amplitude and broadening is increased. From here we choose $\theta = 58°$ in order for the peak structure to coincide with the RAS experimental result, which gives a red-shifted Soret band at $\lambda = 395$ nm [3]. However, the value of p_o is taken to reproduce the same experimental magnitude of $\mathcal{R}$ for $L = 2$ in Fig. 4, since in this figure we present our best agreement with experiment. In the inset of

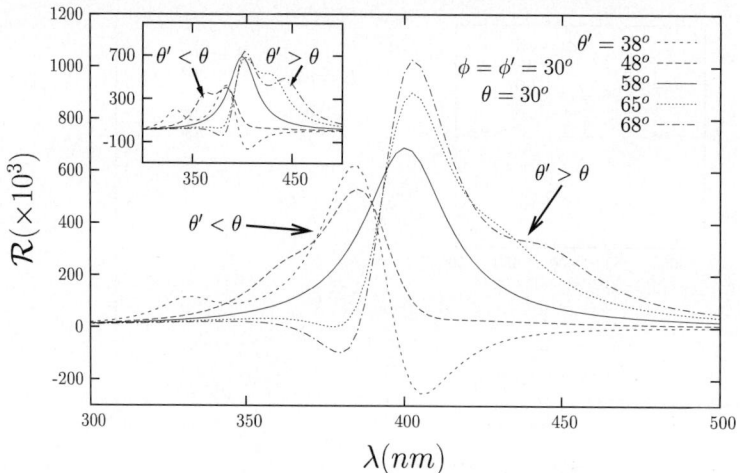

Fig. 3 $\mathcal{R}$ vs. λ for $L = 8$ and a fixed $\phi = 30^\circ$. The lower six layers have $\theta = 58^\circ$, and in the upper two layers θ' varies from 38° to 68°. On the other hand, in the inset, the lower four layers have $\theta = 58^\circ$, and the upper four layers have the same variation in θ' as in the main panel.

Fig. 2 we also show $\mathcal{R}$ as a function ϕ for a fixed $\theta = 58^\circ$. We see that only the amplitude of $\mathcal{R}$ changes and as we go above $\phi = 45^\circ$ its sign is reversed. At $\phi = 45^\circ$ $\mathcal{R} = 0$ because at this particular value of ϕ, $p_x(\ell, \omega) = p_y(\ell, \omega)$. From this figure we learn that a system with all the layers with the same θ and ϕ always presents a peak-like $\mathcal{R}$.

Now, we try to simulate the RAS experiments of Ref. [3]. Fig. 3 shows $\mathcal{R}$ for a porphyrin system of $L = 8$ and $\phi = 30^\circ$, in which the tilt angle θ' of the porphyrins in the last two (four) planes varies from 38° to 68°, and the first six (four) have a fixed value of $\theta = 58^\circ$ (see Fig. 1). We see that the spectrum has a derivative-like shape when θ' is different from θ, and that as the value of θ' in the last two or four planes approaches to the value of 58° the derivative-like shape tends to disappear and we get the peak-like characteristic of a multilayer system of planes with dipoles with the same orientation. For $\theta' < \theta$ the derivative-like shape in $\mathcal{R}$ has the same structure as in the experiment, but as the molecule tilts closer to the substrate, i.e. $\theta' > \theta$ the signs in the line-shape are reversed.

From Fig. 3, we see that $\mathcal{R}$ is controlled by the tilt angle θ and that by allowing some of the top layers to reorient with respect to the underlying ones in such a way that they get closer to the substrate's normal, the experimental trend is qualitatively reproduced. Thus, we do a more detailed analysis and we progressively change θ and ϕ as we increase L. In Fig. 4, we show $\mathcal{R}$, changing L from 2 till 16 in steps of 2, and for each corresponding change in L we decrease θ by 2° and increase ϕ by 1.2°, starting at $\theta = 58^\circ$ and $\phi = 30^\circ$ for $L = 2$. Indeed, as L is increased $\mathcal{R}$ goes from peak-like to derivative-like, indicating that the model qualitatively reproduces in terms of the geometrical reordering of the porphyrin layers the experimental trend as L is increased [3]. In the same figure, we show $\mathcal{R}$ when ϕ is fixed to 30°, where we see that the agreement in the short wavelength region is not as good, and therefore the azimuthal variation that we introduced, also induces a twist in the porphyrins, that is required in order to get agreement with the experiment. For completeness we have checked that the same behavior is found when we change the substrate from a metal to a dielectric, thus changing $\epsilon_2(\omega)$ which controls the image term in Eq. 3. Since the interlayer local field interaction decays exponentially with z, the image's contribution is only marginal. Thus, our model would apply to the films of Ref. [2], grown by the Langmuir-Blodgett technique on a quartz substrate.

The decrease of θ as the number of layers is increased could be understood by the fact that the interaction with the substrate will be screened by the underlying layers, thus producing porphyrins which will try to

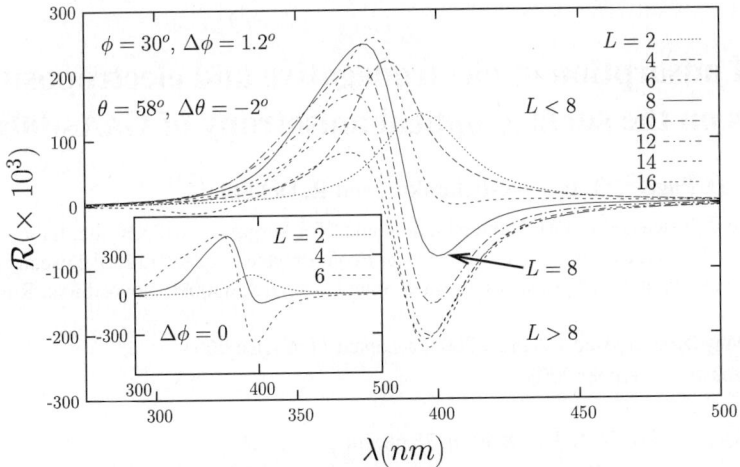

Fig. 4 $\mathcal{R}$ vs. λ for different values of L, θ, and ϕ. In the main panel, θ decreases 2° and ϕ increases by 1.2^o. In the inset $\phi = 30^o$ is fixed and θ varied as in the main panel.

align along the substrate normal. This is in agreement with the reasoning of Goletti et al. [3] that attribute the interaction with the substrate to π orbitals of the porphyrin. However, the twisting comes out of our model, and the increment of ϕ is needed from the fact that the correct magnitude of p_x and p_y is required to reproduce the experimental derivative-like feature in $\mathcal{R}$. Then, as ϕ is increased p_y grows while p_x decreases simply from the geometrical projection along the two perpendicular axis where one measures RAS. Finally, we mention that our model shows that the behavior of the system is ultimately driven by the local field among the porphyrins.

4 Conclusions We have presented a theoretical study for the reflectance anisotropy spectroscopy, RAS, of porphyrin layers grown onto a gold substrate by the Langmuir-Schaefer technique. We have obtained RAS as function of the number of porphyrin layers, and shown that the spectra changes from a peak-like to derivative-like line shape as the tilting and twisting of the molecules is varied. More specifically, at low coverage the molecules tend to be more inclined relative to the substrate, due to stronger interaction with it, but as the number of layers is increased the substrate influence decreases as it is screened by the underlying layers, and the molecules tend to straighten up. In conclusion, with the polarizable dipole model, we are able to explain qualitatively the optical RAS behavior of the growth process of this very fascinating films, and in principle, determine parameters such as the tilt and twist angle of the porphyrin molecules.

Acknowledgements We thank partial support from CONACyT-México (Grant 36033-E) and CONCYTEG-México (C. C. Grant 03-04-k119-049 and R. V. 03-04-k118-039 Anexo 5).

References
[1] For a review see the articles in this volume and *256. WE-Heraeus-Seminar Optical Spectroscopy at interfaces (OSI-2001), Phys.Stat. Sol. (a)* **188** *No.4 (2001)*.
[2] C.Goletti, et al., Surf. Sci. **501**, 31 (2001).
[3] C. Goletti, et al., Langmuir **18**, 6881 (2002).
[4] B.G. Frederick, et al., Phys. Rev. Lett. **80** 4470 (1998).
[5] F.W. de Wette and G.E. Schacher, Phys. Rev. **137**, A78(1965).
[6] W.L. Mochán and R.G. Barrera, Phys. Rev. Lett. **55**, 1192 (1985).
[7] J.H. Weaver, et al., Physics Data Series (1981).

phys. stat. sol. (c) **0**, No. 8, 2976–2981 (2003) / **DOI** 10.1002/pssc.200303838

Effect of adsorption of electronegative and electropositive elements on the surface optical anisotropy of GaAs(001)

C. Hogan[*1], **D. Paget**[2], **O. E. Tereshchenko**[3], and **R. Del Sole**[1]

[1] Dipartimento di Fisica and INFM, Università di Roma "Tor Vergata", 00133 Roma, Italy
[2] Laboratoire de Physique de la Matière Condensée, Ecole Polytechnique, 91128 Palaiseau cedex, France
[3] Institute of Semiconductor Physics, Novosibirsk State University, 630090 Novosibirsk, Russia

Received 30 May 2003, revised 4 August 2003, accepted 11 August 2003
Published online 10 November 2003

PACS 71.15.Qe, 73.20.-r, 78.40.Fy, 78.40.-q, 78.68.+m

Surface optical anisotropy of GaAs(001) is extremely sensitive to adsorption of electropositive and electronegative atoms. The oxygen-induced experimental changes of the reflectance anisotropy spectrum of the As-rich (2×4) surface are interpreted using *ab initio* calculations. We conclude that oxygen adsorption, contrary to previous beliefs, does not quench all surface-related optical transitions, but only those for which the *initial* state is surface-related. Alkali metal (AM) adsorption is found to induce a strong negative signal in the 3.5 eV–5 eV range, weakly dependent on surface reconstruction and AM nature, whose origin is discussed.

1 Introduction

Surface optical techniques, such as Reflectance Anisotropy Spectroscopy (RAS) or Surface Differential Spectroscopy (SDR), possess excellent potential for investigating the adsorption of foreign elements at surfaces. Besides the possibility of non-destructive in-situ analysis, the spectroscopic nature of the techniques offers detailed information on the microscopic mechanism of the adsorption process. Previous works have illustrated the applicability of RAS across a range of semiconductor-adsorbate interfaces. Adsorption of oxygen has been shown to quench RA signals on As-rich [1] and Ga-rich [2] GaAs(001) and, quite generally, its chemisorption is believed to quench surface-related optical signals. For adsorption of alkali metals (AM) on GaAs, such as cesium [3] and sodium and potassium [4] has been investigated with an emphasis on the application of RAS to the study of disorder-order phase transitions under annealing. it has been proposed that AM deposition only weakly affects the main features of the spectrum, which is related to the fact that, in strong contrast with oxygen, AM weakly interact with the solid [5, 6].

In this work, we examine the usefulness of RAS as a technique for monitoring and understanding the process of adsorption of electronegative elements on the As-rich (2×4) surface of GaAs(001). Complete results will be published elsewhere [7]. We also consider adsorption of electropositive elements on both As- and Ga-rich surfaces. For this study we will adopt a more phenomenological approach based on interpretation of experimental data. A theoretical study of Cs adsorption on the $\beta2(2 \times 4)$ As-rich surface will be published independently [8].

2 Oxygen adsorption: Experimental

We define the RA signal by $\Delta R/R = (R_{[1\bar{1}0]} - R_{[110]})/R$, where $R_{[1\bar{1}0]}$ and $R_{[110]}$ are, respectively, the surface reflectivities for light polarized parallel and perpendicular to the dimers. Shown in the top panel of Fig. 1 are experimental spectra of clean $c(2 \times 8)$-like

---
* Corresponding author: e-mail: cdhogan@roma2.infn.it, Phone: +39 06 7259 4503 Fax: +39 06 202 3507

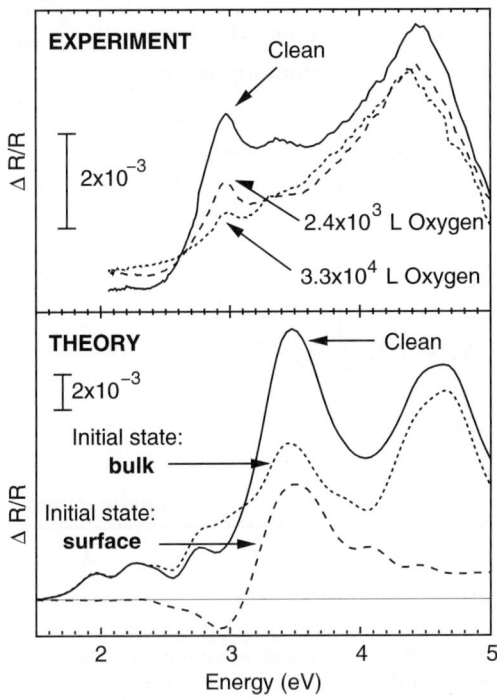

Fig. 1 Top panel: Experimental RAS for clean and oxidized As-rich (2×4) GaAs(001) surface, after coverages of 2.4×10^3 L and 3.3×10^4 L excited oxygen, taken from Ref. [1]. Bottom panel: Selective decomposition of clean surface theoretical spectra.

GaAs(001) and its modification under RT adsorption of atomic oxygen (after Ref. [1]). The spectrum for the clean surface is characterized by positive features near 3 eV and 4.5 eV, the latter lying close to the E_0' bulk critical point. We note three effects due to oxygen adsorption on the spectrum: (i) a significant (75%) quenching of the peak at 2.9 eV, (ii) a slight reduction in the high energy peak at 4.5 eV, and (iii) a slight increase of the signal in the 2.0–2.5 eV range.

The quenching of the peak at 2.9 eV suggests, using so far generally accepted ideas, that this peak has a strong surface contribution, in agreement with tight binding calculations [9]. This result appears to be in contradiction with other studies, including theoretical work [10], *ab initio* calculations [11], and experimental investigations of the peak dependence on GaAs quantum well thicknesses [12] as well as its dependence on the value of x in $Ga_x In_{1-x} As$ compounds [7]. One of the aims of the present paper is to clarify this contradiction.

3 Oxygen adsorption: Interpretation As we will illustrate below, the above controversy has partly arisen due to a somewhat ill-defined use of the terms 'surface' and 'bulk'. True surface states are defined *energetically*, i.e, being those lying inside the gap, and which do not couple with bulk states. In practice, we may instead adopt a definition based on *spatial* criteria, and assume that states that are, e.g, at least 80% localized in the top 2–3 layers are 'surface' (s), while remaining states are 'bulk' (b). This definition of a surface state is more general than the previous one, since it also includes states which are energetically degenerate with bulk ones but which have an increased localization at the surface, i.e., surface resonances. The localization criterion is, however, rather arbitrary, especially in calculations with relatively thin slabs and thick 'surface' regions, such as for the $\beta 2(2 \times 4)$ reconstruction considered here. Based on this spatial definition, we may adopt a *state-state* decomposition of the signal, made up of four combinations (s-s, s-b, b-s, b-b), which is frequently used in theoretical papers. Alternatively, we may decompose the spectra according to the localization of the transition within the surface layer (or any layer, in fact). This yields a *layer-by-layer* decomposition; in its most basic form it reduces to two spectral parts, 'surface layer' and 'bulk layer', which may be sharply [13] or gradually [14] defined. Both decomposition schemes are

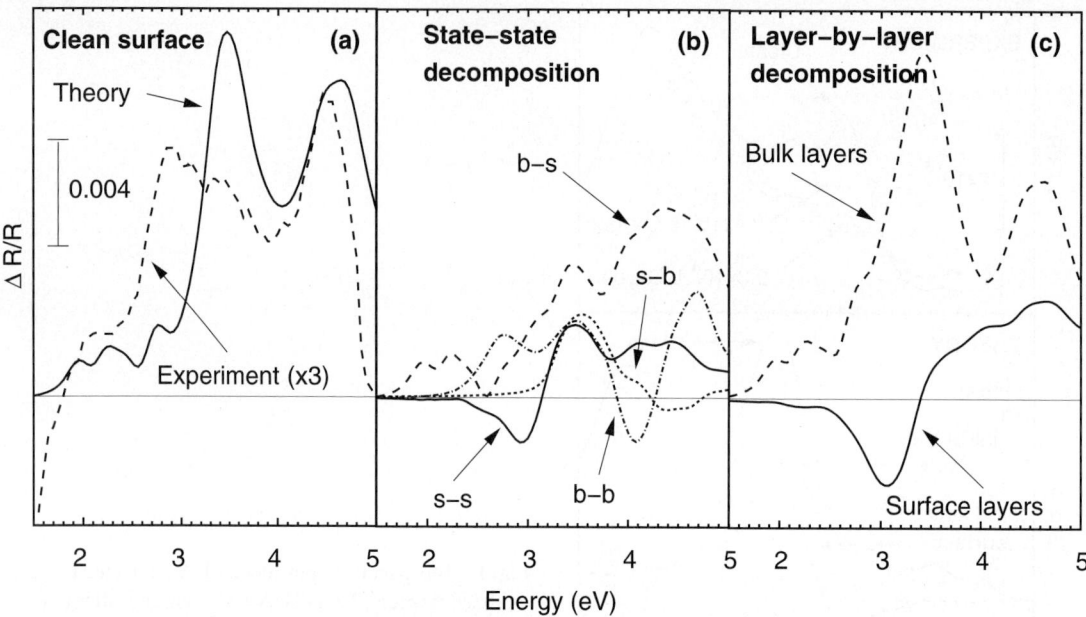

Fig. 2 RA spectra for the clean GaAs(001) (2×4) surface. (a) Experimental and theoretical results. The experimental data has been scaled by a factor of 3. (b) State-state decomposition of the theoretical spectrum in (a). (c) Layer-by-layer decomposition, for surface and bulk layers. A phenomenological broadening parameter of 0.15 eV has been used in all the theoretical spectra.

at best qualitative unless they are accompanied by a detailed microscopic examination of the underlying transitions [13, 15].

The structure of the As-rich $\beta 2(2 \times 4)$ reconstruction of GaAs(001) is well-known and consists of two As dimers at the top layer with a further As dimer situated at the third atomic layer. In the second layer, Ga atoms are planar sp^2 hybridized, and thus have unoccupied p_z type orbitals. Calculations of the optical properties were carried out within a standard *ab initio* density functional theory (DFT-LDA) scheme implemented in our molecular-dynamics Car-Parrinello code. Plane-waves up to a kinetic energy cutoff of 13 Ry were employed along with norm-conserving pseudopotentials of the Hamann type [16]. The surface was modelled using periodically repeating supercells of thin GaAs slabs, 10 layers thick, separated by 10 Å of vacuum. For the optical spectra, we correct DFT-LDA energies using a 'scissor-operator' shift of $+0.8$ eV (the bulk value) [17]. The use of 32 $\boldsymbol{k}$-points in the SBZ was judged to give reasonable convergence. Further details will be given elsewhere [8].

The theoretical spectrum, shown in the bottom panel of Fig. 1 and again in Fig. 2(a), compares reasonably well with the experimental one. The blue-shift in the lower energy peak of about 0.6 eV can be attributed to a combination of the use of the 'scissors-operator' approximation and the neglect of many-body effects (e.g. excitons) in the calculation [11]. In Fig. 2(b) we show the state-state decomposition. We have defined our surface region to be about 3 layers thick. The results suggest that anisotropy arising solely from surface states (the s-s contribution) accounts for only about 25% of the lower energy peak, and even less for the peak at 4.5 eV. In both cases the contribution is positive. The contribution of pure bulk-related transitions (b-b) is also of the order of 25%, so that the signal in fact appears to be largely derived from transitions between surface and bulk states (b-s and s-b curves).

For comparison, we show in Fig. 2(c) the result of the layer-by-layer decomposition for the surface and bulk regions. In this case, the contribution to the anisotropy from the surface layer again appears to be negative, albeit with greater magnitude, up to about 3.4 eV (theoretical energy). This result, consistent with that found by calculations [14] and high resolution electron energy loss spectroscopy [18], occurs in spite of there being only a small negative component in the state-state decomposition, at 2.95 eV. Neither negative signal appears in the total spectrum due to cancellation with other (stronger) bulk-related contributions. Elsewhere [7] we have shown that the negative peak at the s-s contribution results from transitions between dimer backbonds and Ga dangling bonds, i.e., along the [110] direction, which explains its negative sign.

In order to explain in detail the effect of oxygen adsorption, it would be necessary to perform a study of the adsorption sites and bonding mechanisms of oxygen atoms with the surface. Such study is beyond the scope of the present work. Instead, we simulate the effects of adsorption on the RA signal by making the physically reasonable assumption that the highly electronegative oxygen atoms will react most strongly with negatively-charged (occupied) surface states. Such states are uniquely present at the first- and third-layer As dimers. Since free oxygen atoms hybridize into an $[sp^2]^5 p_z^1$ configuration, bonding to the As dimer lone-pairs is not favoured; rather, substitution of topmost As atoms is more likely, as found in cluster calculations [19]. Therefore, we expect oxygen to influence transitions involving occupied surface states (i.e., the s-s and s-b terms) only. As we have shown elsewhere [7], the optically active unoccupied surface states are localized on the second layer Ga atoms, and are hence unperturbed by the oxygen.

In the bottom panel of Fig. 1 we therefore group the state-state contributions of Fig. 2(b) into two sets: those that involve the occupied surface states (s-s and s-b) and those that do not (b-s and b-b). According to our hypothesis, the latter contribution is all that should remain following oxygen adsorption. Indeed, the agreement with experiment is remarkably good, in that it reproduces the three features (i, ii, iii) noted above. It is important to recognize that the only part of the spectral change that can be attributed solely to surface states is the slight increase at 2.5 eV. Furthermore, it is evident that oxygen does not quench all the transitions located in the surface *layer*, responsible for the spectrum shown in Fig. 2(c).

4   Adsorption of electropositive elements   Results were obtained using the same setup as described elsewhere [20]. The undoped sample was first treated by a HCl isopropanol solution before being annealed at increasingly high temperatures in order to reveal the characteristic spectra of the As-rich surface or of the Ga-rich one [21]. RA spectra were taken at room temperature. The samples were subsequently exposed at room temperature to Cs, K and Na, using thoroughly outgassed getters. The Cs coverage [22, 23] and the K and Na coverages were characterized by Auger spectroscopy [20]. We show in Fig. 3(a) the RA spectra for the As-rich GaAs(001) surface after RT adsorption of approximately 0.1 ML Cs, which corresponds to a coverage of one atom per (2 × 4) cell. The main effect on the spectrum is the emergence of a relatively broad and negative signal that extends between 3 eV and 4.5 eV, as seen in the difference spectrum (b). We observe that the signal is extremely sensitive even to slight cesium doses, since a coverage as small as 0.01 ML should produce a detectable change. Based on its broadness, it is likely that the signal is composed of contributions from different sources, such as from different adsorption sites or from a possible quenching of the bulk-related peak at 4.5 eV. Notably, the line at 3 eV is largely unaffected and only experiences a decrease at higher coverages. In addition, a small negative signal is observed near 2 eV in the difference spectrum, which we found to saturate at low coverages.

Interestingly, the signal appears to be rather independent of the surface reconstruction. Fig. 3 (d) and (e) show results from a similar experiment for the Ga-rich (4 × 2) surface, with the difference again given as (f). On comparison with the As-rich result in (c), spectrum (f) differs mostly in the surface-state dominated region below 3 eV, previously proposed to be due to the effect of Cs on gallium dimer states [3]; above this energy, the similarity with the As-rich result is striking, suggesting that signals for both surfaces are of the same nature. Furthermore, we find that the chemical nature of the adatom is not very important. In Fig. 3(g) and (h) we present the experimental spectral differences for K and Na adsorption (atomic radii 2.4 Å and 1.9 Å respectively; for comparison, the As dimer length is 2.4 Å). Similar broad signals are again observed in both spectra. Clearly, it appears that the adsorption-induced signal above 3 eV has a universal character.

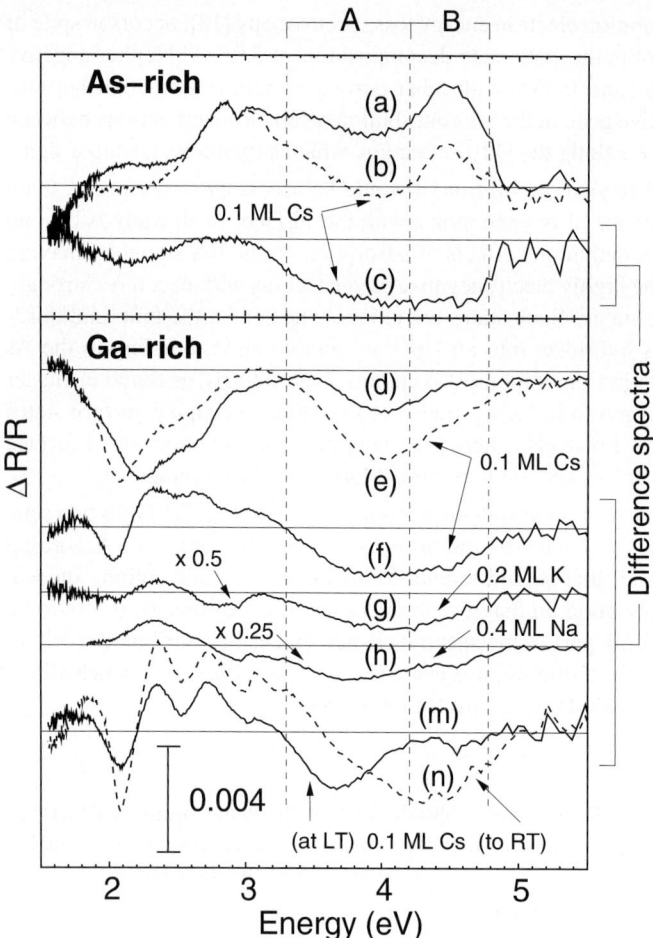

Fig. 3 Effect of alkali atom adsorption on the RA spectrum of GaAs(001). Top: Curves (a) and (b) show spectra of the clean $c(2 \times 8)$ As-rich (2×4) surface, and after RT adsorption of 0.1 ML of cesium, while curve (c) shows their difference [(b)−(a)]. Bottom: Curves (d) and (e) show the corresponding clean and Cs-covered spectra for the Ga-rich $c(8 \times 2)$ surface, and (f) is the difference spectrum. Curves (g) and (h) show the K-induced and the Na-induced difference spectra, normalized with respect to the alkali concentration. Curves (m) and (n) show spectra taken at low temperature, after adsorption at 90K of 0.1 ML of Cs (m), and again following cycling of the sample to RT (n). Regions of the spectra corresponding to different mechanisms of optical anisotropy are marked A and B.

Closer inspection of Fig. 3 reveals some differences as a function of AM nature. In fact, the signal seems to be composed of two parts — particularly noticeable for the case of sodium — which we have marked A and B on Fig. 3. The influence of the adsorption site can be shown from the difference curves at low temperature (90K), for which the features are better resolved. Such adsorption is known from LEED [24] and RAS [3] to create a disordered overlayer, which becomes ordered under annealing to RT. Curve (m) shows the LT Cs-induced change after adsorption of 0.1 ML of Cs on the Ga-rich surface. Initially, the signal is composed of a dominant component near 3.7 eV, with a weak but visible tail up to 5 eV. After annealing to RT, we find a signal very close to curve (e), taken at the same coverage. Cooling to LT does not further modify the adsorbate structure, but induces a change of the spectrum, due to the known temperature effect on surface optical anisotropy. The corresponding signal, shown in curve (n), is mostly composed of the second component. Comparison of curve (m) and of curve (n), taken at the same temperature and at the same coverage, demonstrates that the nature of the adsorption site does have an effect on the *energy* of the AM-induced signal.

The latter difference, although sufficiently large to enable using RAS for monitoring the activation of GaAs to negative electron affinity, is relatively weak in view of the large difference expected for surface-related features: for the $\beta2(2 \times 4)$ As-rich surface, *ab initio* calculations indicate that a strong charge transfer occurs for adsorption near empty Ga dangling bonds, whereas at the negatively charged As dimers, the transfer is minimal [8]. A possible mechanism that unites all these data derives from the presence of a large surface dipole in *all* AM/semiconductor systems, since as shown in Ref. [8], the surface dipole, unlike the transferred charge, seems to be relatively independent on adsorption site. Such reasoning would imply that the signal originates from modification of bulk states, through the polarization of their wavefunctions or shifts in their eigenvalues, and is consistent with the observed relative independence with respect to surface reconstruction, as well as the fact that the signal energies lie well above the band gap. As shown by Schmidt *et al.* [14], an electric field imposed normal to the surface, consistent with the direction of the charge-transfer induced surface dipole, induces negative structures in GaAs(001) in the 3.4-4.5 eV region. Model calculations [25] also predict AM-induced changes occuring near the bulk critical points, i.e., in the B region on Fig. 3. Finally, subsurface layer anisotropy may arise through presence of deep alkali resonances. These are bulk states which add to the anisotropy due to coupling with alkali valence states. Only particular bulk states will couple depending on the location of the alkali atom, and hence contribute to the anisotropy. All these effects will be discussed in a forthcoming publication [26].

5 Summary   We have shown that the surface anisotropy of GaAs(001) is strongly sensitive to adsorption of electronegative and electropositive atoms, and we have discussed the mechanisms for this sensitivity, which differ for the two types of elements. Oxygen adsorption on the As-rich (2×4) surface reduces the peak near 2.9 eV because of quenching of optical transitions for which the initial state is surface-related. Purely surface-related transitions are found to contribute to only 25% of this peak, however. This is completely consistent with the oxygen-induced effect due to the dominance of mixed optical transitions involving both surface and bulk states. Adsorption of alkali metals give rise to a very large (detectable at a coverage as small as 0.01 ML) negative signal in the 3.5–5 eV energy range that is weakly-dependent on surface reconstruction and on the nature of the AM, for which the origin is discussed.

Acknowledgements   C.H. has been supported by the EU through the NANOPHASE Research Training Network (Contract No. HPRM-CT-2002-00167). Computer time was granted by IDRIS (project 544)

References

[1] V. L. Berkovits, P. Chiaradia, D. Paget, A. B. Gordeeva, and C. Goletti, Surf. Sci. **441**, 26 (1999).
[2] C. Novello, V. Emiliani, C. Goletti, V. Berkovits, and P. Chiaradia, in: Proceedings of the 19th Course of the International School of Solid State Physics, Epioptics 2000, ed. A. Cricenti (World Scientific, Singapore, 2000), pp 109–114.
[3] V. L. Alperovich and D. Paget, Phys. Rev. B **56**, R15565 (1997).
[4] O. E. Tereshchenko, D. V. Daïneka, and D. Paget, Phys. Rev. B **64**, 85310 (2001).
[5] F. Bechstedt and M. Scheffler, Surf. Sci. Reports **18**, 145 (1993), and references therein.
[6] M. Prietsch, M. Domke, T. Mandel, C. Xue, and G. Kaindl, Z. Phys. B **74**, 1989 (1989).
[7] D. Paget, C. Hogan, V. L. Berkovits, and O. E. Tereshchenko, Phys. Rev. B (accepted 2003).
[8] C. Hogan, D. Paget, Y. Garreau, M. Sauvage, P. Chiaradia, G. Onida, L. Reining, and V. Corraddini, to be published.
[9] M. Murayama and T. Nakayama, Jpn. J. Appl. Phys. **36**, L268 (1997).
[10] K. Uwai and N. Kobayashi, Phys. Rev. Lett. **78**, 959 (1997).
[11] W. G. Schmidt, F. Bechstedt, K. Fleischer, C. Cobet, N. Esser, W. Richter, J. Bernholc, and G. Onida, phys. stat. sol. (a) **188**, 1401 (2001).
[12] L. F. Lastras-Martinez, D. Ronnow, P. V. Santos, M. Cardona, and K. Eberl, Phys. Rev. B **64**, 245303 (2001).
[13] C. Hogan, R. Del Sole, and G. Onida, Phys. Rev. B. (accepted 2003).
[14] W. G. Schmidt, F. Bechstedt, W. Lu, and J. Bernholc, Phys. Rev. B **66**, 85334 (2002).
[15] W. G. Schmidt, F. Bechstedt, and J. Bernholc, Appl. Surf. Sci. **190**, 264 (2002).
[16] D. R. Hamann, Phys. Rev. B **40**, 2980 (1989).
[17] R. Del Sole and R. Girlanda, Phys. Rev. B **48**, 11789 (1993).
[18] A. Balzarotti, E. Placidi, F. Arciprete, M. Fanfoni, and F. Patella, Phys. Rev. B **67**, 115332 (2003).
[19] S. I. Yi, P. Kruse, M. Hale, and A. C. Kummel, J. Chem. Phys. **114**, 3215 (2001).
[20] O. E. Tereshchenko, V. S. Voronin, H. E. Scheibler, V. L. Alperovich, and A. S. Terekhov, Surf. Sci. **507–510**, 51 (2002).
[21] O. E. Tereshchenko, S. I. Chikichev, and A. S. Terekhov, J. Vac. Sci. Technol. A **17**, 2655 (1999).
[22] B. Kierren and D. Paget, J. Vac. Sci. Technol. A **15**, 2074 (1997).
[23] G. Vergara, L. J. Gomez, J. Capmany, and M. T. Montojo, Surf. Sci. **131**, 278 (1992).
[24] B. Goldstein, Surf. Sci. **47**, 143 (1975).
[25] R. Del Sole and G. Onida, Phys. Rev. B **60**, 5523 (1999).
[26] C. Hogan, D. Paget, O. E. Tereshchenko, L. Reining, and G. Onida, to be published.

phys. stat. sol. (c) **0**, No. 8, 2982–2986 (2003) / **DOI** 10.1002/pssc.200303857

Giant reflectance anisotropy of polar cubic semiconductors in the far infrared

Yu. A. Kosevich*, **J. Ortega-Gallegos, A. G. Rodriguez, L. F. Lastras-Martinez,** and **A. Lastras-Martinez**

Instituto de Investigacion en Comunicacion Optica, Universidad Autonoma de San Luis Potosi, Alvaro Obregon 64, 78000 San Luis Potosi, S.L.P., Mexico

Received 30 May 2003, revised 17 September 2003, accepted 19 September 2003
Published online 10 November 2003

PACS 78.30.Fs, 78.68.+m

We present our measurements and model for the reflectancre anisotropy of the (001) surface of polar cubic zinc-blende semiconductor in the far infrared. We observe that the relative reflectance difference of GaAs(001) in the far infrared can reach the value of twenty percents which is two orders of magnitude higher than the reflectance difference of the GaAs(001) in the near-ultraviolet - visible range. The most strong reflectance anisotropy was observed in the optical phonon Reststrahlbande and its vicinity. We relate the observed reflectance anisotropy with the anisotropy of the optical-phonon and plasma damping constants. Such anisotropy can be caused by anisotropic inhomogeneous broadening of the frequencies of the optical-phonon and plasma oscillations polarized respectively along the [110] and [1$\bar{1}$0] directions. This effect can be understood in terms of the lattice-deformation-induced changes of the optical-phonon force constants and electron-effective-mass tensor components. Anisotropic inhomogeneous strain of the lattice can in turn be induced by anisotropic microscopic short-range ordering of point defects (dopants) and dislocations in near-surface regions of noncentrosymmentric zinc-blende semiconductors. The observed giant reflectance anisotropy can be used as a sensitive tool for the far infrared characterization of zinc-blende semiconductors.

1 Introduction

Reflectance anisotropy (or reflectance-difference) spectroscopy has emerged in the last years as a sensitive probe for the study of surface and surface-induced phenomena in zinc-blende semiconductors [1]. This spectroscopy deals with the measurements of the difference in reflectivity from cubic semiconductor between two mutually orthogonal polarizations of the light and therefore it is specific to the mechanism which breaks the cubic symmetry of the lattice. In particular, surface reconstructions [2,3,4], local field effects [5], surface electric fields [6] and near-surface bulk dislocations [7] provide such mechanisms. Most of the reflectance-anisotropy measurements and corresponding analysis were performed earlier either in the near-ultraviolet - visible (see, e.g. [6,7]) or in the near infrared (see, e.,g., [4]) photon energy ranges respectively. From the other hand, in the literature there were described several mechanisms of the reflectance anisotropy of cubic semiconductors which are specific to the far infrared frequency range and which should manifest themselves in the optical-phonon Reststrahlbande and its vicinity, see, e.g., [8,9].

In this paper we report on measurements and model for the reflectance anisotropy of GaAs(001) in the far infrared. We have observed that the relative reflectance difference of GaAs(001) in the far infrared can reach the value of twenty percents which is two orders of magnitude higher than the reflectance difference

---

* Corresponding author: e-mail: kosevich@cactus.iico.uaslp.mx, Phone: +52 444 825 0183, Fax: +52 444 825 0198

of the same surface in the near-ultraviolet - visible range, cf. [6,7]. The most strong reflectance anisotropy was observed in the optical phonon Reststrahlbande and its vicinity. We ascribe the observed reflectance anisotropy to the anisotropy of the optical-phonon and plasma damping constants. Such anisotropy can be caused by anisotropic inhomogeneous broadening of the frequencies of the optical-phonon and plasma oscillations polarized respectively along the [110] and [1$\bar{1}$0] directions. This effect can be understood in terms of the lattice-deformation-induced change of the optical-phonon force constants and electron-effective-mass tensor components. Anisotropic inhomogeneous strain of the lattice can in turn be induced by anisotropic microscopic short-range ordering of point defects (dopants) and dislocations in near-surface regions of noncentrosymmentric zinc-blende semiconductors. The values of the observed reflectance anisotropy both in semi-insulating and n-doped GaAs(001) are consistent with previous measurements of the change of the far infrared reflectivity caused by mechanical polishing of the samples, see, e.g., [10].

2   Experimental results and the model

We have measured the reflectivity of polarized far-infrared radiation, normally incident on the (001) surface of commercial epi-ready Cr-doped semi-insulating as well as n-doped GaAs substrates at room temperature. The light was polarized either along the [110] or [1$\bar{1}$0] directions. The reflectance measurements were performed with Michelson interferometer (Bruker, model IFS66v).

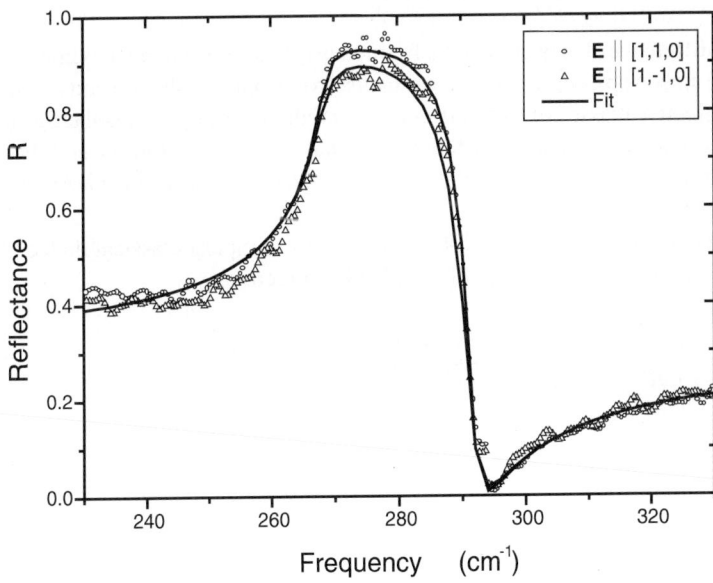

Fig. 1 Reflectance spectra and their fit for semi-insulating Cr-doped GaAs(001) for light polarized along the [110] and [1$\bar{1}$0] directions close to the optical-phonon Reststrahlbande. The fitting parameters are given in the text.

Figure 1 shows the measured reflectance spectra and their fit (see below) for the semi-insulating GaAs(001). We see that the main reflectance difference occurs within the optical-phonon Reststrahlbande, or in the frequency range of maximal reflectivity. The positions of the lower and upper frequency edges of the reflectance table, which are determined by the frequencies of the transverse (ω_T) and longitudinal (ω_L) optical phonons respectively, coincide for the both polarizations and therefore the macroscopic cubic symmetry of the lattice is not broken. Such reflectance anisotropy can be modelled in the main approximation by different optical phonon damping constants for the phonons polarized along the orthogonal [110] and [1$\bar{1}$0] directions. It is known that the reflectance of semiconductor within the optical-phonon Reststrahlbande is very sensitive to the damping of optical phonons, see, e.g., [11].

Figure 2 shows the measured reflectance spectra and their fit (see below) for the n-doped GaAs(001). The Hall measurements give us free-carrier density in the sample of 1.4×10^{18} cm^{-3}. From Fig. 2 we see

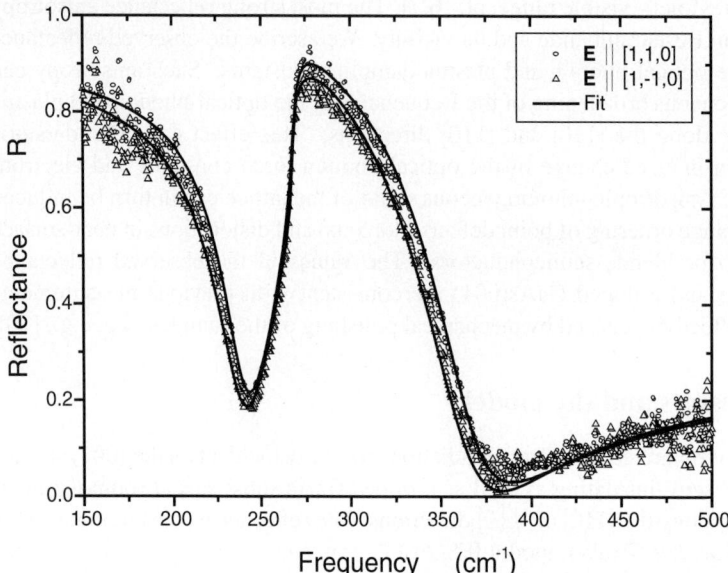

Fig. 2 Reflectance spectra and their fit for n-doped ($N_{el} \approx 10^{18} cm^{-3}$) GaAs(001) for light polarized along the [110] and [1$\bar{1}$0] directions close to the optical-phonon Reststrahlbande. The fitting parameters are given in the text.

that in this case the main reflectance difference also occurs in the frequency range between the maximal and minimal relectivity. Such reflectance anisotropy we can model with anisotropic optical-phonon and plasma damping constants and anisotropic effective masses of free carriers. Essentially in both semi-insulating and n-doped GaAs samples the observed relative reflectance difference reaches the value of twenty and more percents which is two orders of magnitude larger than the relative reflectance difference of the GaAs(001) in the near-ultraviolet - visible range, cf. [6,7].

We model the bulk far infrared optical anisotropy in the (001) plane by the simplest classical dielectric tensor, assuming that its main axes x' and y' are along the [110] and [1$\bar{1}$0] directions:

$$\epsilon_{1,2} = \epsilon_{\infty} \left(1 + \frac{\omega_L^2 - \omega_T^2}{\omega_T^2 - \omega^2 - i\omega\Gamma_{1,2}} - \frac{\omega_{p1,2}^2}{\omega(\omega + i\gamma_{1,2})} \right) = (\eta_{1,2} + i\kappa_{1,2})^2, \tag{1}$$

to calculate the reflectance:

$$R_{1,2} = \frac{(\eta_{1,2} - 1)^2 + \kappa_{1,2}^2}{(\eta_{1,2} + 1)^2 + \kappa_{1,2}^2}, \tag{2}$$

where $1 = x'x'$, $2 = y'y'$. In Eq. (1), ϵ_{∞} is the high-frequency lattice dielectric constant, $\Gamma_{1,2}$ and $\gamma_{1,2}$ are the optical-phonon and free-carrier plasma damping constants for the oscillations polarized in the corresponding direction, $\omega_{p1,2} = \sqrt{4\pi N e^2/m_{1,2}^* \epsilon_{\infty}}$ define the plasma frequencies of the free carriers with density N and effective masses $m_{x'x'}^*$ and $m_{y'y'}^*$. To fit our reflectivity data in the GaAs samples, we use $\epsilon_{\infty} = 11.1$, $\omega_T = 268.2$ cm^{-1} and $\omega_L = 291.5$ cm^{-1}, see [10].

The best fit for the reflectance anisotropy in the semi-insulating GaAs(001) can be done with the use of Eq. (1) with $\omega_p = 0$, $\Gamma_1 = 2.3$ cm^{-1}, $\Gamma_2 = 3.1$ cm^{-1}, see Fig. 3. The obtained optical-phonon damping constants are close to the damping constant $\Gamma = 2.3$ cm^{-1} which was found from the measurements of the reflectivity of unpolarized far infrared radiation from semi-insulating GaAs, see [10]. In Fig. 3 we also plot the spectrum of relative reflectance difference $\Delta R/R \equiv 2(R_1 - R_2)/(R_1 + R_2)$ together with its fit with the above parameters.

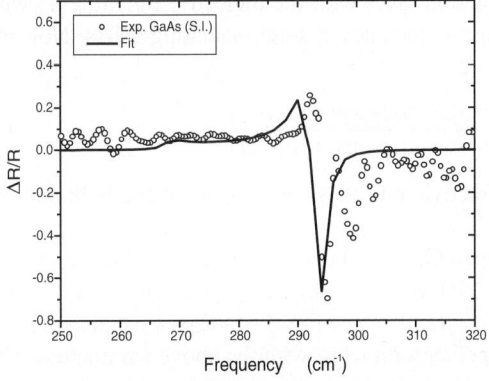 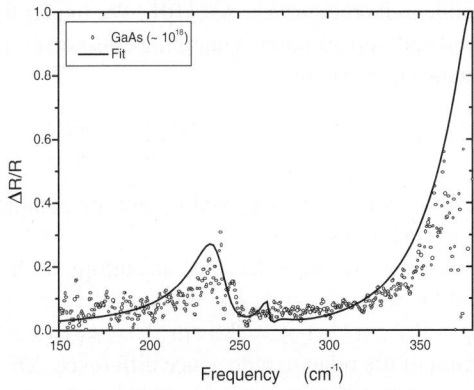

Fig. 3 Reflectance-difference spectra and their fit for the semi-insulating GaAs-(001) for light polarized along [110] and [1$\bar{1}$0] close to the optical-phonon Reststrahlbande. The fitting parameters are given in the text.

Fig. 4 Reflectance-difference spectra and their fit for the n-doped GaAs-(001) for light polarized along [110] and [1$\bar{1}$0] close to the optical-phonon Reststrahlbande. The fitting parameters are given in the text.

In order to understand the change of the optical phonon damping constants, we decompose the inhomogeneous lattice deformation (strain) tensor $u_{i'k'}(\mathbf{r})$ into the two, in general independent, components, namely into the averaged homogeneous $\varepsilon_{i'k'}$ and inhomogeneous $\delta u_{i'k'}(\mathbf{r})$ components:

$$u_{i'k'}(\mathbf{r}) = \varepsilon_{\mathbf{i'k'}} + \delta\mathbf{u_{i'k'}}(\mathbf{r}). \tag{3}$$

The averaged homogeneous deformation $\varepsilon_{i'k'}$ produces the linear-in-strain shift (and splitting) of the optical-phonon frequencies [9,12], while the r.m.s. value (the dispersion) of inhomogeneous deformation $\langle\delta u_{i'k'}(\mathbf{r})\rangle \equiv \zeta_{i'k'}$ causes inhomogenous broadening of the optical-phonon lines.

Since there are no observable splittings either of the transverse or of longitudinal optical-phonon frequencies, see Fig.1, the homogeneous component of the orthorhombic lattice deformation $\varepsilon_{i'k'}$ is zero or at least is negligibly small in comparison with the r.m.s. value of the inhomogeneous strain $\zeta_{i'k'}$. Essentially the anisotropy of the tensor of r.m.s. value of inhomogeneous strain $\zeta_{i'k'}$ will result in *anisotropic inhomogeneous broadening* of the optical-phonon lines and therefore in anisotropy of the optical-phonon damping constants. To explain within the model of anisotropic inhomogeneous broadeding the reflectance anisotropy spectra presented on Fig.1, namely that $\omega_{T1} = \omega_{T2}$ and $\Gamma_1 < \Gamma_2$, we have to assume that $\zeta_2 > \zeta_1 >> (\varepsilon_1, \varepsilon_2)$. A significant change of the broadening of the optical-phonon lines, in the absence of their net shifts, has been reported for the near-surface regions of mechanically polished GaAs [13]. Since the mechanical treatment of the surface introduces point defects and dislocations in the near-surface region, we can relate the observed in our measurements anisotropy of the optical-phonon damping constants with possible anisotropy in microscopic short-range ordering of point defects (dopants) and dislocations in near-surface regions of noncentrosymmentric zinc-blende semiconductor caused by effective local lowering of microscopic symmetry of the lattice, cf. [14]. Therefore our measurements show that anisotropic inhomogeneous lattice deformation with zero average can lower the symmetry of the far infrared surface optical response with respect to the macroscopic symmetry of the lattice via the anisotropic inhomogeneous broadening of the corresponding spectral lines. To restore the macroscopic symmetry of the system, the anisotropy of microscopic short-range ordering of point defects and dislocations close to the opposite equivalent parallel surface of zinc-blende semiconductor slab should have the opposite sign and therefore the reflectance anisotropy at the opposite surfaces of the slab should be sign-reversed.

In order to make a fit for the reflectance anisotropy spectra for the n-doped GaAs(001), one has to take explicitly into account the presence of the depletion layer with the apriori unknown thickness l at the semiconductor surface, see [10]. We model the depletion layer with the dielectric function (1) with $\omega_{pl} = 0$ and optical-phonon damping constants $\Gamma_{d1,2}$ close to the ones of semi-insulating GaAs. Now the reflectance is given by:

$$R = |\, r \,|^2, \; r = \frac{(1 - n_d)(n_b + n_d) + (1 + n_d)(n_d - n_b)\exp(4\pi i\omega n_d l)}{(1 + n_d)(n_b + n_d) + (1 - n_d)(n_d + n_b)\exp(4\pi i\omega n_d l)}, \tag{4}$$

where ω is in cm^{-1} and n_b and n_d are the complex refractive indices ($n = \eta + i\kappa$) of the bulk and the depletion layer, respectively.

The best fit for the reflectance anisotropy in the n-doped GaAs(001) can be done with the use of Eqs. (1) and (4) with $\omega_{p1} = \omega_{p2} = 333$ cm^{-1}, $\Gamma_1 = 4.6$ cm^{-1}, $\Gamma_{d1} = 3.0$ cm^{-1}, $\gamma_1 = 60$ cm^{-1}, $l_1 = 1000$ Å, $\Gamma_2 = 7.7$ cm^{-1}, $\Gamma_{d2} = 3.0$ cm^{-1}, $\gamma_2 = 67$ cm^{-1}, $l_2 = 2150$ Å, see Fig. 2. In Fig. 4 we also plot the spectrum of the relative reflectance difference $\Delta R/R$ together with its fit with the above parameters. The obtained plasma frequency $\omega_p = 333$ cm^{-1} and plasma damping constants $\gamma_1 = 60$ cm^{-1} and $\gamma_2 = 67$ cm^{-1} are close to the plasma frequency $\omega_p \approx 380$ cm^{-1} and plasma damping constant $\gamma \approx 50$ cm^{-1} which were found from the measurements of the reflectance of unpolarized far infrared radiation from n-doped GaAs with a close level of doping, see [10]. The anisotropy of the plasma damping constants we can also relate with the anisotropic inhomogeneous broadening of the frequencies of plasma oscillations polarized respectively along the [110] and [1$\bar{1}$0] directions. This effect can be understood in terms of the linear-in-strain change of the electron-effective-mass tensor components by the lattice deformation, see, e.g., [15]. From the other hand, the obtained from the fit rather large and different thicknesses l_1 and l_2 of the depletion layer should not be considered literally. The above fitting parameters one can use to extract the values of different optical-phonon $\Gamma_{1,2}$ and plasma $\gamma_{1,2}$ damping constants together with the value of the plasma frequency ω_p. Therefore similarly to the case of semi-insulating GaAs(001), we relate the observed giant reflectance anisotropy of the n-doped GaAs(001) with the anisotropy of the optical-phonon and plasma damping constants. The absence of the induced anisotropies of the optical-phonon and plasma frequencies shows the absence of homogeneous orthorhombic lattice deformation $\varepsilon_{i'k'}$ (or at least its smallnesses in comparison with the r.m.s. value of inhomogeneous strain $\zeta_{i'k'}$) and therefore the absence (or nonimportance for the observed effects) of the change of macroscopic symmetry of the cubic lattice of the n-doped GaAs(001) substrate.

We have also performed reflectance-difference measurements on MBE-grown layer of undoped GaAs (001) with presumably very low concentration of point defects and dislocations. These measurements did not show any observable reflectance anisotropy. This probably means that the reflectance anisotropy in the far infrared is sensitive to the surface treatment (mechanical and chemical polishing) which is applied to the epi-ready-wafers but is not applied to the MBE-grown layers. The enhancement of the optical-phonon and plasma damping constants in the damaged surface layer has been directly observed, see, e.g., [10].

References

[1] D. E. Aspnes and A. A. Studna, Phys. Rev. Lett. **54**, 1956 (1985).
[2] Y. C. Chang, S. F. Ren, and D. E. Aspnes, J. Vac. Sci. Technol. A **10**, 1856 (1992).
[3] R. E. Balderas-Navarro et al., Appl. Phys. Lett. **78**, 3615 (2001).
[4] C. Coletti et al., Phys. Rev. B **66**, 153307 (2002).
[5] W. L. Mochan and R. G. Barrera, Phys. Rev. Lett. **55**, 1192 (1985).
[6] S. E. Acosta-Ortiz and A. Lastras-Martinez, Phys. Rev. B **40**, 1426 (1989).
[7] L. F. Lastras-Martinez and A. Lastras-Martinez, Phys. Rev. B **64**, 085309 (2001).
[8] W. L. Mochan and J. Recamier, Phys. Rev. Lett. **63**, 2100 (1989).
[9] Yu. A. Kosevich, Solid State Commun. **104**, 321 (1997).
[10] R. T. Holm, J. M. Gibson and E. D. Palik, J. Appl. Phys. **48**, 212 (1977).
[11] C. F. Klingshirn, *Semiconductor Optics* (Springer, Berlin, 1997), p. 67.
[12] B. A. Weinstein and M. Cardona, Phys. Rev. B **5**, 3120 (1972).
[13] D. J. Evans and S. Ushioda, Phys. Rev. B **9**, 1638 (1974).
[14] T. Hofmann, V. Gottschalch, and M. Schubert, Phys. Rev. B **66**, 195204 (2002).
[15] G. L. Bir and G. E. Pikus, *Symmetry and Strain-Induced Effects in Semiconductors* (Wiley, New York, 1974).

phys. stat. sol. (c) **0**, No. 8, 2987–2991 (2003) / **DOI** 10.1002/pssc.200303843

Model for the strain-induced reflectance-difference spectra of InGaAs/GaAs (001) epitaxial layers

A. Lastras-Martínez[*, 1], **R. E. Balderas-Navarro**[1, 2], **C. I. Medel-Ruiz**[1], **J.M. Flores-Camacho**[1], **A. Gaona-Couto**[1], and **L. F. Lastras-Martínez**[1]

[1] Instituto de Investigación en Comunicación Optica, Universidad Autónoma de San Luis Potosí, Alvaro Obregón 64, San Luis Potosí, S.L.P., México
[2] Facultad de Ciencias, Universidad Autónoma de San Luis Potosí, Alvaro Obregón 64, San Luis Potosí, S.L.P., México

Received 30 May 2003, revised 4 August 2003, accepted 11 August 2003
Published online 10 November 2003

PACS 68.60.Bs, 78.66.Fd, 68.65.Hb, 78.68.+m

We present a model to describe the Reflectance-difference (RD) spectra of InGaAs grown on GaAs (001) at T=500 °C, in the energy range from 2.3-3.5 eV. The model assumes the presence of an orthorhombic strain in the InGaAs epilayer that accounts for an anisotropic process of nucleation of InGaAs islands. We show that the developed model leads to accurate fits to the experimental RD spectra of InGaAs/GaAs for epilayer thickness both below and above of the critical thickness for the 2D-3D growth-mode transition. From the fitting of the theoretical model to the experimental RD line shapes we obtain quantitative information on changes in surface stoichiometry and morphology at the onset of the 2D-3D transition. Our results demonstrate that the RD line shape of InGaAs/GaAs in the 2.3-3.5 eV energy range can be explained entirely in terms of surface-modified InGaAs E_1 and $E_1+\Delta_1$ bulk transitions with no need to invoke surface states.

1 Introduction

Reflectance Difference (RD/RAS) spectroscopy [1] has emerged in the last two decades as a sensitive probe for the study of surface and interface-induced phenomena in zincblende semiconductors [2-4]. RD spectroscopy measures the difference in reflectivity between two mutually orthogonal polarizations and thus provides information on semiconductor regions were the cubic symmetry has been broken by the presence of the surface or interface [1]. One important area of application of RDS is the *in situ*, real time monitoring of the epitaxial growth of zincblende semiconductors, where RDS spectroscopy has been shown to be highly sensitive to changes in surface reconstruction and in general to changes in surface stoichiometry and epilayer morphology during growth [2-5].

In an earlier paper we have shown that the *in situ* RD amplitude at 2.5 eV of $In_{0.3}Ga_{0.7}As$ grown by MBE on GaAs (001) rises sharply at the onset of 3D growth, indicating that the InGaAs islands grow anisotropic [6]. We further showed that the RD line shape after the 3D growth mode transition could be accurately modeled by assuming the $In_xGa_{1-x}As$ islands to be under an orthorhombic strain [6]. In this paper we show that this line shape model can be extended to RD line shapes for thickness below 2D-3D growth transition. Quantitative information on surface morphology changes and In segregation induced by the 2D-3D transition are obtained from the theoretical fit to the RD experimental spectra. Our results demonstrate that the RD line shape of InGaAs/GaAs in the 2.3-3.5 eV energy range can be explained entirely in terms of surface-modified InGaAs E_1 and $E_1+\Delta_1$ bulk transitions with no need to invoke surface states.

---

[*] Corresponding author: e-mail: alastras@cactus.iico.uaslp.mx, Phone: 52 (444) 825 0183, Fax: 52 (444) 825 0198

2   Theoretical line shape model

Let us consider an $In_xGa_{1-x}As/GaAs$ epilayer under an orthorhombic strain with in-plane components given by

$$e_{xx} = e_{yy} = \left(1 - \frac{\gamma}{2}\right)\frac{\Delta a}{a}; \quad e_{xy} = \frac{\gamma}{2}\frac{\Delta a}{a}, \tag{1}$$

where $\Delta a/a$ is the interface strain corresponding to the pseudomorphic layer and γ is a parameter that measures interface relaxation along [110]. The tensor given by Eq. (1) splits the eight-fold, Λ-symmetry interband transitions (along <111>directions in the Brillouin Zone) into two sets of fourth-fold transitions; a first set containing points along $\{[111], [11\bar{1}], [\bar{1}\bar{1}1], [\bar{1}\bar{1}\bar{1}]\}$ (set 1), and a second set containing points along $\{[\bar{1}11], [1\bar{1}\bar{1}], [1\bar{1}1], [\bar{1}1\bar{1}]\}$ (set 2) [7, 8]. In what follows we will label with subscript 1 the splitting energy and interband squared matrix element of the critical points in the first set and with subscript 2 those of the second set. As a result of the critical point splitting the interband transition matrix elements become polarization-dependent.

The difference in dielectric function between $[110]$ and $[1\bar{1}0]$ directions due to the strain given in Eq. (1), <u>for a single critical point i</u> (either E_1 or $E_1+\Delta_1$), may be written as [8-10]

$$\Delta\varepsilon_i = \frac{C}{E^2}\left[\Delta M_1 J\left(E_{0i} + \Delta E_1\right) + \Delta M_2 J\left(E_{0i} + \Delta E_2\right)\right], \tag{2}$$

where $\Delta M_{1,2}$ are the differences in interband matrix elements between $[110]$ and $[1\bar{1}0]$ directions, E_{0i} is the unperturbed interband critical point energy, $\Delta E_{1,2}$ are the perturbation energies, $J(E)$ is the interband joint density of states and C is a constant independent of energy. The RD line shape is given by $\Delta R/R = Re[(\alpha - i\beta)\Delta\varepsilon$, where $\Delta\varepsilon$ is the overall change in dielectric function given by the superposition of the contributions $\Delta\varepsilon_i$ of E_1 and $E_1+\Delta_1$ critical points and α and β are Seraphin coefficients. We note that, due to the fact that the light penetration depth is larger that the epilayer thickness, we calculated α and β on the basis of a three-phase model air-InGaAs-GaAs [11].

$\Delta\varepsilon$ and $\Delta R/R$ may be calculated from the strain tensor given by Eq. (1) through a perturbative approach based on Pikus-Bir strain Hamiltonian [6] and the dielectric function line shapes for E_1 and $E_1+\Delta_1$ critical points [8-10]. Dielectric function line shapes for E_1 and $E_1+\Delta_1$ transitions were obtained by fitting Lorenzian line shapes to the real and imaginary parts of the dielectric function spectra for a 0.5 μm-thick $In_{0.3}Ga_{0.7}As/GaAs$ epilayer measured by spectroscopic ellipsometry, following a procedure described elsewhere [9]. The measured pseudo dielectric function was corrected for the presence of an oxide layer.

3   Results and discussion

In Fig. 1 we show *in situ* RD spectra for an $In_{0.3}Ga_{0.7}As/GaAs$ epilayer with several nominal thickness. RD spectra were measured at T=513 °C. We note that spectra have been displaced vertically for the sake of clarity. Nominal thickness after each growth step are as indicated, with the lowermost spectrum corresponding to the GaAs buffer layer just before the starting of InGaAs growth. Vertical dashed lines correspond to E_1 and $E_1+\Delta_1$ energies, assuming a pseudomorphic fully strained epilayer with the nominal composition x=0.3. There is a continuous evolution of the RD line shape with growth time. We note the step rise in amplitude when a film thickness of about 37 Å is reached. As reported previously, this rise is correlated with a rise in RHEED intensity at a spot corresponding to bulk diffraction, thus indicating that the onset of 3D growth was reached [6].

In what follows we will show that spectra of Fig. 1 can be modeled on the basis of the theory discussed above. Before that, nevertheless, because spectra of Fig. 1 shows a considerable temperature-broadening, we will discuss the modeling of a RD spectrum measured at 150 °C that, as shown in

Fig. 2a, displays a relatively sharp optical structure. We note that a sharp spectrum is critical in order to assess the validity of the developed line shape model.

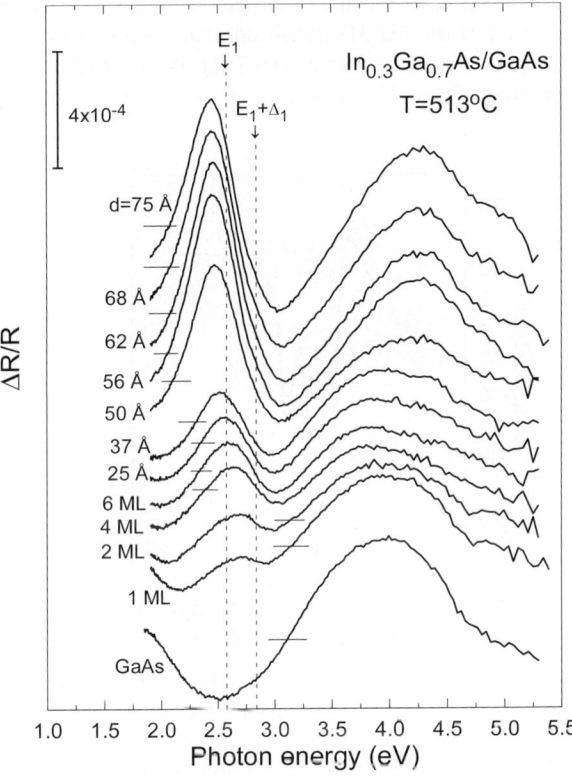

Fig. 1 RD spectra for an $In_{0.3}Ga_{0.7}As/GaAs$ (001) hetero-structure measured at 513 °C. Spectra correspond to a sample that was grown in steps. Epilayer total thickness after each growth step is indicated. Lowermost spectrum corresponds to the c(4x4) GaAs buffer layer just before InGaAs growth initiated. Vertical dashed lines show the position of the E_1 and $E_1+\Delta_1$ energies for a pseudomorphic film with the nominal In composition $x=0.3$.

In Fig. 2a we plot with open circles the RD spectrum for the sample of Fig. 1 corresponding to a thickness (nominal) of 75-Å. The measurement was carried out in situ at a temperature of 150 °C by rapidly cooling down the sample after the end of the growth. Continuos line in Fig. 2 correspond to a fitted line shape based on Eqs. (1)-(3). We obtained the following values for the fitting parameters: $\Delta a/a=0.45$, $\gamma=1$ and $d=230$ Å. We further note that a vertical offset of the calculated spectrum was necessary in order to take account for the non-local contributions to the experimental RD line shape that are not considered in the model.

As we can see from Fig. 2, the theoretical line shape accurately reproduces the experimental spectrum. This is more evident in Fig. 2b where we show the first energy-derivative of the spectrum of Fig. 2a (open circles) along with its fitted line shape (continuous line). We may thus conclude that the RD spectrum of Fig. 2 has a strain origin and that the developed model is adequate to describe it.

Vertical arrows in Figs. 2a and 2b indicate the energies of the perturbed E_1 and $E_1+\Delta_1$ transitions. We note that the splitting energy is much larger for $E_1+\Delta_1$ (0.23 eV) than for E_1 (0.015 eV). In consequence, a richer optical structure is observed around $E_1+\Delta_1$ than around E_1. We note that the mismatch value $\Delta a/a=0.45$ employed in the fitting is considerably larger than that corresponding to the nominal $(x=0.3)$ In composition. The fitted thickness value has to be compared with the average island height average of 170 Å determined by *ex situ* AFM measurements. We further note that, with the exception of the low-energy tail below 2.2 eV, the RD line shape of Fig. 2a remains essentially the same after exposure to atmospheric pressure, thus demonstrating that it is not related to surface reconstruction but to surface-modified bulk states.

Let us now consider high-temperature RD spectra. In Figs. 3a and 3b we reproduce with open circles the RD spectra of Fig. 1 corresponding to nominal thickness $d=37$ Å (before the 2D-3D transition) and $d=56$ Å (after the 2D-3D transition), respectively. Continuous lines in Figs. 3a and 3b are the corresponding line shape fittings. We can see that, as with the spectrum of Fig. 2, the line shape fittings describe accurately the experimental spectra. We conclude that the RD spectra have a strain origin for thickness both below and above the 2D-3D transition. Furthermore, we note that fitting parameters differ considerably in both cases. As a matter of fact, the evolution of these parameters provides information on processes occurring during the 2D-3D transition. Previous to this transition, for a nominal thickness 38 Å, we have $\Delta a/a=0.3$, $\gamma=0.85$ and $d=60$ Å, while after the 2D-3D transition (nominal thickness 56 Å) we obtained $\Delta a/a=0.45$, $\gamma=1$ and $d=230$ Å. Thus, there is a large change in lattice mismatch $\Delta a/a$ that is consistent with well-known In redistribution at the onset of the 2D-3D transition, that results in InGaAs islands with In average compositions considerably higher that the nominal ones [12]. We further see that there is a considerable increase in the film thickness (from 70 to 160 Å) that is of course consistent with the 2D-3D surface change in morphology.

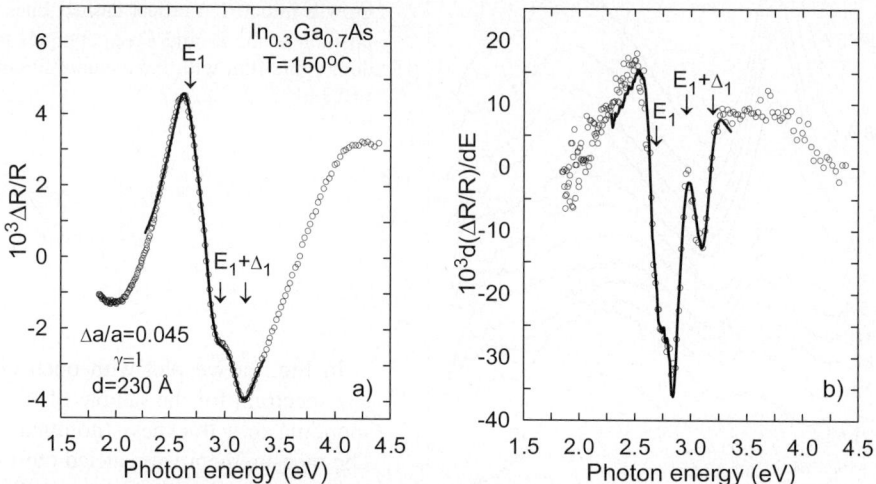

Fig. 2   a) Open circles: experimental RD spectrum at 150 °C for the sample of Fig. 1 (75 Å nominal epilayer thickness). Continuous line: fitted RD line shape. Values of fitting parameters are as indicated. b) Open circles: first energy-derivative of the RD spectrum of Fig. 2a. Continuous line: corresponding line shape fitting. Arrows in both figures indicate the calculated positions of the perturbed E_1 and $E_1+\Delta_1$ transitions. Note that the splitting energy of E_1 is negligible in the scale of the figures.

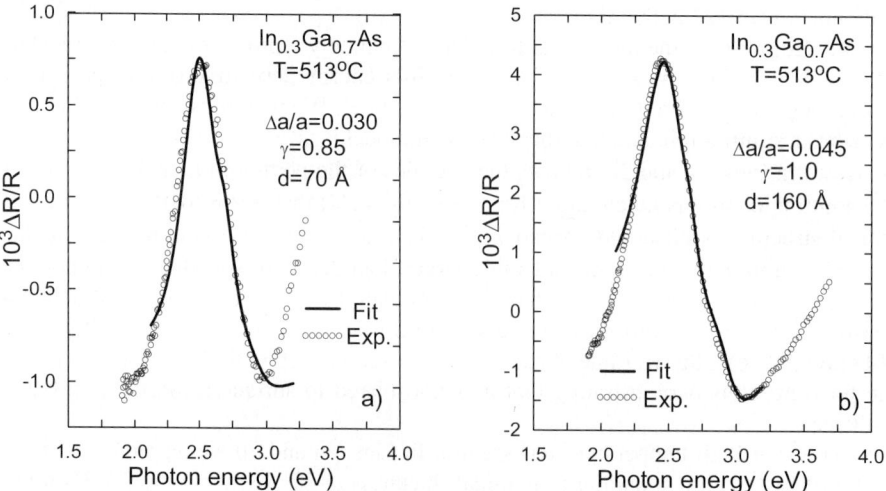

Fig. 3   a) Open circles: experimental RD spectrum at 513 °C for the sample of Fig. 1 (37 Å nominal epilayer thickness). Continuous line: fitted RD line shape. b) Open circles: experimental RD spectrum at 513 °C for the sample of Fig. 1 (56 Å nominal epilayer thickness). Continuous line: fitted RD line shape. Values of the fitting parameters are as indicated.

Results presented here demonstrates the strain origin of the RD spectra of $In_{0.3}Ga_{0.7}As/GaAs$ for energies around E_1 and $E_1+\Delta_1$ transitions. $In_{0.3}Ga_{0.7}As/GaAs$ 3D islands are known to relax the interface strain at their tops, keeping the bases under a tetragonal pseudomorphic strain [13]. For anisotropic shaped islands we expect this inhomogeneous strain to contribute to reflectance anisotropy as the strain relaxation at the top of an island should be anisotropic. We further note that for thickness lower that the critical thickness for the 2D-3D transition an orthorhombic strain may be as well expected due to the surface stress associated to surface reconstruction. Indeed, calculations show that surface reconstruction leads to substantial subsurface anisotropic atomic displacements [14]. Experimental evidence for surface stress induced optical anisotropies have been reported in Ref. [59].

We note that an additional source of orthorhombic strains is the unbalance in the density of dislocations with cores oriented along the two [110] directions, that is known to occur in zincblende semiconductors. More work is however necessary in to order establish whether there is a contribution of dislocations to the InGaAs reflectance anisotropy observed here.

The RD spectra presented in this paper show around E_1 and $E_1+\Delta_1$ the same general features observed in the RD spectra of GaAs (001) under [110] uniaxial strain; i.e., in both cases we observe a maximum and minimum in this energy range. We note, nevertheless, that the energy separation between the maximum and minimum is larger in the case of InGaAs/GaAs due to the large interface strain that leads to an apparent in crease in the Δ_1 energy. We further note that the energy broadening of the InGaAs/GaAs RD band around E_1 is considerably larger than that observed for GaAs. This can be explained on the basis of the inhomogeneous strains expected for the InGaAs islands.

In conclusion, we reported on the development of a line shape model for the RD spectrum of $In_{0.3}Ga_{0.7}As/GaAs$ heterostructures. The line shape model assumes the InGaAs epilayer to be under an orthorhombic strain and is found to describe accurately the in situ RD spectrum of MBE $In_{0.3}Ga_{0.7}As/GaAs$ heterostructures, for thickness both below and above the 2D-3D growth-mode transition. Quantitative information on surface morphology changes and In segregation induced by the 2D-3D transition are obtained from the theoretical fit to the RD experimental spectra. We conclude that the RD spectrum of $In_{0.3}Ga_{0.7}As/GaAs$ for energies around E_1 and $E_1+\Delta_1$ transitions RD spectra is associated to strain-modified bulk states. Results presented here show the high potential of RD spectroscopy as an *in situ* probe for the study of $In_xGa_{1-x}As/GaAs$ growth processes on a quantitative basis.

Acknowledgements Work partially supported by Consejo Nacional de Ciencia y Tecnología of México under contracts No. 32147-E and 485100-5-3397E.

References

[1] D.E. Aspnes and A.A. Studna, Phys. Rev. Lett. **54**, 1956 (1985).
[2] D.E. Aspnes, D.E. Harbison, A.A. Studna, L.T. Florez, and M.K. Kelly, J. Vac. Sci. Technol. B **6**, 1127 (1988).
[3] Z. Sobiesierski, D. I. Westwood, and C.C. Mathai, J. Phys. C **10**, 1 (1998).
[4] V. Emiliani, A.M. Frisch, C. Goletti, N. Esser, W. Richter, and B.O. Fimland, Phys. Rev. B **66**, 085303 (2002).
[5] K. Hingerl, R.E. Balderas-Navarro, W. Hilber, A. Bonanni, and D. Stifter, Phys. Rev. B **62**, 13048 (2000); R.E. Balderas-Navarro, K. Hingerl, A. Bonanni, H. Sitter, and D. Stifter, Appl. Phys. Lett. **78**, 3615 (2001).
[6] C.I. Medel-Ruiz, A. Lastras-Martínez, R.E. Balderas-Navarro, S.L. Gallardo-Cruz, V. Méndez-García, J.M. Florez-Camacho, A. Gaona-Couto, and L.F. Lastras-Martínez, submitted to Appl. Surf. Sci. (2003).
[7] F.H. Pollak and M. Cardona, Phys. Rev. **172**, 816 (1968).
[8] L.F. Lastras-Martínez and A. Lastras-Martínez, Phys. Rev. B **54**, 10726 (1996).
[9] A. Lastras-Martínez, R.E. Balderas-Navarro, L.F. Lastras-Martínez, and M.A. Vidal, Phys. Rev. B **59**, 10,234 (1999).
[10] L. F. Lastras-Martínez, M. Chavira-Rodríguez, A. Lastras-Martínez, and R. E. Balderas-Navarro, Phys. Rev. B **66**, 075315 (2002).
[11] See for instance, O. S. Heavens, in: Optical properties of thin solid films (Dover, New York, 1965), p. 63.
[12] T. Walther, A.G. Cullis, D.J. Norris, and M. Hopkinson, Phys. Rev. Lett. **86**, 2381 (2001).
[13] N. Grandjean and J. Massies, J. Cryst. Growth **134**, 51 (1993).
[14] J.A. Appelbaum and D.R. Hamann, Surf. Sci. **74**, 21 (1978).

phys. stat. sol. (c) **0**, No. 8, 2992–2996 (2003) / **DOI** 10.1002/pssc.200303842

Optical properties of the cleavage InAs(110) surface

X. López-Lozano[1], **Cecilia Noguez**[*2], and **L. Meza-Montes**[1]

[1] Instituto de Física, Universidad Autónoma de Puebla, Apartado Postal J-48, Puebla 72570, México
[2] Instituto de Física, UNAM, Apdo. Postal 20-364, D.F. 01000, México

Received 30 May 2003, revised 4 August 2003, accepted 11 August 2003
Published online 10 November 2003

PACS 73.20.At, 78.20.Ci, 78.68.+m

The electronic and optical properties of the cleavage InAs(110) surface are studied using a semi-empirical tight-binding method which employs an extended atomic-like basis set. We calculate the surface electronic band structure and the Reflectance Anisotropy Spectrum. We describe and discuss the optical properties in terms of the surface electronic states and we compare our results with other theoretical approaches, and with experimental observations.

1 Introduction Most of the theoretical [1–5] studies about InAs(110) do not provide a way for directly compare with experiments [6–14]. The available experimental measurements are not enough to elucidate the atomic structure and electronic properties of InAs(110). Andersson and collaborators [6] found the energies of occupied-surface states at high-symmetry points using photoemission techniques. Six years later, the same group measured again the occupied-surface states at the same high-symmetry points [7], founding a systematic shift of -0.15 eV with their previous measurements [6]. Swantson *et al.* [8] also measured the energies of occupied-surface states at high-symmetry points and they found differences up to 0.5 eV with those reported by Andersson [6, 7]. The interpretation of photoemission spectra has been difficult to do because of the small bulk-band gap of InAs. Theoretically, the electronic structure and atomic positions of InAs(110) were calculated using a quantum-molecular dynamics [2] based in a semi-empirical tight-binding (TB) approach. Almost a decade latter, an *ab initio* quantum-molecular dynamics [1] was performed. The reported atomic structure and electronic surface states differ between TB and *ab initio* calculations, and also differ with the available experimental measurements. This fact is because those semi-empirical calculations were performed using an atomic reconstruction that was not fully relaxed, while *ab initio* calculations were performed using a DFT-LDA with a plane-wave basis set that depend on the cut-off energy to get accurate surface states. In summary, there is no a consensus on the physical properties of the InAs(110) surface. Shkrebtii and collaborators [5] calculated and measured the Reflectance Anisotropy Spectrum (RAS) of InAs(110). Although the authors claimed to elucidate the optical properties of such surface, their calculated RAS is far to resemble the measurements. The discrepancies between their theoretical results and measurements can be associated again to the small optical bulk gap of InAs.

In this work, we present the optical properties of InAs(110) employing a semi-empirical TB formalism [15, 16] and using the atomic coordinates obtained from *ab initio* quantum-molecular dynamics [1]. The use of the fully relaxed slab coordinates guarantees that the calculated electronic properties include all the subtle effects of surface-induced strain and appropriate geometry. Our calculated surface electronic band structure is very similar to that calculated by Shkrebtii and collaborators [5], however, in that work they focused to study the electronic properties of the adsorption of antimony on III-V(110) surfaces, and

---

* Corresponding author: e-mail: cecilia@fisica.unam.mx, Phone: +52 55 562 251 106, Fax: +52 55 561 61 535

only a rough analysis of the clean InAs(110) surface was done. They compared the theoretical and experimental RAS [5], however, their calculations and measurements do not agree.

2 Theoretical method The III-V(110) semiconductor surfaces relax in such a way that the surface cation atom moves inwards the surface into an approximately planar configuration, with a threefold coordination with its first-neighbors anion atoms. The topmost anion atom moves outward to the surface, showing a pyramidal configuration with its three first-neighbors cation atoms [17, 18]. In Fig. 1 (a) we show the top view of a surface unit cell that contains one Indium atom in color pink or large atom (cation), and one Arsenic atom in color green or small atom (anion), per atomic layer. The parameter a_0 is the bulk lattice constant and $d_0 = a_0/2\sqrt{2}$. In Fig. 1 (b) we show a side view of the surface with only the three outermost atomic layers. Here, we define the structural parameters associated to the surface relaxation whose values are taken from Ref. [1]. In Fig. 1 (c) we show the corresponding Two-Dimensional Irreducible Brillouin Zone (2DIBZ). The InAs (110) surface was modeled using a slab of 52 atoms, yield-

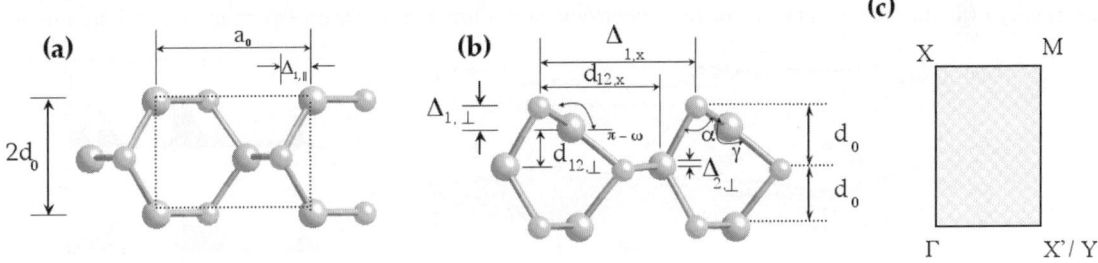

Fig. 1 Model of the atomic geometry of InAs(110). (a) Top view of a surface unit cell. The short side of the surface unit cell is along the [1$\bar{1}$0] crystallographic direction, while the long side is along [001]. (b) Side view of the first three atomic layers of the surface. (c) Two-Dimensional Irreducible Brillouin Zone.

ing a free reconstructed surface on each face of the slab. The thickness of the slab is large enough to decouple the surface states at the top and bottom surfaces of the slab. Periodic boundary conditions were employed parallel to the surface of the slab to effectively model an infinite two-dimensional crystal system. The atomic coordinates were taken from Ref. [1], corresponding to Density Functional Theory (DFT) calculations with an energy cutoff of 18 Ry. We calculate the electronic level structure of the slab using a well known parameterized TB approach with a sp³s* orbital-like basis, within a first-neighbor interaction approach [16]. This basis provides a good description of the valence and conduction bands of cubic semiconductors. This TB approximation has been applied to calculate the electronic and optical properties of a variety of semiconductor surfaces, including other III-V compounds [19]. The TB parameters are taken to be the same as those of Vogl [16] for the bulk but they are scaled by a factor of $(D/d)^2$, where d is the bond length of any two first-neighbor atoms, and $D = \sqrt{3}a_0/4$ [20]. Once the electronic-level structure of the slab has been obtained, we calculate the average slab polarizability, which is in terms of the transition probability between eigenstates induced by an external radiation field. We take an average over 7000 points distributed homogeneously in the irreducible two-dimensional Brillouin zone (2DBZ). The real part of the average polarizability is calculated using the Kramers Kronig relations. Finally, the RAS is calculated as the difference of the Differential Reflectance between the two orthogonal directions in the surface plane, as

$$RAS = \left(\frac{\Delta R}{R_0}\right)_{[1\bar{1}0]} - \left(\frac{\Delta R}{R_0}\right)_{[001]}, \tag{1}$$

where R_0 is the reflectivity calculated with the well known Fresnel formula, and $\Delta R = R - R_0$ is the difference between R_0 and the actual reflection coefficient. The details are fully explained in Ref. [15].

3 Results and Discussion The surface electronic band structure along high-symmetry points of the 2DIBZ of InAs(110) has been recently discuss in a previous paper [21]. Here, we only present a summary of such results to understand the optical properties of InAs(110). We found that our calculation of the electronic structure reproduce the photoemission measurements made by Anderson *et al* [7]. We label the surface electronic states with A_i and C_i for surface anions and cations, respectively.

3.1 *Electronic Structure* Below the Valence Band Maximum (VBM) we found four well-defined occupied surface electronic states denoted by A_5, A_3, A_2 and C_2. The A_5 surface states correspond to the dangling bonds of the As atoms located at the first atomic layer. These A_5 states form a band from the high-symmetry point X to the point X', going through the high-symmetry point M in the 2DIBZ. This band has a minimum at X with an energy of -1.20 eV and disperse upwards towards the Γ point. From X, the band also disperse upwards towards the M point, where the A_5 surface states have an energy of about -0.8 eV. From M to X, this band disperses into the projected bulk band. The A_5 band shows a small dispersion around M given rise to a large contribution on the Local Density of States of the first layer at an energy of about -1 eV. The A_3 surface electronic states are at a lower energy than A_5, and are due to

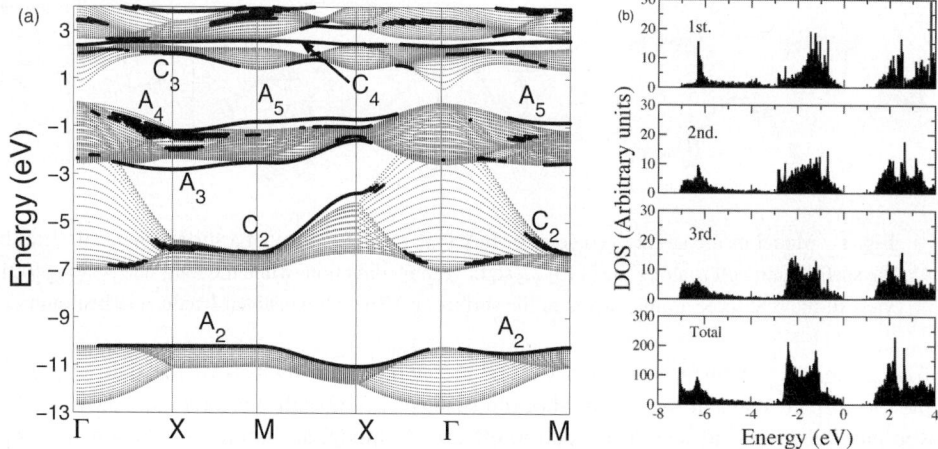

Fig. 2 (a) Electronic band structure of the reconstructed InAs(110) surface. Tiny dots represent the projected bulk states, while black dots represent surface electronic states. (b) Total and the projected local density of states in the first, second and third atomic layers of reconstructed InAs(110).

the backbonds between the anions localized in the first atomic layer and the cations in the second layer. The A_3 band has a minimum in the X high-symmetry point with an energy of -2.8 eV from the VBM. The band reaches its maximum at X' with an energy of -1.6 eV from the VBM. The band shows a dispersion of 1.2 eV, however, around X and M the band is almost flat, contributing to a large density of states in the first and the second layers at energies of about -2.8 eV and -2.5 eV, respectively. This can be observed on the LDOS in Fig. 2(b), where two peaks are found at these energies in the panels showing the LDOS in the first and the second layers. We found surface states with an energy of about -6.0 eV at the high-symmetry point X, that form a band denoted by C_2. This band shows a large dispersion of about 2.5 eV, where the minimum of the band is at M with an energy of -6.3 eV, and its maximum is around X' with an energy of -3.8 eV. These surface states are located at the cation (In) atoms and are due to the bonding between the In and As atoms at the top layer. From X to M, the band shows a small dispersion which is reflected in the LDOS where a large contribution is found at about -6.2 eV in the first layer. From M to X', the C_2 band disperses upwards of about 2.5 eV given rise to a small contribution to the LDOS as shown in Ref. [4]. At lower energies we found another occupied surface electronic band denoted by A_2 which extends almost along all the high-symmetry points in the 2DIBZ. These surface states are located in the anion (As) atoms, and have a s character due to the backbonds between the atoms at the first and second layers, and some contribution is also found from the backbonds between the atoms in the second and third layers. From Γ to M going along X, the band does not show dispersion and is at -10.2 eV. From M to Γ going along X', the band has a dispersion of about 1 eV, showing a minimum around X'. We have also found several localized states at X between Γ and M with an energy from -2 eV to -1 eV. These states are inside the projected bulk band, therefore, they are resonance-like states. These resonance states, denoted by A_4, disperse upwards from X towards both, Γ and M. Most of the A_4 states have a p-character, and are localized at the anion in the first atomic layer. The A_4 states with lower energy, between -1.9 eV and -2.0 eV, show also a s character and are localized at the third layer. Above the VBM we found two unoccupied surface states bands, namely, C_3 and C_4. The C_3 surface states are localized at the cations in the first and third atomic layers. They show a strong p character due to the dangling bonds at cations. At X, we found that C_3 has a maximum with an energy of about 2 eV, and has its minimum value between M and X' with a energy of about 1.4 eV. Finally, at 2.7 eV from VBM we found empty surface states that form a band along all the high symmetry points in the 2DIBZ. This band is denoted by C_4, and shows a very small dispersion along the 2DIBZ.

3.2 *Optical Properties*

We show the RAS for the InAs(110) surface as a function of the energy of the incident light. RAS has contributions of electronic transitions from occupied to empty states that can be labeled as surface to surface (ss), surface to bulk (sb), bulk to bulk (bb), and bulk to surface (bs) transitions. What is different of our calculation to the one did in Ref. [5], is that we averaged the bb transitions coming from the two orthogonal directions in the surface plane to make the bulk optical response isotropic as expected in cubic materials. In Fig. 3(a), we show our calculated RAS, where we obtained that only bb electron transitions are found below 2 eV, and the bulk band gap is about 0.5 eV. In Fig. 3(b), we show the experimental RAS taken from Ref. [5], where also the theoretical RAS was calculated (not shown here), however, our calculated RAS is closer to the experimental one, as we explain below.   At about

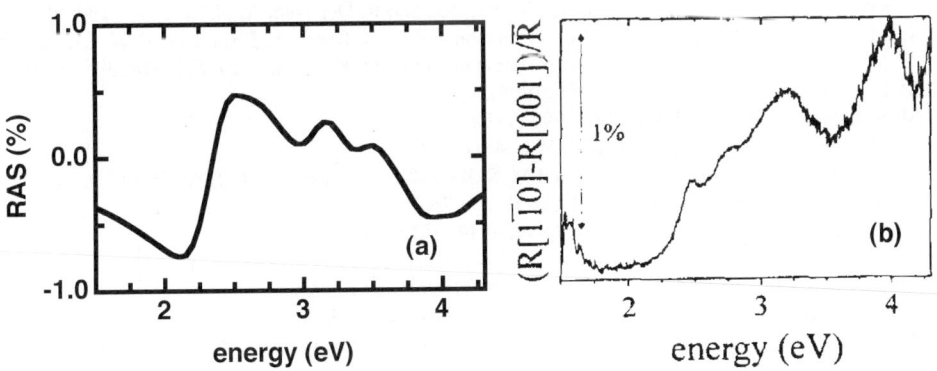

Fig. 3   (a) Theoretical and (b) experimental RAS of InAs(110). The exp. figure was taken from Ref. [5].

2.5 eV, our theoretical RAS shows a peak mainly due to electronic transitions from ss, and sb states, while RAS measurements shows a shoulder. These ss and sb electronic transitions involve dangling-bonds states located at the As and In atoms in the first layer that form atomic chains along the [1$\bar{1}$0] crystallographic direction. Shkrebtii *et al.* [5] associated this peak to ss transitions only. Our theoretical peak has its maximum at 2.5 eV, and is about 0.75 eV wide, because the sb electronic transitions are mainly from the A_5 surface states to empty bulk states along the M to X' high-symmetry points, that show such dispersion. This wide structure is also observed experimentally, where an additional shoulder appears at 2.75 eV which we assign to the sb transitions, while Shkrebtii *et al.* [5] associated it to modifications of bulk states at the E_1 critical point. The theoretical RAS shows another peak at 3.2 eV that comes from ss electron transitions between A_4 occupied and C_3 empty states, and A_5 occupied and C_4 empty states. Again, this peak was associated to bb electronic transitions by Shkrebtii *et al.* [5]. We found another shoulder at 3.5 eV, that comes from sb and bs electron transitions. At 3.8 eV, our theoretical RAS shows a local minimum that corresponds to the minimum observed in the experimental RAS at 3.5 eV. In conclusion, we found a RAS spectrum that better resembles the experimental measurements. We assign the main features in the optical spectrum between 2 eV to 4 eV to the surface atomic and electronic structure, while previous calculation did to bulk electronic states.

4 Summary

4 Summary   We performed a TB calculation using a fully relaxed atomic geometry to study the electronic structure and optical properties of the clean InAs(110) surface. We found a good agreement between our calculations and photoemission experimental data for the surface electronic states. We also found a good agreement between our calculations and RAS measurements. In conclusion, we found that fully relaxed atomic positions from *ab initio* methods in combination with our semi-empirical tight-binding calculation, better resembles the surface electronic states of cleaveage InAs(110) samples. Although the agreement between our results and other theoretical and experimental data is good, we conclude that more experimental studies are necessary to clearly elucidate the atomic relaxation and the electronic properties of InAs(110).

Acknowledgements   We acknowledge the partial financial support from DGAPA-UNAM grant No. IN104201, CONACyT grants No. 36651-E and 36764-E.

References

[1] J.L. Alves, J. Hebenstreit and M. Sheffler, Phys. Rev. B **44**, 6188 (1991).
[2] C. Mailhiot, C. B. Duke and D. J. Chadi, Phys. Rev. B **31**, 2213 (1985).
[3] R. P. Beres, R. E. Allen and J. D. Dow, Phys. Rev. B **26**, 5207 (1982).
[4] B. Engels, P. Richard, K. Schroeder, S. Blügel, Ph. Ebert and K. Urban, Phys. Rev. B **58**, 7799 (1998).
[5] A. I. Shkrebtii, N. Esser, M. Köpp, P. Haier, W. Richter, and R. Del Sole, Appl. Surf. Sci. **104/105**, 176 (1996).
[6] C. B. M. Andersson, J. N. Andersen, P. E. S. Persson and U. O. Karlsson, Phys. Rev. B **47**, 2427 (1992).
[7] C. B. M. Andersson, J. N. Andersen, P. E. S. Persson and U. O. Karlsson, Surf. Sci. **398**, 395 (1998)
[8] D.M. Swantson, *et al.*, Surf. Sci. **312**, 361 (1994).
[9] D.M. Swantson, *et al.*, Can. J. Phys **70**, 1099 (1992).
[10] H. W. Richter, *et al.*, J. Vac. Sci. Technol. B **4**, 900 (1986).
[11] H. Carstensen, R. Claessen, R. Manzke, and M. Skibowski, Phys. Rev. B **41**, 9880 (1990).
[12] W. Drube, D. Straub and F. J. Himpsel, Phys. Rev. B **35**, 5563 (1987).
[13] W. Gudat and D. E. Eastman, J. Vac Sci. Technol. **13**, 831 (1976).
[14] J. van Laar, A. Huiser and T. L. van Rooy, J. Vac Sci. Technol. **14**, 894 (1977).
[15] C. Noguez and S.E. Ulloa, Phys. Rev. B **53**, 13138 (1996); and references therein.
[16] P. Vogl, H. P. Hjalmarson and J. D. Dow, J. Phys. Chem. Solids **44**, 365 (1983).
[17] See for example, F. Bechstedt and R. Enderlein, *Semiconductor Surfaces and Interfaces*, (Akademie-Verlag, Berlin, 1988); A. Zangwill, *Physics at Surfaces*, (Cambridge University Press, New York, 1992); H. Lüth, *Surfaces and Interfaces of Solids*, (Springer Verlag, Berlin, 1993).
[18] T. J. Godin, J. P. LaFemina and C. B. Duke, J. Vac. Sci. Technol. A **10**, 2059 (1992).
[19] C. Noguez, Phys. Rev. B **58**, 12641 (1998).
[20] *"Electronic Structure and the Properties of Solids"*, Walter A. Harrison, (Dover Publications, Inc., New York, 1980).
[21] X. López-Lozano, C. Noguez, and L. Meza-Montes, submitted to Phys. Rev. B.

phys. stat. sol. (c) **0**, No. 8, 2997–3001 (2003) / **DOI** 10.1002/pssc.200303850

Ab-initio study of the adsorption of acetylene on Si(001) surface

Olivia Pulci[*,1], **Pier Luigi Silvestrelli**[2], **Maurizia Palummo**[1], **Francesco Ancilotto**[2], and **Rodolfo Del Sole**[1]

[1] INFM, Department of Physics University of Rome Tor Vergata, Via della Ricerca Scientifica 1, I-00133 Rome, Italy
[2] INFM (Udr Padova and DEMOCRITOS National Simulation Center, Trieste, Italy) and Dipartimento di Fisica "G. Galilei", Università di Padova, via Marzolo 8, I-35131 Padova, Italy

Received 30 May 2003, revised 4 August 2003, accepted 11 August 2003
Published online 10 November 2003

PACS 71.15.Mb, 73.20.At, 78.68.+m

We present the results of a first principles study of the adsorption of acetylene on Si(001). Several adsorption sites are investigated, and the electronic structures and reflectance anisotropy spectra (RAS) are calculated for the energetically most favorable geometries. We find that the end-bridge adsorption sites, both at 0.5 ML and at 1ML C_2H_2 coverage do not give metallic surfaces, at odd with other calculations. First principles study of optical properties for different configurations of acetylene adsorbed on Si(001) show a strong change of the spectra, thus suggesting that RAS experiments may give a definitive answer on the adsorption sites and saturation coverage.

1 Introduction   The adsorption of simple hydrocarbon molecules on Silicon surfaces is very important in the study of the growth of SiC, a strategic semiconductor material. Acetylene, in particular, thanks to its triple bond is a very reactive molecule and hence represents a promising carbon source for the heterogrowth of silicon carbide on silicon. For this reason it has been extensively studied both theoretically and experimentally. Even if acetylene is a simple molecule, still many aspects of its adsorption on Si(001) surface are not clarified, and results are often contradictory.

Many experimental techniques have been used to investigate the adsorption of C_2H_2 on Si, ranging from STM, to LEED, HREED, Auger, photoemission, and photoelectron diffraction [1–10], with results often leading to different conclusions. What seems to be clarified is the fact that at room temperature, and below, the adsorption of acetylene is not dissociative, and that upon annealing at 879 K the C–H bonds break and the hydrogen leaves the surface.

The saturation coverage is still under debate, ranging from 0.5 ML [4] (corresponding to one hydrocarbon molecule each second Silicon dimer) to 0.83 ML [3], interpreted as an effective 1ML coverage when Si defect-free surfaces are considered. Still, even the adsorption site of acetylene is under debate, and what happens to the surface silicon dimers upon C_2H_2 adsorption is controversial.

Theoretical studies are also numerous [10–17] and two main classes of possible adsorption geometries, the di-sigma and the tetra-sigma, have been discussed because they are compatible with experimental findings. The di-sigma class contains the adsorption sites where one hydrocarbon molecule bonds with two silicon atoms; the tetra-sigma class where one hydrocarbon molecule bonds to four silicon atoms. The tetra-sigma class is higher in energy: theoretical results have, in fact, shown that

* Corresponding author: e-mail: olivia.pulci@roma2.infn.it, Phone: +39 06 725 945 03, Fax: +39 06 202 3507

this configuration is thermodynamically unfavorable, so we concentrate our analysis on the di-sigma class. Within this class, two are the adsorption sites mainly under debate: the 'on-top' site, where acetylene adsorbs on top of a silicon dimer (Fig. 1a), and the 'end bridge' site, where the acetylene bridges two Si dimers within the same row (Fig. 1b). These two configurations for a coverage of 0.5 ML have been discussed as the most favorable ones, but recently Miotto and coworkers [16] have stated that the end-bridge structure gives a metallic surface and hence would be Peierls unstable.

For higher coverage, namely 1 ML coverage, it has been agreed that the most stable geometry is the 1ML end-bridge structure, with two acetylene molecules per 2×2 cell bridging two silicon dimers within the same row (Fig. 1d). However, also this structure has been recently discarded by Miotto et al. because of its metal character. In this paper, we present our first-principles calculation in order to shed some light on the still controversial problem of the adsorption site and saturation coverage of $Si(001) : C_2H_2$.

2 Scheme of the calculations

We have performed ab-initio Density Functional Theory calculations on Si(001) clean and upon C_2H_2 adsorption. The exchange correlation Functional has been evaluated within the Local Density Approximation and also within the Generalized Gradient Approximation (GGA, PW91 [18]). A repeated Silicon slab consisting of 12 atomic layers in a 2×2 geometry has been used, with the bottom surface saturated with hydrogen. A set of 8 k-points in the Irreducible part of the Brillouin Zone (IBZ) and an energy cutoff of 40 Ry was used for the geometry optimizations and for the calculations of the adsorption energy. This has been evaluated as:

$$E_{ads} = E[Si : C_2H_2] - E[Si(001) \, 2 \times 2] - E[C_2H_2] \tag{1}$$

where E is the energy of the Silicon surface covered with C_2H_2, of the clean Silicon surface, and of the isolated C_2H_2 molecule, respectively.

Optical properties have been computed using a set of 28 k-points in the IBZ. The Reflectance Anisotropy has been calculated following Del Sole [19] as:

$$\frac{R_x(\omega) - R_y(\omega)}{R_0(\omega)} = \frac{4\omega d}{c} \, \text{Im} \, \frac{4\pi(\alpha_{xx}^{hs}(\omega) - \alpha_{yy}^{hs}(\omega))}{\epsilon_b(\omega) - 1} \, , \tag{2}$$

with ϵ_b bulk dielectric function and α equal to the half slab polarizability. c is the light speed and d the half slab thickness. Here we have taken x as the direction of the dimers of the clean Si(001) surface.

3 Results

The geometries studied, ranging from 0.5 ML (one acetylene molecule per 2×2 silicon cell) to 1 ML coverage (2 acetylene molecules per 2×2 cell) are shown in Fig. 3. In the on top geometry (Fig. 1a) the Silicon dimer bonded to the C_2H_2 is not buckled, whereas the Si dimer not bonded with acetylene preserves its buckling as in the clean Si(001) surface.

Interestingly, we have found two possible equilibrium bridge configurations for 0.5 ML, one almost symmetric (Fig. 1b), the other more stable and asymmetric (Fig. 1c). The asymmetry, not visible in

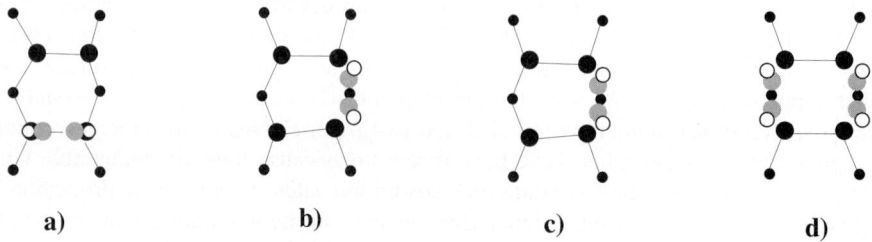

     a)   **b)**   **c)**   **d)**

Fig. 1 (online colour at: www.interscience.wiley.com)   Top view of the geometries studied: a) 'on-top', b) end-bridge symmetric 0.5 ML, c) end-bridge asymmetric 0.5 ML, d) end-bridge 1 ML coverage. Black circles: Si atoms; Gray circles: C atoms; white circles: H atoms.

Table 1 Adsorption energy per molecule, E_{ads}, and structural parameters for acetylene on Si(001). Si–Si(1) and Si–Si(2) indicate the Si dimers to which the acetylene molecule is bonded in the end-bridges configurations; in the case of the on-top structure the molecule is bonded only to the Si–Si(1) dimer. The buckling angles relative to these Si dimers are also given. Data have been computed using the gradient-corrected PW91 functional.

	on-top	end-bridge 0.5 ML asym.	end-bridge 0.5 ML symm.	1 ML end-bridge
E_{ads} (eV/C$_2$H$_2$)	−2.75	−2.62	−2.54	−2.87
Si–Si(1) (Å)	2.33	2.42	2.39	2.41
buckling(1) (deg)	0.5	5.2	2.5	0.0
Si–Si(2) (Å)	2.29	2.38	2.39	2.41
buckling(2) (deg)	17.4	10.0	2.5	0.3
Si–C (Å)	1.88	1.90	1.90	1.90
C–C (Å)	1.34	1.35	1.35	1.35
C–H (Å)	1.09	1.09	1.09	1.09

the top view, is caused by the fact that the dimers are asymmetrically buckled in c), and symmetrically buckled in the b) geometry.

For 1 ML coverage (that is, 2 molecules of C$_2$H$_2$ on a Si(001) 2 × 2 cell), our calculated geometry shows that in the 1 ML end-bridge configuration the dimers are not buckled.

The geometrical details are listed in Tab. 1. In all cases considered, the silicon dimers do not break upon C$_2$H$_2$ adsorption.

3.1 Adsorption energies and electronic band structures
The adsorption energies per C$_2$H$_2$ molecule, calculated according to Eq. 1, are shown in Tab. 1: for 0.5 ML coverage, the most stable structure is the on top one with an adsorption energy of −2.75 eV/C$_2$H$_2$. The corresponding band structure is shown in Fig. 2a. The asymmetric end-bridge is just 0.13 eV higher in energy and, at odd with the result of Miotto et al., turns out to be semiconducting (see Fig. 2b), though the minimum gap is only 0.35 eV. The two surface states within the gap are related to the 'uppermost' and 'the most down' Si surface atoms. The symmetric end-bridge geometry is higher in energy than the asymmetric one. Moreover, this structure turns out to be metallic. Our results hence show that the surface undergoes a spontaneous structural distortion from the symmetric end-bridge to the asymmetric end-bridge, lowering its energy by about 0.08 eV and becoming semiconducting.

Finally, for 1ML coverage, the end-bridge geometry is without doubt semiconducting (see Fig. 3): the Silicon dimers are not buckled anymore, the empty (full) surface states typical of the Si(001)

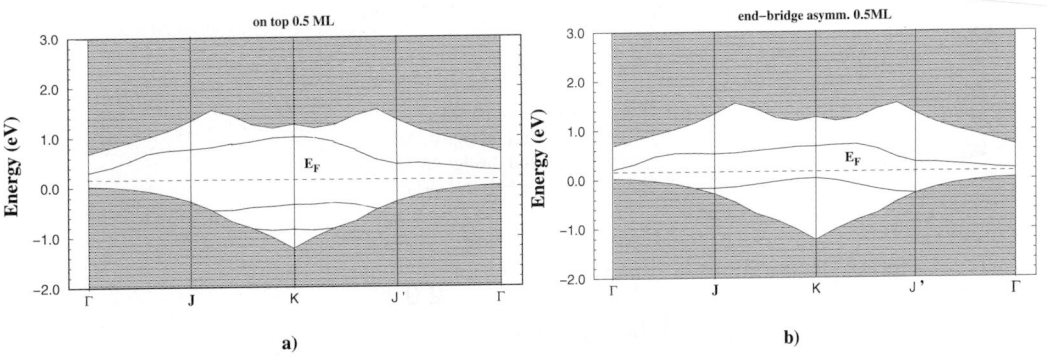

Fig. 2 a) Band structure for the on top band structure. Analysis of the surface states reveals that the filled (empty) surface state in the gap is related to the up (down) silicon atom belonging to the Si dimer not bonded to acetylene. b) Band structure for the end-bridge asymmetric structure 0.5 Ml coverage. See text for details.

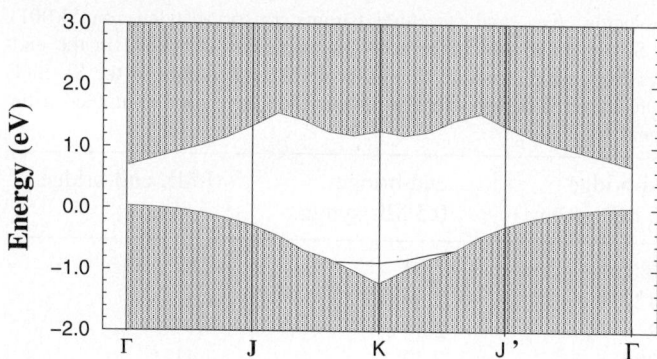

Fig. 3 Band structure for the end-bridge geometry at 1 ML coverage.

surface, connected to the 'down' (up) silicon atoms of the clean surface, are saturated, and the gap becomes empty (Fig. 3).

3.2 Optical properties Figure 4 shows our calculated Reflectance Anisotropy Spectra (RAS) obtained for the clean Si(001) surface and upon C_2H_2 adsorption. As it can be seen, at low photon energy (around 1.0 eV) there is a strong reduction of anisotropy for the on-top configuration with respect to the clean Si(001) surface, while in the end-bridge 0.5 ML configuration there is an opposite behavior with an increase of the negative peak. This negative peak in the RAS of the clean surface [20] (experimentally [21] a peak is found at 1.6 eV, the shift between theory and experiment being due to the well known underestimation of the energy gaps by DFT) is due to optical transitions involving silicon dimer states, and its sign shows that, at low energy, the transitions are favored for light polarized along the dimer rows, thus implying a strong interaction among the dimers along the same row. Therefore, the fact that this peak is reduced in the on-top configuration can be understood as a reduction of the interaction among the dimers, due to the presence of the chemisorbed acetylene molecule; instead, in the end-bridge configuration (at 0.5 ML coverage), since the axis of the molecule is perpendicular to the Si dimers, there is an increase of transitions for light polarized along this direction. Finally, for 1ML coverage, the anysotropy disappears because the acetylene saturates all the surface silicon bonds and the corresponding states are pushed out of the electronic gap.

Our RAS calculations suggest that a direct way to investigate the adsorption sites of C_2H_2 on Si(001) would be to perform reflectance anisotropy experiments in the near infrared regions. At these

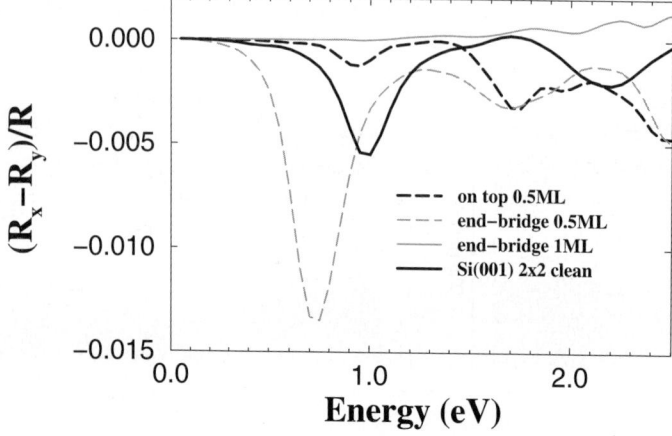

Fig. 4 (online colour at: www.interscience.wiley.com)   Theoretical RAS for Si(001) 2×2 clean (thick solid line), on top (thick dashed line), end-bridge asymmetric 0.5ML (gray dashed line) and end-bridge at 1ML coverage (gray solid line).

energies a dramatic change in the spectra with respect to the clean Si(001) surface is predicted, due to the changes in the electronic structure within the silicon surface gap.

4. Summary and Conclusion We have presented an ab-initio calculation of the geometry, electronic structure and optical properties of acetylene on Si(001) surfaces. We found that the on-top geometry is, for 0.5 ML coverage, the most stable one, but the end-bridge is just 0.13 eV/C_2H_2 higher in energy and not metallic, caused by a spontaneus distortion (asymmetric buckling) that lowers the energy with respect to the symmetric buckled configuration, and makes the surface semiconducting. Moreover, the adsorption energy is lowest for the end-bridge 1 ML, thus suggesting that the saturation coverage could be one full monolayer. This structure, discarded in the past because claimed to be metallic, is in our calculations semiconducting. Finally, the theoretical RAS shows characteristic changes upon acetylene adsorption, indicating that experimental reflectance anisotropy spectra would help in discriminating the adsorption sites and saturation coverage.

Acknowledgements Computer time at CINECA from the INFM 'Iniziativa Trasversale Calcolo Parallelo' is gratefully acknowledged. This work has been supported by INFM-PRA 1MESS, MIUR COFIN 2002, INFM-PAIS-Celex and by the EU through the NANOPHASE Research Training Network (Contract No. HPRN-CT-2000-00167).

References

[1] J. Yoshinobu, H. Tsuda, M. Onchi, and M. Nishijima, Chem. Phys. Lett. **130**, 170 (1986); M. Nishijima, J. Yoshinobu, H. Tsuda, and M. Onchi, Surf. Sci. **192**, 383 (1987).

[2] C. Huang, W. Widdra, X. S. Wang, and W. H. Weinberg, J. Vac. Sci. Technol. A **11**, 2250 (1993).

[3] P. A. Taylor, R. M. Wallace, C. C. Cheng, W. H. Weinberg, M. J. Dresser, W. J. Choyke, and J. T. Yates, Jr., J. Am. Chem. Soc. **114**, 6754 (1992); C. C. Cheng, P. A. Taylor, R. M. Wallace, H. Gutleben, L. Clemen, M. L. Colaianni, P. J. Chen, W. H. Weinberg, W. J. Choyke, J. T. Yates, Jr., Thin Solid Films **225**, 196 (1993).

[4] L. Li, C. Tindall, O. Takaoka, Y. Hasegawa, and T. Sakurai, Phys. Rev. B **56**, 4648 (1997).

[5] F. Matsui, H. W. Yeom, A. Imanishi, K. Isawa, I. Matsuda, T. Ohta, Surf. Sci. **401**, L413 (1998); Phys. Rev. B **62**, 5036 (2000).

[6] R. Terborg, P. Baumgärtel, R. Lindsay, O. Schaff, T. Giessel, J. T. Hoeft, M. Polcik, R. L. Toomes, S. Kulkarni, A. M. Bradshaw, and D. P. Woodruff, Phys. Rev. B **61**, 16697 (2000); R. Terborg, M. Polcik, J. T. Hoeft, M. Kittel, D. I. Sayago, R. L. Toomes, and D. P. Woodruff, Phys. Rev. B **66**, 085333 (2002).

[7] H. W. Yeom, S. Y. Baek, J. W. Kim, H. S. Lee, and H. Koh, Phys. Rev. B **66**, 115308 (2002).

[8] S. H. Xu, M. Keeffe, Y. Yang, C. Chen, M. Yu, G. J. Lapeyre, E. Rotenberg, J. Denlinger, and J. T. Yates, Jr., Phys. Rev. Lett. **84**, 939 (2000).

[9] S. Mezhenny, I. Lyubinetsky, W. J. Choyke, R. A. Wolkow, and J. T. Yates Jr., Chem. Phys. Lett. **344**, 7 (2001).

[10] W. Kim, H. Kim, G. Lee, J. Chung, S.-Y. You, Y.-K. Hong, and J.-Y. Koo, Surf. Sci. **514**, 376 (2002).

[11] Y. Tanida and M. Tsukada, Appl. Surf. Sci. **159–160**, 19 (2000).

[12] J.-H. Cho, L. Kleinman, C. T. Chan, and K. S. Kim, Phys. Rev. B **63**, 073306 (2001).

[13] D. C. Sorescu and K. D. Jordan, J. Phys. Chem. B **104**, 8259 (2000).

[14] P. L. Silvestrelli, F. Toigo, and F. Ancilotto, J. Chem. Phys. **114**, 8539 (2001).

[15] Y. Morikawa, Phys. Rev. B **63**, 033405 (2001).

[16] R. Miotto, A. C. Ferraz, and G. P. Srivastava, Phys. Rev. B **65**, 075401 (2002).

[17] P. L. Silvestrelli, O. Pulci, M. Palummo, R. Del Sole, and F. Ancilotto, submitted to Phys. Rev. B.

[18] J. P. Perdew and Y. Wang, Phys. Rev. B **45**, 13244 (1992).

[19] for more details, see: R. Del Sole, *Reflectance spectroscopy-theory* in *Photonic probes of surfaces*, edited by P. Halevi (Elsevier, Amsterdam 1995), p. 131, and references therein.

[20] M. Palummo, G. Onida, R. Del Sole, and B. Mendoza, Phys. Rev. B **60**, 2522 (1999).

[21] R. Shioda and J. van der Weide, Phys. Rev B **57**, R6823 (1998).

phys. stat. sol. (c) **0**, No. 8, 3002–3006 (2003) / **DOI** 10.1002/pssc.200303854

Structure and magneto optical properties of ferromagnetic Ni films grown on Cu(110)

M. Wahl*, **Th. Herrmann, N. Esser,** and **W. Richter**

[1] Institut für Festkörperphysik, Sekr. PN 6-1, Hardenbergstr. 36, 10623 Berlin, Germany

Received 30 May 2003, revised 4 August 2003, accepted 11 August 2003
Published online 10 November 2003

PACS 75.50.Cc, 75.70.Ak, 78.20.Ls, 78.66.Bz, 78.68.+m

The structure and magnetism of several monolayer (ML) thick ferromagnetic nickel films evaporated on Cu(110) was studied in-situ under ultra- high-vacuum conditions. Optical-, magnetic- and structural properties of the layers are correlated using simultaneous Reflectance Anisotropy Spectroscopy (RAS) and Magneto Optical Kerr Effect Spectroscopy (MOKE-S) as well as STM at certain coverages. Using oxygen as a surfactant for the nickel/copper system, it is possible to obtain pseudomorphic layer growth up to 18 ML. These films turn out to be nearly single domain and show complete out-of-plane magnetization after a spin-reorientation transition from in-plane at 6 ML.

1 Introduction

Most experimental work about nickel on copper was performed on the Cu(001) surface because of the pseudomorphic growth and out of plane magnetization of the Ni film. In contrast, the growth of Ni on Cu(110) has been rarely investigated yet. Additionally, the usability of surfactants on the Ni/Cu(110) system has not been reported so far. In order to study such films we developed a dedicated optical setup for measuring Reflectance Anisotropy Spectroscopy (RAS) and Surface Magneto Optical Kerr Spectroscopy (MOKE-S) simultaneously. RAS has been shown to be very useful for surface analysis of metals [1, 2, 3]. In contrast to MOKE at a single photon wavelength, Magneto Optical Kerr Spectroscopy is rarely used. However it can give useful information on the modification of the electronic structure[5, 6]. Thus RAS and MOKE-S are an ideal combination for in-situ analysis and control of ferromagnetic layer growth. Information about electronic and magnetic properties can be obtained simultaneously with the same equipment.

2 Experimental details and sample preparation

The optical measurements were done in-situ, using a RAS-MOKE-spectrometer developed at the TU Berlin. Our spectrometer measures optical anisotropy and Magneto Optical Kerr Effect in polar configuration within an energy range from 0.75 eV to 6.5 eV. Furthermore we can measure hysteresis loops and RAS/MOKE transients at any desired energy within the energy range. RAS measures the difference of the complex reflectivity for polarization along two perpendicular axes (in our case: [1$\bar{1}$0] and [001]) within the surface. RAS is surface related for cubic crystals like Cu and Ni. MOKE measures the difference of the reflectivity for left- and right-circular polarized light, which depends on the sample's magnetization and can be obtained with the same optical modulation technique [10]. Since MOKE is linearly dependent

---

* Corresponding author: e-mail: mwahl@gift.physik.tu-berlin.de, Phone: +49 30 314 22079, Fax: +49 30 314 21769

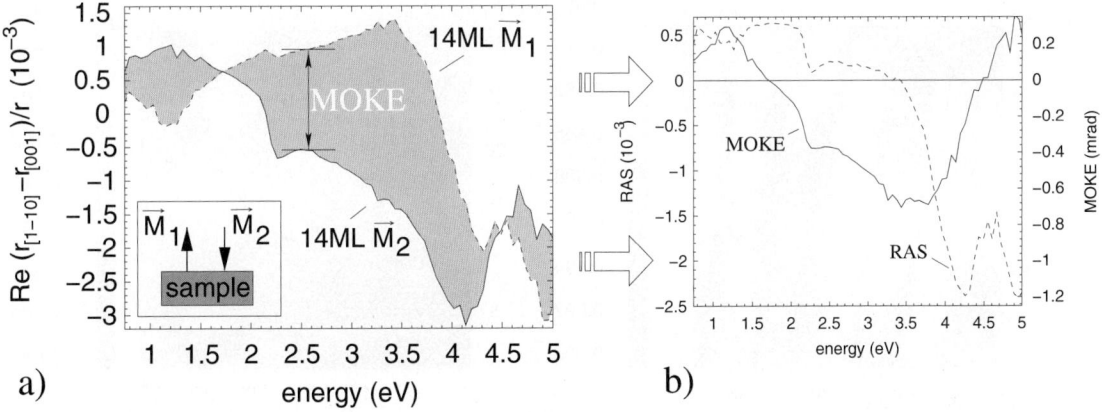

Fig. 1 RAS/MOKE measurement procedure. a) Two RAS/MOKE spectra of the same film, done at different magnetizations $\vec{M}_1$ and $\vec{M}_2$ ($\vec{M}_1 = -\vec{M}_2$). b) The average of the two spectra corresponds to RAS, the difference to MOKE

on the magnetization, RAS and MOKE can be separated. The polarization states of the reflected light are measured after magnetizing the sample in two opposite directions perpendicular to the surface (see inset figure 1a)). The Kerr rotation is derived by taking the difference, whereas the RAS signal is given by the average of the two measurements. MOKE transients during the growth were taken with and without the magnetic field to obtain field dependent (non remanent) and remanent Kerr signals. The experiments were performed in an ultra-high vacuum system, equipped with STM, AES, RAS, MOKE and LEED. After cycles of Ne^+ ion bombardment (1 keV, 10 min at room temperature) followed by subsequent annealing at 850 K a well ordered Cu(110)-(1×1) surface was obtained and no impurities could be detected with AES. The clean surface was exposed to 10 Langmuir of oxygen at room temperature to form the Cu(110)-(2×1)-O "added-row" structure [8, 9]. Ni was evaporated at room temperature with a flux rate of 0.25 monolayers per minute. Layer thicknesses were estimated by AES, by analysing the relative heights of the Ni and Cu Auger intensities.

3 Experimental results

3.1 Structure and layer morphology

Figure 2 compares STM images of Ni films between 9 ML and 10 ML on the clean and oxygen terminated Cu(110) surfaces. On the clean substrate the surface is rather rough with approx. 5 nm broad row like island structures extending along the $[1\bar{1}0]$ direction. On the Cu(110)-(2×1)-O surface, the morphology of the Ni film is very different. The surface turns out to be almost flat with monoatomic steps and structures pointing towards the [001] direction. Ni-O rows can be clearly seen on top of the film, forming the (2×1) surface reconstruction observed by LEED. Additional AES measurements prove that oxygen always stays on top of the film, even for thicknesses above 9 ML. Thus, we conclude that oxygen acts as a surfactant for the Ni/Cu(110) system and enhances the film quality dramatically.

3.2 Optical Anisotropy

To monitor the different growth modes optically, the change in the optical anisotropy was measured at 2.1 eV during Ni evaporation (Figure 3a)). A structured RAS transient is only found during the growth on the Cu(110)-(2×1)-O surface, showing a change in shape at 0.9 ML and pronounced minima at 1.9 ML and 4.9 ML. Compared to the growth of Cobalt on Cu(110)-(2×1)-O the structured RAS transient is an

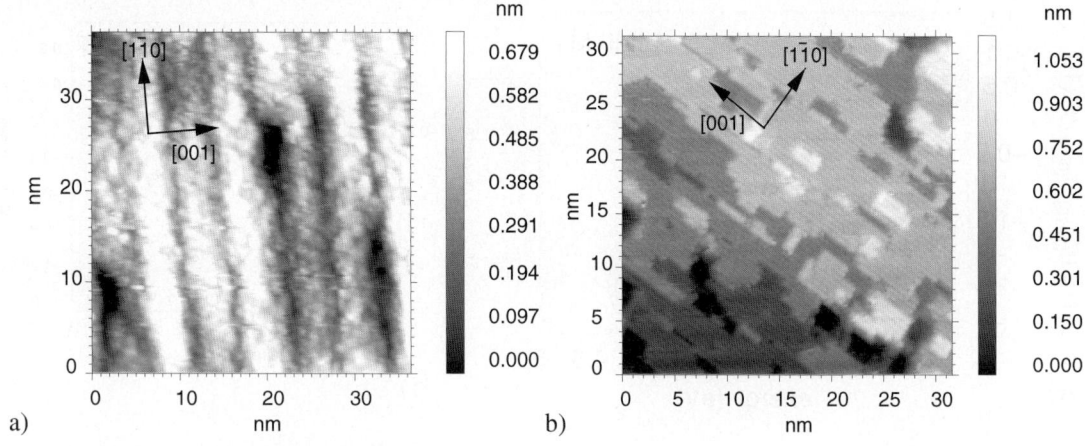

Fig. 2 STM images of a) 10 ML Ni on Cu(110) and b) 9 ML Ni on Cu(110)-(2×1)-O. Crystal directions are indicated in the figures.

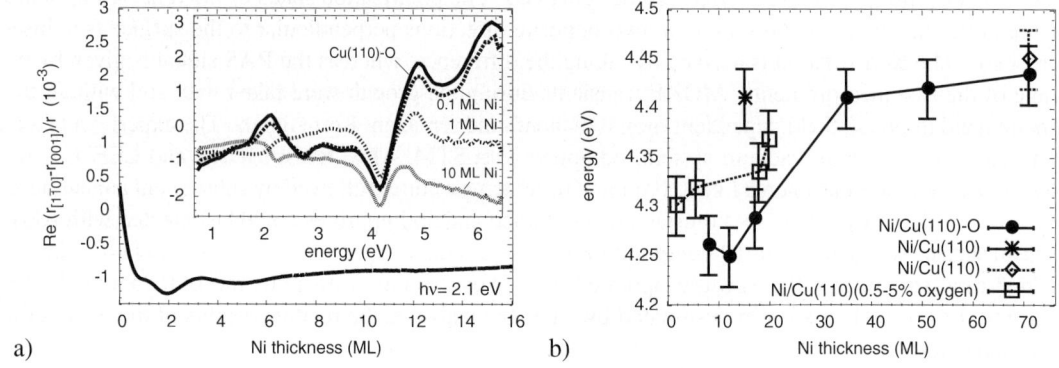

Fig. 3 Optical anisotropy of Ni on Cu(110). a) Thickness dependent optical anisotropy at 2.1 eV obtained during growth. The insets show RAS spectra of various film thicknesses. b) Thickness dependent energy position of the 4.3 eV RAS structure for different Ni films on Cu(110). There are significant differences in energy position below 12 ML for different amounts of oxygen on the substrate.

indication of pseudomorphic layer growth [7]. The inset of Figure 3a) shows RAS spectra for different film thicknesses. There is a significant change in line-shape around 2 eV for Ni thicknesses above 1 ML, which is absent for the clean substrate. A similar behaviour has been found for Cobalt on Cu(110)-(2×1)-O [7]. Accordingly we assume that at low coverage Ni, Cu and O intermix and form an intermediate interface structure similar to Co on this substrate. A closer look at the energy positions of the minimum around 4.3 eV for different film thicknesses is shown in Figure 3b). For Ni films below 12 ML on the Cu(110)-(2×1)-O substrate the peak position is around 4.25 eV, then increases its energy position between 12 ML and 36 ML and finally saturates at 4.3 eV for higher coverages. On clean or slightly oxygen contaminated Cu(110) substrates the peak positions are at higher energies for all thicknesses. Since layer by layer growth only occurs on the pre-oxidised surface we propose that the different peak positions are related to the film quality.

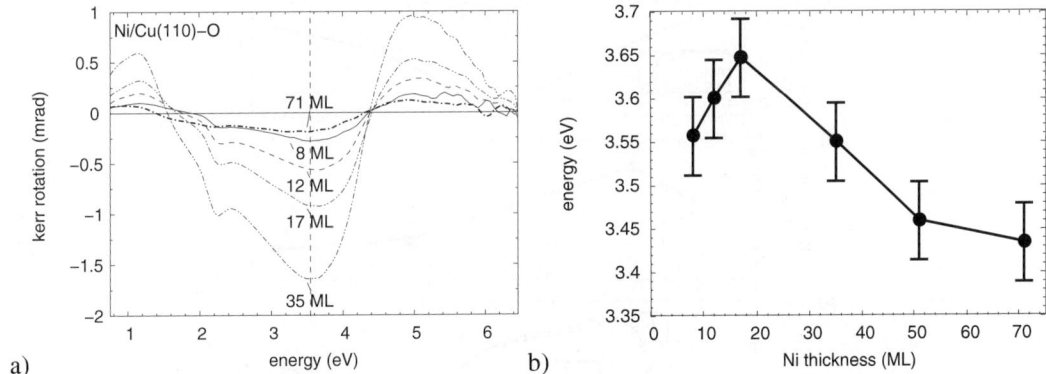

Fig. 4   Spectral magneto optical properties of Ni on Cu(110)-(2×1)-O for different film thicknesses. a) polar MOKE spectra measured in remanence. b) Thickness dependent energy position of the MOKE minimum specified by the dashed line in a). The peak shift direction changes between 18 ML and 36 ML.

3.3   Magnetic- and magneto optical properties

Kerr rotation spectra of Ni on Cu(110)-(2×1)-O are shown in 4a). The spectra were measured in remanence (no external magnetic field applied) from 8 ML to 71 ML. A linear increase of the magneto optical Kerr effect is observed until 35 ML, as is expected from the linear thickness dependence of MOKE within the thin film limit[4]. In the case of thin films the polar Kerr effect is, to a good approximation, proportional to the out of plane magnetization of the sample. A smaller remanent Kerr effect is observed for 71 ML. Here, the magnetic easy axis is already tilted towards the film plane, decreasing the amount of the out of plane magnetization. Complete out of plane magnetization is only obtained on the pre-oxidised Cu(110) substrate. The spectral structures exhibit a red-hift with increasing film thickness. A closer look on the peak shift of the MOKE minimum at 3.6 eV is shown in figure 4b) for different film thicknesses. It exhibits an increase of the energetic position until 18 ML followed by a decrease. The change of the peak shift direction cannot be described within a simple layer model as introduced in Ref[4]. Thus, there must be a difference in the optical or magneto optical components of the dielectric tensor between the thicknesses below 18 ML and above 36 ML.

Figure 5 resembles the magnetic properties of Ni on Cu(110)-(2×1)-O at different film thicknesses. Figure 5a) shows the remanent polar Kerr rotation at 2.1 eV obtained during the growth. Additionally the remanent and non-remanent Kerr Signals at low Ni coverages are shown in the inset. We observe a magneto optical signal above 6 ML, marking the onset of ferromagnetism. Between 6 ML and 7.5 ML the magnetization is oriented within the film plane, which is clear from the absence of a remanent polar signal. The spin reorientation transition to out of plane takes place at 7.5 ML. On further deposition no in plane component of magnetization is found up to 36 ML. For thicknesses above 36 ML the magnetization is slowly tilting back in plane. Figure 5c) shows the coercivity versus the film thickness obtained from hysteresis loops. To give an impression of the loop shapes, several scaled loops are shown in Figure 5c). A strong increase in coercivity is found above 10 ML which saturates at 18 ML, followed by a slow decrease. Although this behaviour is not yet fully understood, we assume that the increase is related to the growing dislocation density within the Ni film. Dislocations can lead to domain wall pinning and therefore flipping of magnetization by domain wall movement becomes more energy consumptive. For comparison, the peak position of the spectral MOKE minimum at 3.6 eV is shown again in Figure 5b). Although additional MOKE spectra should be analysed between 18 ML and 36 ML, it seems to be obvious that the changing coercivity and the differences in peak positions are related to each other, probably governed by changes in the film quality.

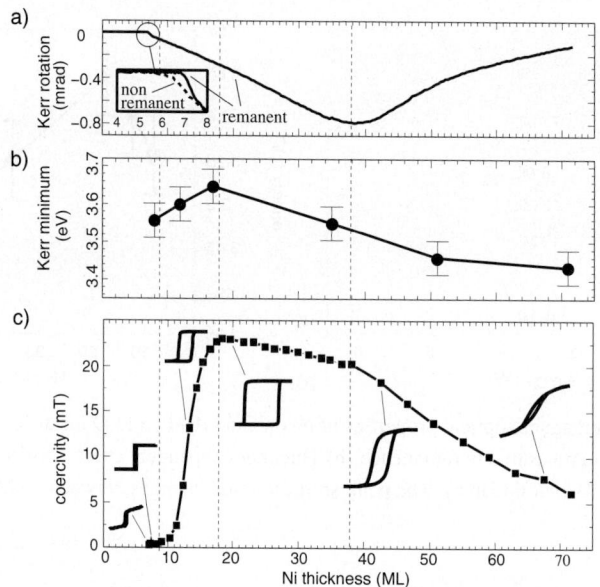

Fig. 5 Compilation of thickness dependent magnetic and magneto optical properties of Ni on Cu(110)-(2×1)-O. a) Remanent polar Kerr rotation obtained during growth. The inset shows the remanent and non remanent Kerr signal at low coverages. b) Spectral position of the MOKE minimum around 3.6 eV obtained from Kerr spectra. c) Coercivity versus film thickness obtained from hysteresis loops during growth interruptions. Some typical hysteresis loops are shown as insets.

4   Summary

We have successfully combined RAS and MOKE spectroscopy to investigate the growth properties of Ni/Cu(110) films. It was shown that oxygen acts as a surfactant on Ni/Cu(110). The pseudomorphically grown films show out of plane magnetization in a thickness range from 7.5 ML to 35 ML, while the magnetization of Ni on the clean Cu(110) surface is in plane. We find that energy positions of features in MOKE-S and RAS spectra are related to layer quality. The results on oxygen terminated substrates, in conjunction with a strong change in the coercivity, imply changes of the film structure around 12 ML.

Acknowledgements   This work was supported by the SFB 290 of the Deutsche Forschungsgemeinschaft (DFG)

References

[1] K. Stahrenberg, Th. Herrmann, N. Esser, and W. Richter, Phys. Rev. B **61**, 3043 (2000).
[2] Ph. Hofmann, K. C. Rose, V. Fernandez, A. M. Bradshaw, and W. Richter, Phys. Rev. Lett. **75**, 2039 (1995).
[3] J.-K. Hansen, J. Bremer, and O. Hunderi, Surf. Sci. **418**, L58 (1998).
[4] J. Zak, E.R. Moog, C. Liu, and S.D. Bader, J. Magn. Magn. Mater. **89**, 107 (1990).
[5] Š. Višňovský, M. Nývlt, V. Prosser, R. Lopušník, R. Urban, J. Ferré, G. Pénissard, D. Renard, and R. Krishnan, Phys. Rev. B **52**, 1090 (1995).
[6] K. Nakajima, H. Sawada, T. Katayama, and T. Miyazaki, Rev. B **54(22)**, 15950 (1996).
[7] Th. Herrmann, K. Lüdge, P. Poulopoulos, J. Lindner, K. Baberschke, N. Esser, and W. Richter, Phys. Rev. B **64**, 184424 (2001).
[8] D. J. Coulman, J. Winterlin, R. J. Behm, and G. Ertl, Phys. Rev. Lett. **64**, 1761 (1990).
[9] K. Kern, H. Niehus, A. Schatz, P. Zeppenfeld, J. George, and G. Comsa, Phys. Rev. Lett. **67**, 855 (1991).
[10] K. Sato, Jpn. J. Appl. Phys. **20(12)**, 2403 (1981).

phys. stat. sol. (c) **0**, No. 8, 3007–3011 (2003) / **DOI** 10.1002/pssc.200303835

PbSe(100) surface electronic states studied by surface differential reflectivity

S. Colonna[1], **F. Ronci**[1], **C. Ottaviani**[1], **A. Szczerbakow**[2], **B. J. Kowalski**[2], **B. A. Orlowski**[2], and **A. Cricenti**[*, 1]

[1] CNR Istituto di Struttura della Materia, via Fosso del Cavaliere, 100, 00133 Roma, Italy
[2] Institute of Physics, Polish Academy of Sciences, Al. Lotnikow 32/46, 02-668 Warsaw, Poland

Received 30 May 2003, revised 4 August 2003, accepted 11 August 2003
Published online 10 November 2003

PACS 73.20.At, 78.40.Fy, 78.68.+m

Using Surface Differential Reflectivity (SDR) we have studied the optical properties of a PbSe(100) surface obtained by cleaving a PbSe sample in ultra-high-vacuum environment. SDR showed the existence of a gap of approximately 1.6 eV. The derived surface dielectric function showed, in the energy range of the present study (1.2 – 3.6 eV), three optical transitions observed at energies of 1.59, 2.01 and 2.39 eV. By putting together angle resolved photoemission data and theoretical calculations, these transitions have been, tentatively, assigned to different points in the surface Brillouin zone.

1 Introduction

Compounds of the IV-VI family are narrow gap semiconductors which crystallize in the rock-salt structure. Among these compounds lead salts (PbS, PbSe) have been the subject of a lot of studies during the last years due to their technological applications as infrared radiation detectors and laser diodes[1-6]. These semiconductors have the peculiarity of a small bulk gap so that no surface states occur inside such gap, thus making their surface electronic properties quite different from the well-known zinc-blend-type semiconductors. Regarding the PbSe compound, several works addressing bulk optical properties and band structure, both experimentally and theoretically, have appeared in literature [7-10].

In this paper, the optical properties of a PbSe(100) surface, obtained by cleaving a PbSe sample in ultra-high-vacuum (UHV) environment, are investigated by Surface Differential Reflectivity (SDR). SDR showed the existence of a gap of approximately 1.6 eV. By taking into account the bulk optical properties, we derived the surface dielectric function obtaining, in our energy range (1.2 – 3.6 eV), three optical transitions at energies of 1.59, 2.01 and 2.39 eV. By putting together angle resolved photoemission data and theoretical calculations, these transitions have been, tentatively, assigned to different points in the surface Brillouin zone (SBZ).

2 Experimental set-up

The sample was grown as a 10x5x5mm^3 parallelepiped with the (100) cleavage plane parallel to the 5x5 plane at the Institute of Physics Polish Academy of Science by the horizontal version of the process defined as Self-Selecting Vapor Growth SSVG [11] . The PbSe(100) surface was prepared by cleaving the sample using the double wedge technique in a vacuum chamber (base pressure 10^{-10} mbar). A sharp LEED pattern evidenced a clean unreconstructed surface. The SDR experiment consists of shining light at normal incidence onto the sample surface in UHV conditions and measuring the intensity of the re-

---

* Corresponding author: e-mail: cricenti@ism.cnr.it, Phone .+39 0649934143, Fax +39 0649934153.

flected light normalized to a dummy sample. The light emitted from a 100 W lamp passes through a lens and is divided into two beams by a semitransparent CaF_2 beam splitter. One beam (I) is focused through a second lens on the sample surface in the UHV chamber while the reference beam (I_0) is focused onto an aluminum mirror. The intensities of the two reflected beams then enter an EG&G PARC half-meter monochromator and are measured with a computer controlled Optical Multi-channel Analyzer (OMA). The values of the ratio ($R = I/I_0$) between the reflectivity of the cleaved surface and the reflectivity of the dummy sample, normalized with respect to the background signal, were obtained by using four computer controlled shutters. The SDR signal was obtained by measuring the reflectivity at normal incidence on the freshly cleaved (100) surface and repeating the same measurement after oxidation by molecular oxygen at a pressure of 10^{-5} mbar. The reflectivity spectra were recorded while the ion gauge was still on in order to get excited molecular oxygen. The SDR spectrum consists of the difference between the two measurements   normalized   to   the   reflectivity   spectrum   of   the   oxidized   surface   [12]
$$\Delta R/R = \left(R_{clean} - R_{ox} \right)/R_{ox} \ .$$

3 Results

Figure 1 shows $\Delta R/R$ measured for a PbSe(100) surface. The different curves were obtained after different oxidation time. The stability of the experimental setup was tested before the oxidation process by repeating the reflectivity measurement of the clean surface and calculating the differential signal. This test showed a stability better than 0.1%, clearly smaller than the features observed in the SDR spectra.

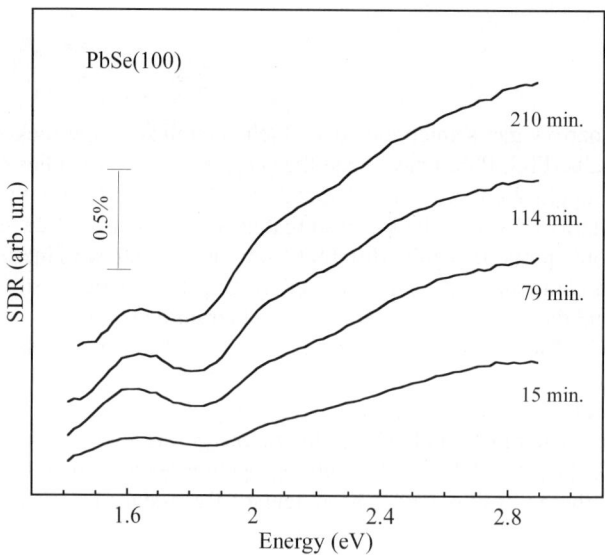

Fig. 1   SDR measurements for a PbSe(100) surface after different oxidation times.

The SDR signal can be written in linear approximation, in the case of small thickness ($d \ll \lambda$) of the region interested by the surface states, as a function of the surface and bulk dielectric functions [12-14]:

$$\frac{\Delta R}{R} = d\left[A(\omega)\,\varepsilon_S'' - B(\omega)\left(\varepsilon_S' - \varepsilon_{ox}' \right) \right] \tag{1}$$

where $A(\omega)$ and $B(\omega)$ are functions of the bulk dielectric constants, ε'_{ox} is the real part of the dielectric function of the film formed after the oxidation process. In the case of a non-absorbing bulk in the experimental range the B factor is negligible and the SDR spectrum is proportional to ε''_S. Due to the very narrow gap (0.17 eV at 4 K) of the PbSe, the surface transitions result resonant with bulk states in our experimental energy range (1.2-3.6eV) because the bulk is optically absorbing. In this case the B term must be taken into account (see Fig. 2) so that the SDR signal contains also the contribution from the real part of the surface dielectric function. The two contributions can be separated in the integral equation by making use of the Kramers-Kronig relation between the real and the imaginary part of the surface dielectric function. The obtained integral equation was solved by transforming the integral in a summation over the experimental points and inverting the kernel matrix of the integral equation. From this analysis we obtained the ε''_S function, shown in Fig. 3, using the 210 min. spectrum when the oxidation process has saturated and the surface states are completely removed.

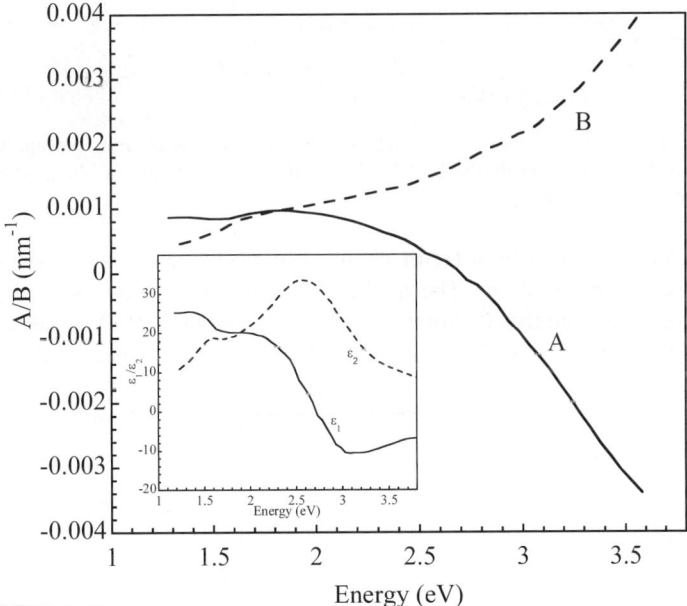

Fig. 2 A and B coefficients used in the analysis of the experimental data. The coefficients were calculated from the bulk dielectric functions published in ref. [9] and reported in the inset.

The PbSe bulk dielectric function [9] was used for the determination of the surface dielectric function. A value of 2 for the ε_{ox} was assumed and the thickness d of the surface layer was fixed to 1 Å. It is worthwhile to note that changing ε_{ox} and d values has only small variations in the calculated surface dielectric function. In any case, when reasonable values for d and ε_{ox} are used, only the intensity of the ε''_S features are affected without strong differences in their energy position. The so obtained surface dielectric function is shown in Fig. 3 and was fitted using several Gaussian contributions: from this fitting procedure three different optical transitions were evidenced at 1.59 eV, 2.01 eV and 2.39 eV.

In order to tentatively assign these transitions to the band structure we show in Fig. 4 the two highest valence band structures measured in ref. [7] combined with the lowest calculated conduction band from ref. [10].

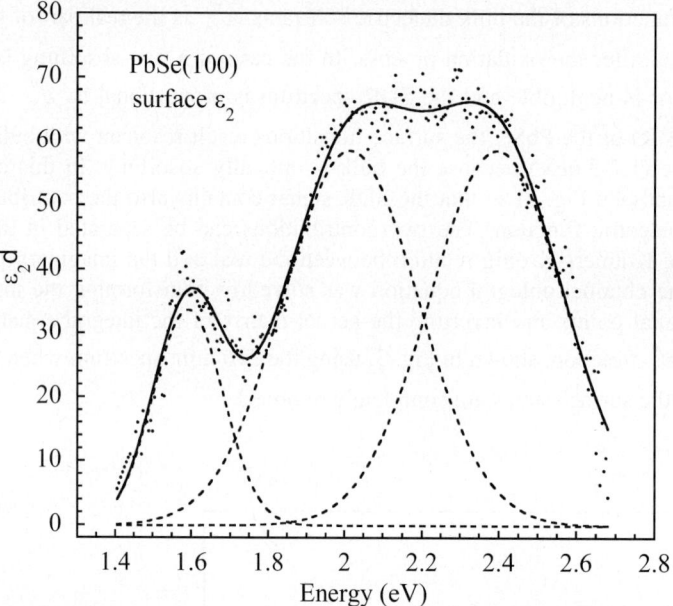

Fig. 3 Imaginary part of the PbSe(100) surface dielectric function obtained from the analysis of the spectrum measured after 210 min. of oxidation (dots). The continuous line is the fit obtained using three Gaussian curves, shown as dashed lines.

In Fig. 4 the conduction band was scaled by a factor obtained by normalizing the dispersion of the calculated valence band with the experimental one. The smallest transition was fixed to 1.59 eV, as the ε_2 lowest optical transition. We can single out that the transitions at 1.59 and the 2.01 eV occur around the middle of Γ-X direction in the Surface Brillouin Zone, while the 2.39 eV transition occurs at Γ and at the

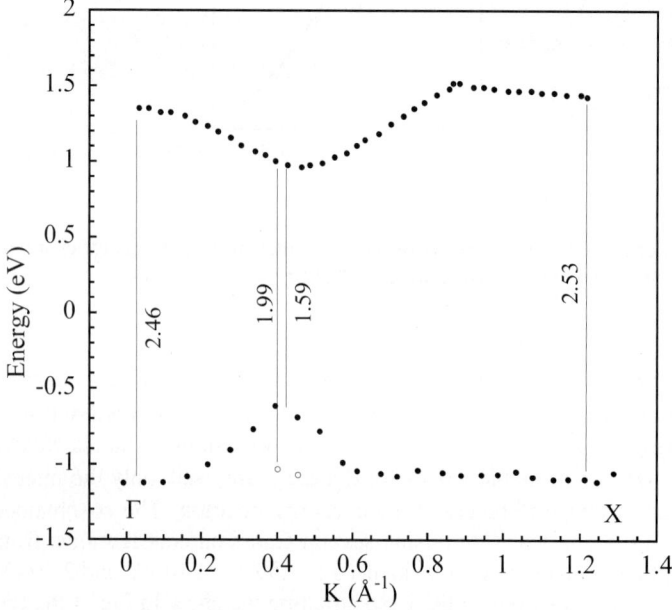

Fig. 4 Measured valence band [7] and theoretical conduction band [10] along the Γ-X direction. The possible transitions are in agreement with the contributions in ε_2 shown in Fig. 3.

border of the SBZ (X point). In this picture the empty band needs to be mapped all along the symmetry lines in order to fully assign the optical transitions to precise points of the SBZ and, moreover, to check if excitonic effects are present in the SDR spectra.

4 Conclusions

Using Surface Differential Reflectivity we obtained the dielectric function of the PbSe(100) surface. This function is dominated by three optical transitions located at 1.59, 2.01 and 2.39 eV. From a comparison of experimental and theoretical band structure it was possible to assign these transitions to different points of the reciprocal space along the Γ-X direction of the surface Brillouin zone.

Acknowledgements This work was partly funded by grant PBZ/KBN/044/P03/2001 and SPUB-M/DESY/P-03/DZ-213/2000. We thank S. Selci and M.F. Righini for useful discussions.

References

[1] F.R. McFeely, S. Kowalczyk, L. Ley, R.A. Pollak, and D.A. Shirley, Phys. Rev. **7**, 5228 (1973).
[2] S.E. Kohn, P.Y. Yu, Y. Petroff, Y. R. Shen, Y. Tsang, and M.L. Cohen, Phys. Rev. **8**, 1477 (1973).
[3] R. Bottner, S. Ratz, N. Schroeder, S. Marquardt, U. Gerhardt, R. Gaska, and J. Vaitkus, Phys. Rev. B **53**, 10336 (1996).
[4] J.A. Leiro, K. Laajaletho, I. Kartio, and M.H. heinonen, Surf. Sci. **412/413**, L918 (1998).
[5] H. Kanazawa and S. Adachi, J. Appl. Phys. **83**, 5997 (1998).
[6] A. Cricenti, M. Tallarida, C. Ottaviani, B. Kowalski, E. Gutievitz, A. Szczerbakow, and B.A. Orlowski, Surf. Sci. **482-485**, 659 (2001).
[7] T. Grandke, L. Ley, and M. Cardona, Phys. Rev. B **18**, 3847 (1978)
[8] G. Allan, Phys. Rev. B **43**, 9594 (1991).
[9] N. Suzuki, K. Sawai, and S. Adachi, J. Appl. Phys. **77**, 1249 (1995).
[10] E.A. Albanesi, C.M.I. Okoye, C.O. Rodriguez, E.L. Peltezer y Blanca, and A.G. Petukhov, Phys Rev. B **61**, 16589 (2000).
[11] A. Szczerbakow, J. Cryst. Growth **82**, 709 (1987).
[12] J.D.E. McIntyre and D.E. Aspnes, Surf. Sci. **24**, 417 (1971).
[13] S. Selci, F. Ciccacci, G. Chiarotti, P. Chiaradia, and A. Cricenti, J. Vac. Sci. Technol. **5**, 327 (1987).
[14] S. Nannarone and S. Selci, Phys. Rev. B **28**, 5930 (1983).

phys. stat. sol. (c) **0**, No. 8, 3012–3016 (2003) / **DOI** 10.1002/pssc.200203846

Surface dynamics during MBE growth of GaAs(001) monitored by *in-situ* reflectance difference spectroscopy

C.I. Medel-Ruíz[1], **R.E. Balderas-Navarro**[*, 1, 2], and **A. Lastras-Martínez**[1]

[1] Instituto de Investigación en Comunicación Optica, Universidad Autónoma de San Luis Potosí, Alvaro Obregón 64. San Luis Potosí, S.L.P. 78000 México
[2] Facultad de Ciencias. Universidad Autónoma de San Luis Potosí, Alvaro Obregón 64. San Luis Potosí, S.L.P. 78000 México

Received 30 May 2003, revised 4 August 2003, accepted 11 August 2003
Published online 10 November 2003

PACS 68.35.Bs, 78.40.Fy, 78.66.Fd, 81.15.Hi, 81.70.Fy

We have used *in-situ* reflectance difference spectroscopy to observe changes in GaAs(001) surface reconstructions during homoepitaxy. With a constant As_4 flux of 5×10^{-6} Torr and growth rates of 1.1 ML/s, the starting (2×4) surface exhibits a sudden decrease in the RD signal corresponding to a (3×1) reconstruction as simultaneously evidenced by RHEED observations. Furthermore, by varying Ga fluxes so as to control changes in surface reconstructions between c(4×4) and (2×4), we traced reproducible features that lead us to propose a possible physical origin based on strained islands, induced by surface reconstruction, in order to explain the experimental observations. This last assumption was tested by depositing some atomic layers of InGaAs on a (2×4) GaAs (001) where the change in strain is evident.

1 Introduction

The (001) surface of GaAs shows a variety of reconstructed structures depending on surface stoichiometries [1]. Surface reconstructions change from the most As-rich c(4×4) phase trough the 2×4, 3×1, and 4×6 phases, and finally to the Ga-rich 4×2 phase as the As coverage is decreased. Among these reconstructions, the 2×4 phase has been the most extensively studied because it is the starting surface for several optoelectronic devices and therefore is crucial to have a reliable control of this surface during several growth conditions. In this regard, a continuing goal of epitaxial crystal growth is the production of high quality material and atomic control of surfaces. Optical in-situ monitoring techniques, such as reflectance difference (RD) spectroscopy, are non-destructive and non-invasive and are compatible with most of the growth techniques and therefore well applicable for monitoring and control [2]. RD spectroscopy measures the difference in reflectivity between two mutually orthogonal polarizations and thus provides information on semiconductor regions were the cubic symmetry has been broken by the presence of the surface or interface [2]. Although there is not a conclusive explanation on the physical origin for the reconstruction-induced RD spectra, some theoretical and experimental reports have pointed out that the optical anisotropy related to reconstructions comprises at least two components, namely surface- and bulk- related, corresponding to dimmer resonances and surface strain-induced bulk anisotropy, respectively [3-6]. Nevertheless, because RD is extremely sensitive to surface conditions, it allows to be used in parallel with standard techniques to assess surface processes during epitaxial growth.

---

[*] Corresponding author: e-mail: rbn@cactus.iico.uaslp.mx

In this paper we focus our attention on reconstruction changes between (2×4) and (4×4) observed during GaAs homoepitaxy by modifying the Ga flux arriving onto the GaAs surface. We performed kinetic RD measurements on (001)-oriented GaAs during growth and simultaneously characterized the change by high energy electron diffraction (RHEED) under As$_4$ flux of about 5×10^{-6} Torr. Under high Ga fluxes, so as to obtain a growth rate as high as 1.1 ML/s, the RD response exhibits the larger change corresponding to an As-deficient growing (3×1) surface, that eventually allows to track the Ga-rich (4×6) phase formation and its evolution prior recovering (2×4) reconstruction. Furthermore, kinetic RD data is presented during tuning surface conditions in order to switch between several reconstructions at low and high substrate temperatures, that leads us to propose a possible physical model, based on strained islands induced by surface reconstruction, to account for the experimental changes in kinetic RD.

2 Experimental details

The experiments were carried out in a Riber 32 molecular beam epitaxy (MBE) chamber equipped with effusion cells containing Ga and As and RHEED. We used a semi-insulating GaAs(001) substrate that was cleaned at 600 °C under As$_4$ flux, and subsequently a 1 μm thick GaAs buffer layer was deposited at the same temperature. The substrate temperature (T_s) was monitored with a pyrometer. The beam equivalent pressure for the As$_4$ was 5×10^{-6} Torr during all the experiments and the Ga cell temperature was varied in order to achieve the desired growth rates as calibrated by RHEED oscillations. The normalized reflectance difference between [1$\bar{1}$0] and [110] directions, $\Delta R/R = \mathrm{Re}(\Delta r/r) = 2(R_{[1\bar{1}0]} - R_{[110]})/(R_{[1\bar{1}0]} + R_{[110]})$, was measured by the standard technique similar to the arrangement reported by Aspnes, et. al. [2].

3 Results and discussion

Figure 1a shows real time kinetic RD and RHEED data following surface stoichiometry changes caused by switching ON and OFF Ga beam at a substrate temperature of 590 °C. The Ga cell temperature was set to 920 °C for a growth rate of 1.1 ML/s. Kinetic RD corresponds to a photon energy of 2.5 eV, whereas RHEED corresponds to specular intensity at [110] azimuth. Just prior the first opening of Ga, the surface exhibits a well defined (2×4) reconstruction that changes abruptly to a (3×1) pattern as soon as the Ga shutter is opened. The RHEED oscillations are damped fast. Once the growth is interrupted RD intensity is recovered in a faster time scale than RHEED, but the surface shows again a (2×4) reconstruction that eventually gets sharper within several seconds, and does not lead to a further increase in the RD signal.

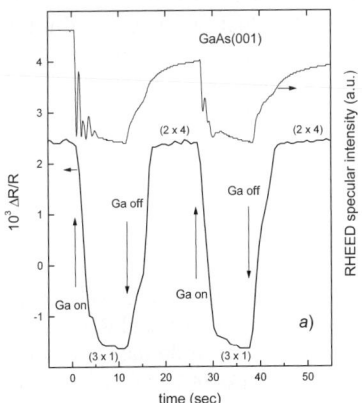 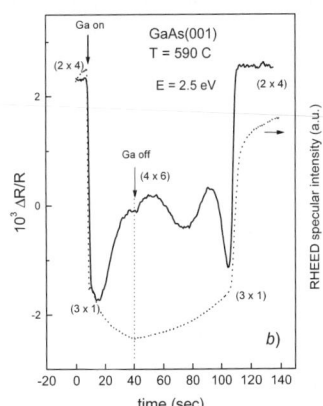

Fig. 1 a) Kinetic RD and RHEED data recorded at 590 °C during GaAs deposition with a growth rate of 1.1 ML/sec. b) Kinetic RD and RHEED data taken at 590 °C during formation of (4×6) reconstruction and its evolution after closing Ga flux while kept it browsed under As$_4$ flux. In both spectra the photon energy used for RD transientsis 2.5 eV. The ON and OFF of Ga shutter is indicated in each case and the corresponding reconstructions.

We now proceed to discuss the evolution of the surface after Ga shutter is left opened for longer time. This evolution is shown in Fig. 1b. In the experiment, the Ga is OFF at t = 40 sec. At this point both RD and RHEED intensities are not recovered immediately but evolve into a transient that develops into a Ga-rich (4×6) pattern followed by a simple oscillation (kinetic RD in time span between 60 and 100 seconds in Fig. 1b), reversing to (3×1) reconstruction and resulting in a sharp raise that finally recovers the (2×4) pattern. It is interesting to note that the RD evolution is more complex than the corresponding RHEED evolution after interrupting the Ga flux. We further note that both, RHEED and RD signals, rise sharply at the same time at the end of the experiment. In the case of RHEED this rise is understood as a smooth-ness of the islands promoted by Ga diffusion. Those islands are anisotropic in shape. Starting with a 4×n geometry, the surface is oversaturated with Ga and because the surface is maintained in an As$_4$ flux, the in-excess Ga atoms migrate anisotropically with different diffusion coefficients along [110] and [1$\bar{1}$0] directions towards the islands edges [7]. Once Ga atoms arrive at the edges, they eventually gather with the impinging As molecules to form GaAs, which, either enters to the lattice or desorbs from the surface. As lesser Ga atoms are left on the top of the islands they favour the (1×3) reconstructions and the (2×4) is finally recovered.

The question arises regarding the physical origin of such changes observed when modifying surface conditions. In the following, we discuss a possible physical mechanism that accounts for the measured optical anisotropy. As pointed out before in refs. [5] and [6] surface reconstruction induces a strain field on top of the islands that propagates through several atomic monolayers into the bulk, thus giving rise to a RD response at energies around the bulk critical points, mainly at E$_1$ and E$_1$ + Δ_1 interband transitions. There exist some reports on the measurements of such a strain field. Men *et. al*, measured the dimmeriza-tion-induced strain in Si(001) by directly applying mechanical stress along [110] direction at elevated temperatures, which lowers the free energy and thus increases the area of terrace with (2×1) reconstruc-tion with respect terraces with (1x2) reconstruction. Recently, Silveira and Briones used the optical de-flection technique on [110] and [1$\bar{1}$0] oriented cantilevers fabricated on thinned GaAs(001) wafers by mechanical polishing. By changing As and Ga dimmer coverage on the surface at high and low tempera-tures, for c(4×4), 2×4, 3×1 and 4×2 reconstructions, they observed different amounts in the deflections of the cantilevers. Prompted by their interesting observations, we performed kinetic RDS measurements under similar MBE conditions used by Siveira and Briones in order to investigate the possible influence of the anisotropic strain associated to reconstructed islands on the RD signal.

Results are shown in Fig. 2. The surface starts with (2×4) reconstruction kept under As flux at T$_s$ = 590 °C. When the As flux is interrupted, the kinetic RD clearly tracks the arsenic desorption during the evolution of the surface into a (3×1) reconstruction. This drop is attributed to a reduction of surface com-pressive stress in the [1$\bar{1}$0] direction, as expected for a progressive reduction of As-As dimmers density [9]. Subsequently, when one ML of Ga is deposited the surface becomes oversatured of Ga atoms and

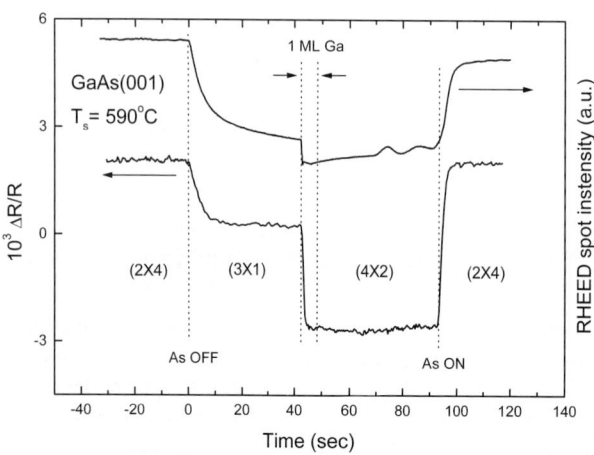

Fig. 2 *In-situ* kinetic RD data at a photon energy of 2.5 eV, and RHEED specular spot intensity taken during As desorption from a (2×4) GaAs(001) surface at 590 °C, and its subsequent Ga saturation by opening Ga cell in order to obtain (4×2) reconstruction. The OFF and ON of As and Ga cells are indi-cated.

attains a (4×2) reconstruction. The situation is reversible to (2×4) when the As shutter is opened again. Interestingly, this experiment exactly resembles the optical deflection measurements reported by Silveira and Briones [9]. Therefore, in our case the same situation occurs: surface reconstruction builds up an anisotropic strain on the islands thereby splitting the E_1 and $E_1+\Delta_1$ interband transitions and thus giving rise to an optical anisotropy. We note that the RD amplitude should depend linearly on the effective strain.

We also performed similar experiments on a c(4×4) GaAs surface at $T_s = 490$ °C, as shown in Fig. 3. Upon starting growth, the initial increasing RD transient corresponds to an increase of surface stress along [1$\underline{1}$0] as reconstruction changes to (2×4). Upon closing the Ga shutter the stress is decreased towards its initial value. Next, if the As flux is interrupted leaving the surface in vacuum, RD and RHEED start to recover their intensities corresponding to (2×4). Eventually, if Ga is supplied to the surface alone, it yields RHEED and RD raisings followed by a linear decrease that finally drives the surface to a (4×2) reconstruction with same signal levels to that of c(4×4). This last observation supports our strain argument: different reconstructions (c(4×4) and (4×2)) yields almost the same RD level as neither of them contain As-dimmers along [1$\underline{1}$0] [9].

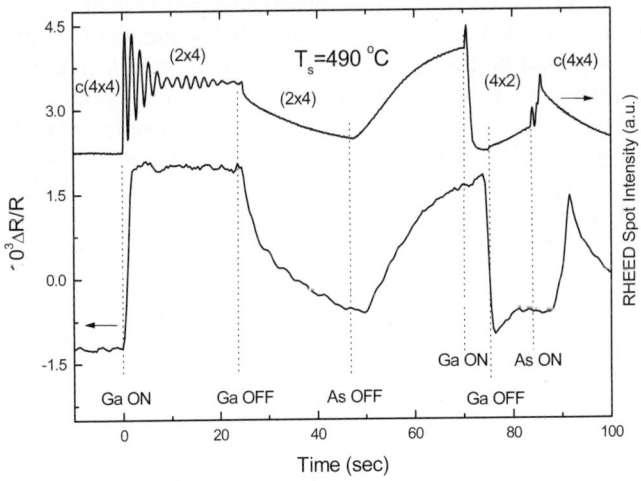

Fig. 3 Kinetic RD and RHEED intensity of specular spot at substrate temperature of 490 °C during changes through c(4×4) to (2×4) and (4×2) during MBE growth.

The stress anisotropy would be sensitive to the top surface layer in such a way that a change in reconstruction should lead to a strain field rotated by 90 degrees [10]. To test this idea we carried out simultaneous RHEED and RD experiments during the growth of a $In_{0.3}Ga_{0.7}As$/GaAs heterostructure at $T_s = 480$ °C . The result of this experiment can be seen in Fig. 4. Initially, when only Ga and As are arriving onto the surface, we can clearly observe up to three RD oscillations that are in phase with the corresponding RHEED oscillations. In contrast, when the In shutter is opened and the reconstruction changes to (1×3), it is evident that RHEED and RD oscillations turn out of phase by 180 degrees, indicating that the anisotropic strain is rotated in sign with respect to the (2×4) reconstructed surface. We can also observe in Fig. 4 that after In is supplied, the oscillating RD response lasts for about three complete periods. After this, oscillations vanish and the RD intensity starts to increase (at point A in Fig. 4). We atributted this last phenomenon to the strain-driven migration of atoms from the edges on top of initial 2D growth islands for their conversion into 3D ones. Further experiments should clarify this issue; for instance by using atomic force microscopy. Furthermore, it would be interesting to investigate with RD the influence that different reconstructions would have on self-assembling of quantum dots by starting with a (2×4) surface and compare it with that starting on an As-rich c(4×4) [11].

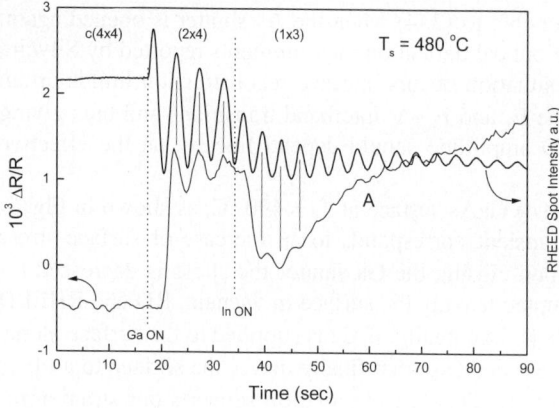

Fig. 4 *In*-situ kinetic RD data at a photon energy of 2.5 eV and RHEED specular spot intensity taken during deposition of few monolayers of GaAs on c(4×4) GaAs(001) surface to produce a (2×4) reconstruction, and its subsequent deposition of a $In_{0.3}Ga_{0.7}As$ film at substrate temperature T_s = 480 °C. Note the one-to-one correspondence in oscillations between RD and RHEED, as indicated with solid vertical lines during the early stages of the growths.

4 Summary

We have carried out simultaneously RD and RHEED measurements for GaAs(001) reconstructed surfaces under MBE conditions. There is a strong correlation between the observed RHEED patterns, the published reconstruction-induced surface strain and the kinetic RD changes at a photon energy of 2.5 eV. Due to the fact that the observed RD features reported in this paper are close to the E_1 interband transition, a bulk origin, besides of a component associated to purely electronic surface states, is thus suggested. To this end, we proposed a possible physical mechanism that leads to a cubic symmetry breakdown of the bulk isotropy, which is related to anisotropically shaped islands that are strained by surface dimers at the top. Because the RD intensity changes should be linear with the corresponding surface strain, we therefore expect that the observed changes in the experiments reported in the present work should be related to those observed by the optical deflection technique applied on thin substrates as reported in Ref. [9].

Acknowledgements Valuable discussions with K. Hingerl and L.F. Lastras-Martínez are greatly acknowledged. This work is partially supported by CoNaCyT (México) under contract No. 32147-E, and FAI-UASLP.

References

[1] P. K. Larsen, J. H. Neave, J. F. van der Veen, P. J. Dobson, and B. A. Joyce, Phys. Rev. B **27**, 4966 (1983).
[2] Itaru Kamiya, D. E. Aspnes, L. T. Florez, and J. P. Harbison, Phys. Rev. B **46**, 15894 (1992).
[3] W. G. Schmidt, F. Bechstedt, K. Fleischer, C. Cobet, N. Esser, W. Richter, J. Bernholc, and G. Onida, phys. stat. sol. (a) **188**, 1401 (2001).
[4] A. Balzarotti, E. Placidi, F. Arciprete, M. Fanfoni, and F. Patella, Phys. Rev. B **67**, 115332 (2003).
[5] K. Hingerl, R. E. Balderas-Navarro, A. Bonanni, and D. Stifter, Phys. Rev. B **62**, 13048 (2000).
[6] L. F. Lastras-Martínez, D. Rönnow, P. V. Santos, M. Cardona, and K. Eberl, Phys. Rev. B **64**, 245303 (2001).
[7] H. Yang, V. P. LaBella, D. W. Bullock, and P. M. Thibado, J. Vac. Sci. Technol. B **17**, 1778 (1999).
[8] F. K. Men, W. E. Packard, and M. B. Webb, Phys. Rev. Lett. **61**, 2469 (1988); and references therein.
[9] J. P. Silveira and F. Briones, J. Cryst. Growth **201/202**, 113 (1999); and references therein.
[10] C. Ratsch, Phys. Rev. B **63**, 161306 (2001)
[11] V. Bressel-Hill, S. Varma, A. Lorke, B. Z. Nosho, P. M. Petroff, and W. H. Weinberg, Phys. Rev. Lett. **74**, 3209 (1995).

phys. stat. sol. (c) **0**, No. 8, 3017–3021 (2003) / **DOI** 10.1002/pssc.200303862

In situ optical monitoring of the interface strain relaxation of InGaAs/GaAs grown by molecular-beam epitaxy

C.I. Medel-Ruiz[**, 1], **A. Lastras-Martínez**[*, 1], and **R. E. Balderas-Navarro**[1, 2]

[1] Instituto de Investigación en Comunicación Optica, Universidad Autónoma de San Luis Potosí, Alvaro Obregón 64, San Luis Potosí, S.L.P., México
[2] Facultad de Ciencias, Universidad Autónoma de San Luis Potosí, Alvaro Obregón 64, San Luis Potosí, S.L.P., México

Received 30 May 2003, revised 4 August 2003, accepted 11 August 2003
Published online 12 November 2003

PACS 68.60.Bs, 78.66.Fd

The evolution of the surface morphology during the MBE growth of $In_xGa_{1-x}As$ on GaAs (001) is studied by *in situ* Reflectance-difference (RD) spectroscopy and reflection high-energy electron diffraction (RHEED). $In_xGa_{1-x}As$ layers with $x = 0.3$ nominal compositions were growth under different arsenic over-pressures. For arsenic-rich conditions RHEED measurements show that the growth mode is largely two-dimensional (2D). In contrast, under As-deficient conditions a 2D-3D growth morphology transition is observed upon closing the In and Ga shutters. Concurrently with this transition, the RD intensity at 2.5 eV shows a sharp increase, indicating the formation of anisotropic $In_xGa_{1-x}As$ islands. We further show that low growth rates (0.3 ML/s) result in higher RD amplitudes and thus in InGaAs islands with enhanced anisotropy.

1 Introduction

The strained system $In_xGa_{1-x}As$/GaAs has attracted much interest from both practical and basic points of view. Besides of being a convenient system to study hetero-interface strain relaxation processes, the heterostructures $In_xGa_{1-x}As$/GaAs have found many important applications to the fabrication of semiconductor quantum devices [1]. The study of the surface processes taking place during the epitaxial growth of $In_xGa_{1-x}As$/GaAs is therefore an issue of considerable interest. For In compositions $x > 0.25$ it is well established that $In_xGa_{1-x}As$ layers grown on GaAs (001) first follow a two dimensional (2D) growth mode, until a critical thickness is reached where a morphological transition to a three dimensional (3D) growth mode takes place as a mechanism to reduce interface elastic energy [2,3]. At larger film thickness, interface mismatch dislocations are introduced as an additional strain-relieving mechanism [4].

Given the complexity of the phenomena associated to the growth of $In_xGa_{1-x}As$/GaAs, the development of *in situ* techniques to characterize the growth process is a matter of considerable importance. Reflectance Difference (RD) spectroscopy offers for this purpose distinctive advantages due to their simplicity and real time capabilities [5-8]. This spectroscopy measures the difference in reflectivity between two mutually orthogonal polarizations and thus provides information on semiconductor regions were the cubic symmetry has been broken by the presence of the surface or interface.

Elsewhere, we reported on *in situ* RD measurements in $In_xGa_{1-x}As$ grown by MBE that show a sharp increase in RD amplitude at 2.5 eV at the onset of the 2D-3D growth transition [9,10]. This, as a result of the formation of anisotropic 3D islands. In this paper we report on the results of a study aimed to investi-

---

[*] Corresponding author: e-mail: alastras@cactus.iico.uaslp.mx, Phone: 52 (444) 825 0183, Fax: 52 (444) 825 0198
[**] CONACyT fellow.

gate the role of growth rate and of As-overpressure on the evolution of the reflectance anisotropy of $In_{0.3}Ga_{0.7}As$/GaAs films grown by MBE. We show that at low enough As-overpressures interrupting the $In_{0.3}Ga_{0.7}As$ growth leads to an increase in reflectance anisotropy at an energy of 2.5 eV as a result of the formation of a 3D surface morphology. In contrast, at high As-overpressure interrupting growth does not lead to a 2D-3D transition and thus no increase in reflectance anisotropy is observed. Furthermore, we present results showing that epilayer growth rate influences the reflectance anisotropy associated to $In_{0.3}Ga_{0.7}As$. Results presented here give additional evidence for the correlation between surface 3D morphology and the amplitude of the RD signal and demonstrates the potential of RD spectroscopy as a probe for the *in situ* characterization of the interface relaxation processes of zincblende heterostructures.

2   Experimental

$In_{0.3}Ga_{0.7}As$ samples were grown on semi-insulating GaAs (001) substrates in a solid-source MBE system (RIBER 32) equipped with a 12 KeV RHEED gun. Prior to the deposition of the hetero-epitaxial film a 0.5 μm thick GaAs buffer was grown at 590 °C. The substrate temperature was then lowered to a temperature in the range from 480-513 °C to start $In_{0.3}Ga_{0.7}As$ growth. During buffer layer growth the surface showed a (2x4) reconstruction. This reconstruction changed to c(4x4) when the temperature was lowered to initiate $In_{0.3}Ga_{0.7}As$ growth. The growth rate and In composition of the InGaAs epilayers were determined from RHEED oscillation periods. In Table I we show the growth parameters for the studied samples. RD measurements were carried out with a setup similar to the one described elsewhere [11]. It allows for spectroscopic measurements in the energy range from 1.5-5.5 eV. To study the evolution of the surface during epitaxial growth, both, RHEED pattern and RD signal at a photon energy of 2.5 eV were simultaneously monitored.

Table 1   Growth parameters for $In_{0.3}Ga_{0.7}As$ samples.

Sample No.	Growth temp.	Growth rate	As BEP
1	480 °C	1 ML/s	$2x10^{-5}$ torr
2	480 °C	1 ML/s	$5x10^{-6}$ torr
3	500 °C	1 ML/s	$9x10^{-6}$ torr
4	513 °C	0.3 ML/s	$9x10^{-6}$ torr

3   Results and discussion

In Figs. 1a and 1b we show the evolution of the intensity of the specular RHEED spot during the growth of samples 1 and 2, respectively. As indicated in Table I, samples 1 and 2 were grown with arsenic overpressures of 2 x 10^{-5} torr and 5 x 10^{-6} torr, respectively. Substantial differences are observed for the damping of the RHEED oscillations in both samples. Fig. 1a, shows that for sample 1 these oscillations were preserved all along the 26 MLs InGaAs growth, indicating that the epilayer grew in a 2D layer-by-layer mode. At the same time, the RHEED pattern remained streaky showing a (1x3) reconstruction. In contrast, as shown in Fig. 1b, RHEED oscillations for sample 2 lasted only for the first 4MLs of InGaAs growth. After this, although the RHEED pattern remained streaky all along the InGaAs growth, no RHEED oscillations were observed. This behavior is similar to that reported in Ref. [12] for the MBE growth of $In_{0.27}Ga_{0.73}As$/GaAs (001) at an As overpressure of 1.2 x 10^{-5} torr. We note that the damping out of RHEED oscillations correlates with a change of surface reconstruction from (1x3) to (4x2). from (1x3) as indicated in Fig. 1b. The damping out of RHEED oscillations and the fact that the RHEED pattern remains streaky, do suggest that the growth mode is step flow.

Upon closing the In and Ga shutters at the end of the InGaAs epitaxial growth we did observe further significant differences between sample 1 and sample 2. Namely, while the RHEED pattern for sample 1 remained streaky after closing In and Ga shutters, preserving the (1x3) reconstruction with the specular spot intensity slowly recovering, the corresponding intensity for sample 2 sharply increased at the same time that the RHEED pattern became spotty.

Let us now consider the RD data. In Figs. 1c and 1d we show kinetic RD measurements at a photon energy of 2.5 eV for samples 1 and 2, respectively. In agreement with previous observations [9], upon starting growth both samples exhibit a sharp increase in RD amplitude, that reverses the negative sign of the measured RD intensity of the starting GaAs(001)-c(4x4) surface. As it is well known, this sharp rise is associated to the change in surface reconstruction from c(4x4) to (1x3) when InGaAs initiates [13-15]. As InGaAs growth progresses, the RD amplitude for sample 1 reaches a maximum at a film thickness of about 10 ML and then shows a slow monotonic decrease. No noticeable change in RD amplitude is observed upon closing the In and Ga shutters.

The evolution of the RD amplitude for sample 2 is more complex and shows a clear correlation with the evolution of the corresponding RHEED pattern. Indeed, after the initial increase associated to the change in reconstruction, the RD amplitude decreases up to the point where does occur the change in surface reconstruction form (1x3) to (4x2). At this point the RD signal partially recovers and then saturates up to the end of the 26 MLs InGaAs growth. Furthermore, in sharp contrast to sample 1, upon closing the In and Ga shutters, we do observe for sample 2 a step rise in the RD amplitude in correlation with the sharp rise in RHEED specular spot intensity. This last result indicates that upon closing In and Ga shutters a substantial redistribution of mass took place that changed the epilayer morphology from 2D to 3D. Results further show that at low values of As-overpressures the 2D-3D transition is kinetically limited.

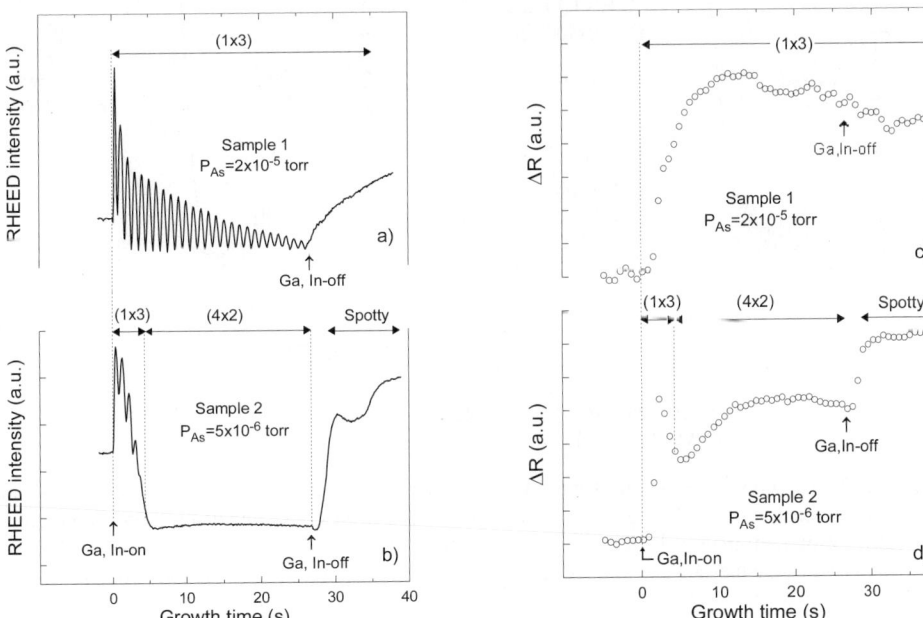

Fig. 1 Evolution of the RHEED specular beam intensity and RD signal amplitude at 2.5 eV during the growth of $In_{0.3}Ga_{0.7}As$ on GaAs for two different conditions of As flux. a): RHEED intensity under As-rich conditions ($P_{As}=2x10^{-5}$ Torr, sample 1) leading to layer-by-layer growth. b): RHEED intensity under As-deficient conditions (($P_{As}= 5x10^{-6}$ Torr, sample 2). c) and d): corresponding RD amplitudes for sample 1 and sample 2, respectively. The growth rate for both samples was 1 ML/s. The observed surface reconstructions are indicated. The final epilayer thickness is of about 26 ML in both cases.

We will discuss now the dependence of the RD amplitude on growth rate. To determine this dependence we carried out simultaneous RHEED and kinetic RD measurements at 2.5 eV at two growth rates, 1 ML/s (sample 3) and 0.3 ML/s (sample 4). Results are shown in Figs. 2a and 2b.

Let us first consider RHEED data. Continuous lines in Figs. 2a and 2b show the RHEED intensity evolution in the [110] azimuth at a position corresponding to a Bragg spot for samples 3 and 4, respectively. As it can be seen, the growth rate has a relatively low influence on the RHEED intensity evolution; i.e., in both cases this intensity shows oscillations for the first 8 ML's and then rises sharply indicating the onset of 3D growth.

The corresponding RD evolution for samples 3 an 4 is shown with open circles in Figs. 2a and 2b, respectively. As we can see from these figures, in contrast to RHEED intensity, the RD kinetics shows a substantial dependence on growth rate. At high deposition rate, Fig. 2a shows that after a delay of about 1 ML, upon starting growth the RD intensity rises steeply and reaches a peak at about 3 MLs. Beyond this point the RD amplitude increases monotonically. At low deposition rate, on the other hand, Fig. 2b shows that the RD amplitude shows a sharp peak just after starting growth and then a monotonic rise. We can also see that upon the occurrence of the 2D-3D transition both samples 3 and 4 show a an increase in the rate of rise of RD amplitude. This rate is, nevertheless, considerably higher for sample 4 than for sample 3. We further note that upon closing In and Ga shutters after 26 MLs deposition, we do observed for sample 3 an increase in RD amplitude that is not found for sample 4. Note that the increase in RD amplitude for sample 3 at the end of the expitaxial growth is similar to that observed for sample 2a (Fig. 1d).

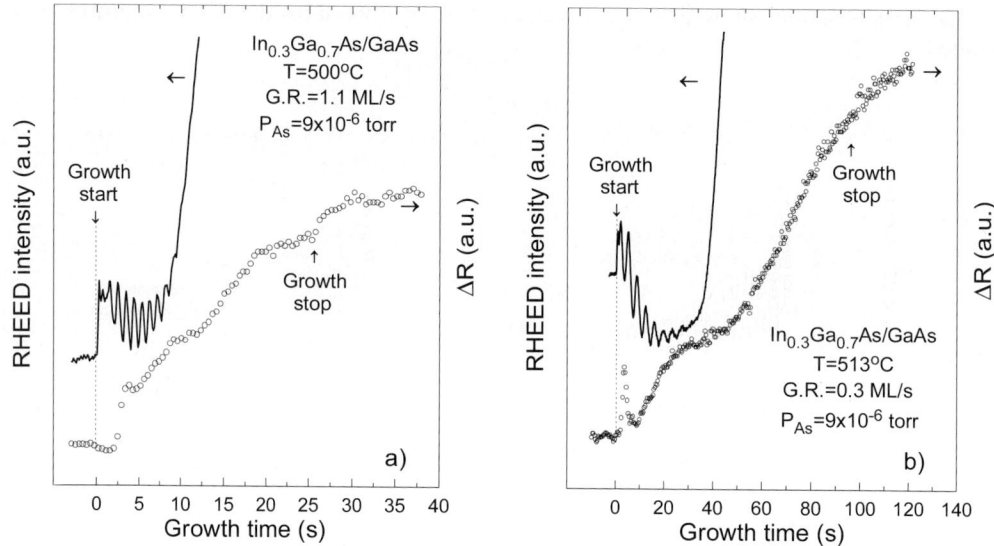

Fig. 2 Evolution of the RHEED intensity at a Bragg spot (continuous lines) and of the RD amplitude at 2.5 eV (open circles) during the growth of $In_{0.3}Ga_{0.7}As$ on GaAs under an As overpressure $P_{As}=9x10^{-6}$ torr, for two different growth rates. a): G.R.=1 ML/*s*. b): 0.3 ML/*s*. We note that the vertical RD scales in both figures are the same so that RD amplitudes for both samples can be directly compared.

The above results establish a correlation between the 3D surface morphology of the InGaAs/GaAs epilayers and the amplitude of the RD signal. Figs. 1 and 2 show that the transition to a 3D growthmode, whether it occurs during growth or at the end of it, induces an increase in the amplitude of the RD signal at 2.5 eV. We may thus conclude that the RD signal in the 3D growth comprise a substantial component originated in InGaAs islands that grow anisotropic. In agreement with this conclusion, elsewhere we have shown that the RD line shape of 3D $In_{0.3}Ga_{0.7}As$/GaAs (001) films in the energy range from 2.4-3.1 eV can be accurately modeled assuming that the InGaAs islands on the epilayer are under an orthorhombic strain [9,10]. This strain may be originated by the well-known anisotropy in the diffusion coefficients of the atomic species on the growth surface that may lead to InGaAs islands with anisotropic shapes [16] and thus with an anisotropic elastic strain relaxation. Further, comparison of Figs 2a and 2b allow us to conclude that the build up of the anisotropic component of the interface strain is enhanced at low growth rates. The results reported here show that the evolution of the RD amplitude provides information on the development of anisotropic strains during InGaAs growth and on the dynamics of 3D island formation and subsequent evolution.

Detailed studies of the RD kinetics at 2.5 eV for the system InAs/GaAs have been reported previously [13-15]. We note, however, that the evolution of the RD amplitude at the 2D-3D transition onset as reported in references 13-15 is different from that discussed here for $In_{0.3}Ga_{0.7}As$/GaAs; i.e., rather than increasing at the 2D-3D onset as reported in this paper, the RD amplitude for InAs/GaAs reaches a maximum at this onset, slightly decreasing afterwards [13-15].

4   Conclusions

We report on the results of a study carried out in order to determine the role of growth-rate and of As-overpressure on the evolution of the reflectance anisotropy of $In_{0.3}Ga_{0.7}As$/GaAs films grown by MBE. We carried out simultaneous RHEED and single-wavelength RD measurements in order to correlate the changes in epilayer surface morphology with changes in the RD intensity. For the study four 26 MLs-thick $In_{0.3}Ga_{0.7}As$/GaAs samples were grown under As-overpressures in the range form $5x10^{-6}$-$2x10^{-5}$ torr. In agreement with previous observations, we found that the value of the As-overpressure establish the conditions for the occurrence of the 2D-3D morphological transition in the InGaAs epilayers. We observed that for a low As overpressure ($5x10^{-6}$ torr) a 2D-3D transition occurs upon closing the Ga and In shutters at the end of growth. In contrast, at an As-overpressure of $2x10^{-5}$ torr this transition is inhibited. We further show that concurrently with the occurrence of the 2D-3D transition there is an increase in the RD amplitude at 2.5 eV. The evolution of the RHEED intensity during the growth of two $In_{0.3}Ga_{0.7}As$/GaAs films with growth rates of 1 and 0.3 ML/s and with an intermediate arsenic pressures of $9x10^{-5}$ torr, show that a 2D-3D transition occurs during growth in both films at an epilayer thickness of about 9 MLs. The corresponding evolution of the RD intensity at 2.5 eV shows significant differences between the two samples. From these differences we conclude that the growth of anisotropic InGaAs islands that contribute to RD intensity is influenced by both As overpressure and growth rate.

Acknowledgements   Work partially supported by Consejo Nacional de Ciencia y Tecnología of México under contracts Nos. 32147-F, and 485100-5-3397E.

References

[1] D. L. Huffaker, G. Park, Z. Zou, O. B. Schekin, and D. G. Deppe, Appl. Phys. Lett. **73**, 2564 (1999).

[2] N. Grandjean and J. Massies, J. Cryst. Growth **134**, 51 (1993).

[3] See for instance, V. A. Shchukin, and D. Bimberg, Rev. Mod. Phys. **71**, 1125 (1999).

[4] C. Lavoie, T. Pinnington, E. Nodwell, T. Tiedje, R. S. Goldman, K. L. Kavanagh, and J. L. Hutter, Appl. Phys. Lett. **67**, 3744 (1995)

[5] D. E. Aspnes and A. A. Studna, Phys. Rev. Lett. **54**, 1956 (1985).

[6] Itaru Kamiya, D. E. Aspnes, L. T. Florez, and J. P. Harbison, Phys. Rev. B **46**, 15894 (1992).

[7] Z. Sobiesierski, D. I. Westwood, and C. C. Mathai, J. Phys. C **10**, 1 (1998).

[8] J. T. Zettler, Prog. Cryst. Growth Charact. Mater. **35**, 27 (1997).

[9] C. I. Medel-Ruiz, A. Lastras-Martínez, R. E. Balderas-Navarro, S. L. Gallardo-Cruz, V. Méndez-García, J. M. Florez-Camacho, A. Gaona-Couto, and L. F. Lastras-Martínez, to be published in Appl. Surf. Sci. (2003).

[10] A. Lastras-Martínez, R. E. Balderas-Navarro, L. F. Lastras-Martínez, and C. I. Medel-Ruiz, submitted to this conference.

[11] D. E. Aspnes, J. P. Harbison, A. A. Studna, and L. T. Florez, J. Vac. Sci. Technol. A **6**, 1327 (1988).

[12] N. Chokshi, M. Bouville, and J. Mirecki Millunchick, J. Cryst. Growth **236**, 563 (2002).

[13] E. Steimetz, J. E. Zettler, F. Schienle, T. Trekt, T. Whetkamp, W. Richter, and I. Sieber, Appl. Surf. Sci. **107**, 203 (1996).

[14] E. Steimetz, J.-T. Zetter, W. Richter, D. I. Westwood, D. A. Woolf, and Z. Sobiesierski, J. Vac. Sci. Technol. B **14**, 3058 (1996).

[15] T. Kita, O. Wada, T. Nakayama, and M. Murayama, Phys. Rev. B **66**, 195312 (2002).

[16] Qi-Kun Xue, Y. Hasegawa, H. Kiyama, and T. Sakurai, Jpn. J. Appl. Phys. **38**, 500 (1999).

phys. stat. sol. (c) **0**, No. 8, 3022–3026 (2003) / **DOI** 10.1002/pssc.200303852

RDS investigation of adsorption
and surface ordering processes on Cu(110)

L. D Sun[1], **M. Hohage**[1], **P. Zeppenfeld**[*,1], and **R. E. Balderas-Navarro**[2]

[1] Institute of Experimental Physics, Johannes Kepler University Linz, A-4040 Linz, Austria
[2] Facultad de Ciencias and IICO, Universidad Autonoma de San Luis Potosi, San Luis Potosi,
SLP 78000, Mexico

Received 30 May 2003, revised 4 August 2003, accepted 11 August 2003
Published online 10 November 2003

PACS 68.43.Mn, 73.20.At, 78.68.+m

The optical anisotropy of Cu(110) has been studied at low temperatures by Reflectance Difference Spectroscopy. Distinct anisotropy features similar to those at room temperature are observed. Taking the second derivative of the measured spectrum, the main feature at 2.1 eV can be clearly separated into two peaks located at 2.0 eV and 2.2 eV, respectively. Adsorption experiments with different gases allow to attribute these two peaks to a surface state transition and a surface modified d-band transition according to their characteristic response to the adsorbates. Additionally, another modified bulk optical transition at 4.4 eV is found to be particularly sensitive to phase transformations such as the $(3 \times 1) \rightarrow (2 \times 1)$ phase transition of CO on Cu(110). We argue that the coverage dependent adsorbate induced surface stress is responsible for this observation.

1 Introduction

Reflectance Difference Spectroscopy (RDS) is widely used to study semiconductor surfaces and thin film growth in various environments. More recently, RDS has also been applied to metal surfaces such as Cu(110) [1, 2, 3, 4, 5], Ag(110) [6] and Au(110) [7]. In particular, the RDS spectrum of Cu(110) shows several distinct features, which are related to a surface state transition at 2.1 eV, a surface resonance transition at 4.12 eV and two surface modified bulk state transitions at 2.2 eV and 4.4 eV [2]. Due to their different physical origin, different responses of each feature to the electronic or morphological changes and to the surface stress induced upon adsorption are expected. In the present report, we will show that the resulting spectra can be used as a 'fingerprint' for different adsorbates and structures. Furthermore, the evolution of the RDS signal with time or adsorbate coverage allows to study the adsorption kinetics, surface ordering and phase transitions in real time.

2 Experimental

The RDS measurements were performed in a UHV chamber with a base pressure lower than 10^{-10} mbar. The apparatus is equipped with facilities for Auger Electronic Spectroscopy (AES) and Low Energy Electron Diffraction (LEED), a home-made Beetle type STM and a Quadruple Mass Spectrometer (QMS) for Temperature Programmed Desorption (TPD) analysis. A high quality Cu(110) sample is fixed on the sample holder which is connected to a liquid helium cryostat via a Cu

[*] Corresponding author: e-mail: zeppenfeld@exphys.uni-linz.ac.at, Phone: +43 732 2468 8510, Fax: +43 732 2468 8509

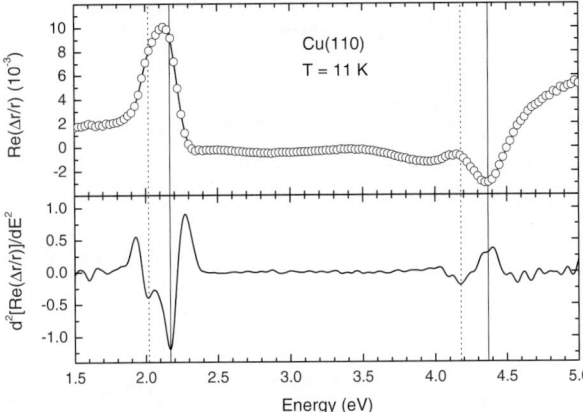

Fig. 1 Real part of the RDS signal of Cu(110) recorded at 11 K (upper panel) and its second derivative (lower panel). The dotted and solid vertical lines indicate the energy positions of the surface state and bulk related optical transitions, respectively.

braid. This allows the sample to be cooled rapidly down to 10 K. An electron impact heater fixed at the back side of the sample can heat the sample up to 1000 K. A K-type thermocouple clamped to the crystal is used to measure the sample temperature with an error smaller than ± 1 K. A programmable temperature controller is used to regulate the sample temperature and to perform well defined heating/ cooling profiles. The sample is cleaned in situ by 900 eV Ar^+ ion sputtering and subsequent annealing at 800 K for 2 minutes. After this sample preparation no surface impurity can be detected by AES, additionally, LEED shows a sharp p(1 × 1) pattern of Cu(110). An RDS spectrometer of the Aspnes type [8] is attached to the chamber via a strain-free optical window. The normalized reflectance difference defined as:

$$\frac{\Delta r}{r} = 2 \frac{r_{[1\bar{1}0]} - r_{[001]}}{r_{[1\bar{1}0]} + r_{[001]}} \qquad (1)$$

can be measured in the energy range between 1.5 eV and 5.5 eV with the current setup. Here $r_{[1\bar{1}0]}$ and $r_{[001]}$ denote the complex reflectance for light polarized along the $[1\bar{1}0]$ and [001] direction, respectively.

3 Results and discussion

The RDS spectrum of Cu(110) recorded at 11 K is shown in the upper part of Fig. 1. Similar to what has been reported previously [2, 9], the spectrum reveals a positive peak at 2.13 eV as well as a negative peak at 4.3 eV surrounded by some smaller features. The energies of these main features are shifted with respect to the room temperature data [10, 11]. The second derivative of this spectrum is displayed in the lower part of Fig. 1. From this it is quite obvious that the feature at 2.13 eV is actually composed of two peaks located at 2.0 eV and 2.2 eV, respectively. Based on the optical transition energies, these two contributions can be ascribed to the surface states transition at the $\overline{Y}$ point of the surface Brillouin zone and the surface modified d-band transition at the X point of bulk Cu, respectively. From their adsorption experiment, Stahrenberg et al. [2] have also come to the conclusion that there are two contributions to the 2.1 eV peak, but the transition energy of the surface state transition could not be determined directly. Two other important features showing up in Fig. 1 are the small peak at 4.2 eV involving a surface resonance at $\overline{X}$ [2] and a negative peak at 4.4 eV which corresponds to the d-band transition in the vicinity of the bulk symmetry point L [11].

Different kinds of adsorbates, from physisorbed rare gases to chemisorbed CO, lead to quite different changes of the optical anisotropy of Cu(110) as can be seen in Fig. 2. Owing to their closed-shell

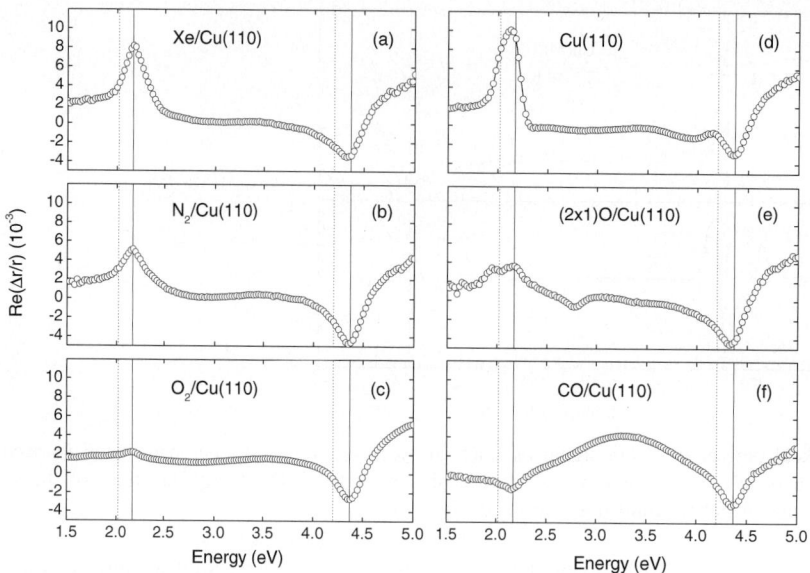

Fig. 2 RDS spectra after adsorption of Xe (a), N$_2$ (b), O$_2$ (c), (2 × 1) O (e) and CO (f) on Cu(110).
As a reference, the spectrum of the bare Cu(110) surface is shown in (d). All the spectra were recorded
at 11 K. Dotted and solid vertical lines indicate the surface state and bulk related optical transitions on
Cu(110), respectively.

electronic structure, the interaction of physisorbed Xe and N$_2$ with Cu(110) are relatively weak
(Figs. 2a, 2b). Therefore, they mainly modify the surface state related intensity at 2.0 eV and 4.2 eV.
Especially, the surface related RDS intensity at 2.0 eV is quenched and the peak maximum is shifted
to 2.2 eV, i.e., the energy position of the bulk transition. The situation for oxygen adsorption at 11 K
is a bit complex, since the oxygen molecules are partially dissociated upon adsorption, therefore, both
atomic and molecular oxygen exists on the surface [12, 13]. For the chemisorbed atomic species a
stronger interaction with the surface electronic structure is expected. Indeed, Fig. 2c shows a spectrum
with very low intensity at 2.2 eV. From Fig. 2a–2c, one may infer a gradually increasing interaction
between adsorbates and the Cu(110) surface as revealed by the vanishing of the intensity first at
2.0 eV and later at 2.2 eV. Fig. 2e shows the RDS spectrum recorded from the Cu(110)–(2 × 1) O
surface, obtained after oxygen adsorption at room temperature followed by annealing to 600 K. In this
case, all the oxygen adsorbates are dissociated and chemisorbed, additionally, the Cu(110) surface is
reconstructed. Besides the complete quenching of the surface state related transition of Cu(110), two
new features appear at 1.9 eV and 2.75 eV, respectively. These two contributions are attributed to the
transitions of oxygen derived surface states [2, 14]. As last of this series, the RDS spectrum after CO
adsorption on Cu(110) is shown in Fig. 2f. The CO molecules experience a strong, chemical interaction with Cu(110) and the RDS intensity at 2.2 eV is even reversed. A more detailed investigation
shows that the reversal of signal is related to the formation of the (2 × 1) superstructure of CO on
Cu(110) [10]. Besides, a broad peak centered around 3.3 eV is observed. It can be attributed to the
electronic transition from the Fermi level to the 2π-derived empty band of CO which shows a very
similar energetic position and width in inverse photoemission [15].

The above examples demonstrate that RDS is sensitive to the details of surface adsorption and that
the resulting spectrum strongly depends on the electronic state of the adsorbate, the adlayer structure
and the possible adsorption induced surface reconstruction. Moreover, the RDS signal at 2.1 eV is
extremely sensitive to surface defects due to the large lateral extent of the quasi-free electron nature
of the metal surface states involved in the corresponding optical transition. Indeed, a cross section as
large as 1000 Å^2 for a single isolated CO molecule adsorbed on Cu(110) has been determined by RDS
[5]. This means that the initial quenching of the RDS signal at 2.1 eV is about 100 times larger than

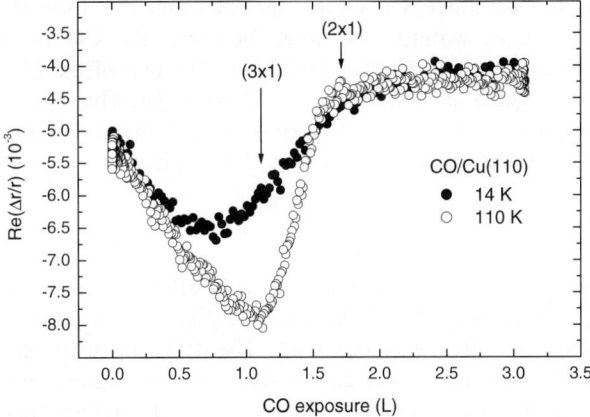

Fig. 3 RDS intensity at 4.4 eV during CO exposure of Cu(110) at 110 K (open circles) and 14 K (filled circles). The CO exposures corresponding to the completion of the (3×1) and (2×1) adlayer phases (as determined by He scattering) are also indicated.

expected from the geometric size of a single molecule. As a result, coverages of the order of 10^{-4} of a monolayer can be easily detected.

On the other hand, it has been pointed out that the RDS intensity at 4.4 eV is due to the optical transition $E_F \rightarrow L_1$ from the Fermi-level to the s-electron like band at the L symmetry point [11]. This transition has been shown to be very sensitive to an externally applied strain [16]. Likewise, this transition is also affected by the presence of an intrinsic or adsorbate induced surface strain. As an example, Fig. 3 shows the RDS signal at 4.4 eV recorded during CO exposure on the Cu(110) surface at 110 K (open circles). The intensity decreases linearly up to 1.1 L (1 L = 10^{-6} Torr · s), then it increases again and saturates at about 1.7 L. These two characteristic exposures (1.1 L and 1.7 L) coincide exactly with the completion of the (3×1) and (2×1) phases observed by He atom diffraction under similar conditions [17]. This observation suggests that the CO phase transition is responsible for the observed variation of the RDS intensity. The detailed physical mechanism which leads to the change of the RDS signal is not clear at the moment, but it should be related to the coverage dependent interaction between the adsorbed CO molecules and the associated adsorbate induced surface strain. In fact, the interaction between CO molecules on neighboring lattice sites on Cu(110) is attractive along the [001] direction but repulsive along [1$\bar{1}$0] [17, 18]. As a result, CO chains running along the [001] direction are formed [17, 18]. The repulsive interaction along [1$\bar{1}$0] keeps the CO chains away from each other as observed by STM [18] and He scattering [17]. At large distances between the chains (such as in the (3×1) phase) the interaction will be small or might even become slightly attractive by 7–10 meV [19]. The (3×1) phase can thus be considered as the limiting case, where the distance between neighboring CO rows in the [1$\bar{1}$0] direction is larger than or equal to three times the Cu distance and the interaction between the rows is attractive, whereas the interaction switches into repulsive when the distance becomes shorter than in the (3×1) phase. This point is reached at a CO coverage of 1/3, i.e., at the completion of the (3×1) phase after an exposure of 1.1 L. For higher coverages, additional CO chains can only be accommodated by compression of the adlayer and formation of energetically unfavorable (2×1) domains. The turning point of the RDS intensity at a CO exposure of 1.1 L could thus be assigned to the switching of the inter-chain interaction from attractive to repulsive. Exposure of CO at 14 K rather than 110 K yields a different dependence of the RDS intensity at 4.4 eV on the CO dose (filled circles in Fig. 3). The decrease of the RDS intensity at low coverage is not as pronounced as at 110 K, and the curve reaches its turning point already after a CO exposure of about 0.6 L. The increase of the RDS intensity beyond this point is much slower as compared to the 110 K case. Finally, the curve also saturates at a later point ($\sim$2 L) than at 110 K. This behavior can be attributed to the limited mobility of the CO molecules on Cu(110) at 14 K. Indeed, if the mobility of CO molecules is high enough, the repulsive interaction along [1$\bar{1}$0] leads to the formation of chains separated by at least three Cu lattice spacings and the formation of any (2×1) domains at a coverage $\leq 1/3$ will be avoided. At 14 K, however, the CO molecules are 'frozen' on the surface, therefore, the local density of CO can be higher than that of the (3×1) phase and compressed (2×1) domains will be formed at relatively low CO coverage. This effect will shift the turning point of the RDS signal at 4.4 eV to lower CO exposure and smear out the intensity increase from this turning point up to the saturation coverage of the (2×1) phase.

4   Conclusion

A careful analysis of the RDS spectrum of Cu(110) recorded at 11 K allows to separate the two contributions located at 2.0 eV and 2.2 eV, respectively. A different response of these two contributions upon adsorption of different kinds of gases was observed. The RDS intensity at 2.0 eV is strongly modified even by physisorbed Xe, in contrast, the 2.2 eV signal is only significantly affected by chemisorbed species such as CO and O. These distinct properties clearly demonstrate that the 2.0 eV signal is related to the surface state transition, whereas the intensity at 2.2 eV is contributed by the surface modified bulk transition. Furthermore, the RDS signal at 4.4 eV has been monitored during CO exposure of Cu(110) at 110 K and 14 K. By comparison with the CO phase transition discovered by He scattering, the observed variation of RDS intensity at 4.4 eV can be clearly related to the coverage dependent interaction between the adsorbed CO molecules and the associated adsorbate induced surface strain. In conclusion, we have further substantiated in this report that RDS can be used as a powerful analytical tool in surface science such as for studies of adsorption and structural phase transitions.

Acknowledgements   This work was financially supported by the Austrian Science Foundation (FWF) under contract numbers 12317-NAW and 15963-N08. One of the authors (R.E.B.-N.) is thankful for financial support from CONACYT (Mexico) and Profactor (Austria).

References

[1] Ph. Hofmann, K. C. Rose, V. Fernandez, and A. M. Bradshaw, Phys. Rev. Lett. **75**, 2039 (1995).
[2] K. Stahrenberg, Th. Herrmann, N. Esser, and W. Richter, Phys. Rev. B **61**, 3043 (2000).
[3] J. K. Hansen, J. Bremer, and O. Hunderi, Surf. Sci. **418**, L58 (1998).
[4] D. S. Martin, A. M. Davarpanah, S. D. Barrett, and P. Weightman, Phys. Rev. B **62**, 15417 (2000).
[5] L. D. Sun, M. Hohage, P. Zeppenfeld, R. E. Balderas-Navarro, and K. Hingerl, Phys. Rev. Lett. **90**, 106104 (2003).
[6] K. Stahrenberg, Th. Herrmann, N. Esser, W. Richter, S. V. Hoffmann, and Ph. Hofmann, Phys. Rev. B **58**, R10207 (1998).
[7] V. Mazine and Y. Borensztein, Phys. Rev. Lett. **88**, 147403 (2002).
[8] D. E. Aspnes, J. Vac. Sci. Technol. B **3**, 1498 (1985).
[9] J. Bremer, J. K. Hansen, and O. Hunderi, Appl. Surf. Sci. **142**, 286 (1999).
[10] L. D. Sun, PhD thesis, University Linz, 2002.
[11] L. D. Sun, M. Hohage, P. Zeppenfeld, R. E. Balderas-Navarro, and K. Hingerl, Surf. Sci. **527**, L184 (2003).
[12] B. G. Briner, M. Doering, H.-P. Rust, and A. M. Bradshaw, Phys. Rev. Lett. **78**, 1516 (1997).
[13] A. Hodgson, A. K. Lewin, and A. Nesbitt, Surf. Sci. **293**, 211 (1993).
[14] Th. Herrmann, K. Ludge, W. Richter, N. Esser, P. Poulopoulos, J. Lindner, and Baberschke, Phys. Rev. B **64**, 184424 (2001).
[15] J. Rogozik, H. Scheidt, V. Dose, K. C. Prince, and A. M. Bradshaw, Surf. Sci. **145**, L481 (1984).
[16] U. Gerhardt, Phys. Rev. **172**, 651 (1968).
[17] J. Goerge, PhD thesis, Jül-Bericht Nr. 3198 (1996).
[18] B. G. Briner, M. Doering, H. P. Rust, and A. M. Bradshaw, Science **278**, 257 (1997).
[19] D. H. Wei, D. C. Skelton, and S. D. Kevan, Surf. Sci. **326**, 167 (1995).

phys. stat. sol. (c) **0**, No. 8, 3027–3031 (2003) / **DOI** 10.1002/pssc.200303829

Angular correlation properties of 2D-nano-roughness-induced speckle patterns for silica-on-silicon wafers

A. Bony[*, 1, 2], **Y. Takakura**[3], **K. Satzke**[2], and **P. Meyrueis**[1]

[1] Laboratory of Photonic Systems, Parc d'innovation Bld S. Brant, 67400 Illkirch, France
[2] Alcatel Research and Innovation Center in Photonics, Lorenzstrasse 10, 70435 Stuttgart, Germany
[3] TRIO / LSIIT (UMR 7005), Parc d'innovation Bld S. Brant, 67400 Illkirch, France

Received 30 May 2003, revised 4 August 2003, accepted 11 August 2003
Published online 10 November 2003

PACS 78.35.+c, 78.68.+m, 78.90.+t

Experimental observation of angular speckle correlations for two-dimensional nano-rough silica layers on silicon substrate are presented. The set-up makes use of a digital CCD camera to record the speckle patterns as a function of illumination and scattering angular conditions. $C^{(1)}$ correlations corresponding to the optical memory and time-reverse memory effects are reported for three different samples. $C^{(10)}$ correlation does not presently seem to be observable.

1 Introduction When light coming from a laser illuminates a sample, surface roughness in the range from nanometer to micrometer (or beyond) results in the formation of far-field speckle patterns, arising from interference between wavelets scattered by different surface elements. For samples composed of a transparent layer on a reflecting substrate, these speckle patterns may be modulated by interference fringes [1, 2]. The way the speckle pattern obtained in a given scattering configuration evolves with the system parameters, i.e. incident and scattering angles, is calculated through the degree of correlation between two speckle patterns. Such a technique was early used to assess surface roughness in the case of single scattering [3]. Feng et al. [4] initiated considerable efforts in the prediction and observation of speckle correlation in the case of volume scattering [5], and then for scattering by rough surfaces[6].

The angular speckle correlations for a transparent layer on a reflecting substrate have been theoretically studied by McGurn and Maradudin [7], and Sánchez-Gil [8], considering mono-dimensional roughness at the air-dielectric interface. The contributions to these spatial correlations are divided into terms of short-range, $C^{(1)}$ and $C^{(10)}$, long-range, $C^{(1.5)}$ and $C^{(2)}$, and infinite-range correlation, $C^{(3)}$. Experimental observation of $C^{(1)}$ [9] with mono-dimensional roughness and $C^{(10)}$ [10] with 2D-roughness were reported for the case of a dielectric layer over a glass substrate, with the sample being wedge-shaped to avoid reflection from the substrate to lie in the plane of incidence.

The purpose of this study is to experimentally investigate short-range angular intensity speckle correlations for samples constituted of 2D-nano-rough silica layers on silicon wafer. Next section describes the samples used in this study, and the experimental set-up developed to record the images of speckle and compute the intensity correlation. In section 3, results obtained for the memory and time-reverse memory effects are presented and discussed. Investigations concerning another type of intensity correlation, namely $C^{(10)}$, are reported. The present study does not deal with theoretically demonstrated higher-order correlations [7] or so-called degenerate optical memory effect [8].

2 Experimental In this section we detail the nature of the samples investigated in this work, and describe the experimental set-up employed for recording the speckle patterns.

---

* Corresponding author: e-mail: Alex_Bony@alcatel-research.de

2.1 Samples   Three samples are investigated: sample A is a silicon wafer covered by a 325nm thermal oxide; in comparison with sample A, sample B contains an additional 5.8μm-thick silica layer deposited using a flame hydrolysis deposition (FHD) apparatus [11]; sample C is a Si wafer covered by a thermal oxide and a FHD layer of respective thickness 15.8μm and 5.7μm. The refractive index of FHD layer for λ=633nm is n=1.46, and 1.44 for thermal oxide. The roughness at the air-silica interface was inspected with a scanning electron microscope (SEM) LEO Gemini 1550, with the sample normal making a 45° angle with the incoming electron beam. This instrument has a resolution of 2nm for an operating voltage of 1.5kV. The roughness was found to be constituted of residual defects of the order of a few tens of nanometers, which could be considered as 2-dimensional. Due to the small roughness of the samples, the scattered light pattern contains reflected components and diffused speckle. Additionally, white light interferometry measurements indicated long range (over several mm) thickness variations [12] of the order of 40nm for sample B and C, and 10nm for sample A. Angle-resolved light scattering measurements carried out on sample B and C revealed so-called "diffused fringes" [1]. Thickness extracted from the positions of their maxima of intensity lead to values of 6.4 and 21μm respectively, which can be assimilated to the sum of the values for the corresponding thermal oxide and FHD layers. Due to the small difference in refractive index between the transparent layers, the samples may reasonably be assimilated to a dielectric over a reflector (silicon), with different thicknesses for the dielectric layers, and slight 2-dimensional roughness at the air-dielectric interface.

2.2 Set-up   A 10mW Helium-Neon laser with random output polarization emits a beam of 1mm diameter at a wavelength λ=633nm that is linearly polarized before illuminating the investigated surface. The sample is mounted on a stage with two independent stepper rotation motors, thus enabling selection of the angle of incidence and of the scattering direction of observation. A second polarizer is fixed on the rotating arm carrying the camera. The speckle pattern is imaged with a scientific digital CCD camera from Hamamatsu (C4742-95-12-NRB), which outputs 1,024x1,280 pixel images with 12-bit resolution, the detector array covering an area of 8.58mm x 6.86mm at a distance of 40cm from the sample. The resulting speckle grains cover approximately 20-30 pixels, and each image has consequently more than 3,000 grains. The motors and the camera are controlled by a computer via a program running under LabVIEW software. Integration time with the camera is varied to optimize image contrast and limit pixel saturation. A typical image is displayed in figure 1, along with the histogram of pixel intensity values, the dark noise of the CCD camera resulting in an upward shift in pixel values from the expected theoretical curve [13].

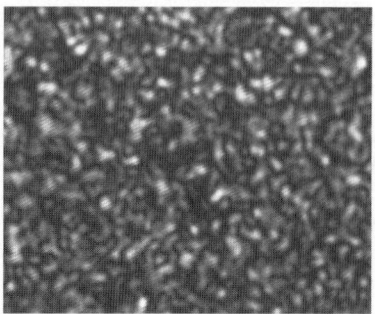 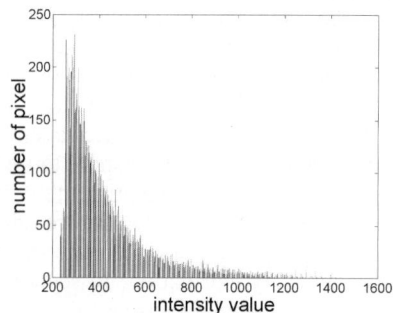

Fig. 1   Typical speckle pattern and histogram of pixel intensity data.

The speckle pattern is obtained by direct illumination of the CCD camera, without the use of an optical objective. This approach experimentally differs from previous similar experiments [9, 10]. As summarized in [14], every pixel composing the speckle pattern in our case is an average of scattering contributions from all the surface elements illuminated by the laser. When imaging the surface with an optical objective, every pixel is formed by scattering contribution of a unique surface element. It has been shown

[13] that when the number of surface scatterers involved is important enough, the statistical behaviors of both speckle patterns are identical.

3 Results and discussion When a first speckle pattern $I(\theta_{i1},\theta_{s1})$ is recorded using illumination and observation angles θ_{i1} and θ_{s1} respectively, its correlation coefficient with a second image $I(\theta_{i2},\theta_{s2})$ obtained for angles θ_{i2} and θ_{s2} is given by [9]:

$$C(\theta_{i1},\theta_{s1};\theta_{i2},\theta_{s2}) = \frac{<I(\theta_{i1},\theta_{s1})\ I(\theta_{i2},\theta_{s2})> - <I(\theta_{i1},\theta_{s1})><I(\theta_{i2},\theta_{s2})>}{\left[\left(<I(\theta_{i1},\theta_{s1})^2> - <I(\theta_{i1},\theta_{s1})>^2\right)\left(<I(\theta_{i2},\theta_{s2})^2> - <I(\theta_{i2},\theta_{s2})>^2\right)\right]^{\frac{1}{2}}} \quad (1)$$

where the symbols $<...>$ imply averaging over the pixels composing the images. In what follows, this parameter C is represented as a function of θ_{s2} for fixed values of the other angular parameters. A maximum in the correlation function is observed when the following angular condition is satisfied [7,8]:

$$\sin\theta_{i1} - \sin\theta_{s1} = \sin\theta_{i2} - \sin\theta_{s2} \quad (2)$$

which leads to the observation of short-range intensity correlations. The memory and time-reversed memory effects are the two main manifestations of these correlations.

3.1 Time-reversed memory effect In the particular case where $\theta_{i2} = -\theta_{s1}$, a peak of correlation is detected in the direction of observation for $\theta_{s2} = -\theta_{i1}$, in agreement with (2). This peak arises from the reciprocal scattering condition. Taking values of $\theta_{i1} = 10°$, $\theta_{s1} = 0°$, $\theta_{i2} = 0°$, and scanning θ_{s2} around the value of -10° expected from (2), a peak of correlation is observed for all the three samples both for s-s and p-p polarization configurations. However, all configurations lead to maximal values for the peak of correlation ranging from 0.5 to 0.8 depending on sample and polarization. In [10], the difference between the correlation peak value and unity was attributed to experimental alignment. A technique similar to the one presented in [15] was used to account for angular misalignment by numerically matching the two speckle patterns considered to obtain the maximum value of correlation achievable. These values remained yet well (0.8) below unity. When changing θ_{i1} to $\theta_{i1} = 15°$, 20°, 30°, the maximal value of the correlation peak decreased and the peak tended to slightly broaden.

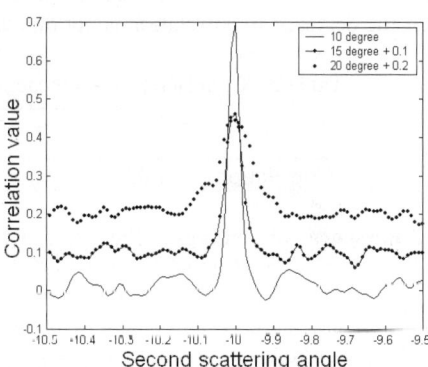

Fig. 2 Correlation in the reciprocal scattering configuration as a function of θ_{s2} for fixed θ_{i2} (initially 10°). (Curves at 15 and 20° are superimposed and shifted for clarity)

Figure 2 illustrates this observation for sample A, with s-s polarization. Comparison of the two speckle images that should present the maximum of correlation for $\theta_{i1} = 30°$ revealed a change in the size of the speckle grains and in the global scale of the pattern. As the observation system is in a non-imaging configuration (no optical objective, nor lens, before the receiver), the average size of the speckles is approximately the one of an Airy task produced by a circular aperture having the diameter of the illuminated area on the sample [14]. Consequently, when increasing the angle of incidence, the illuminated area becomes larger, and the speckle grains smaller than at normal incidence. Nevertheless, the two speckle patterns considered are still visually similar, thus confirming the presence of the time-reversed memory effect, even for large incident angles. However, due to the direct point-to-point correlation calculation, the correlation values are found to decrease with growing angles of incidence.

3.2 Memory effect and general $C^{(1)}$ angular condition When the incident angle θ_{i2} is closed to θ_{i1}, the speckle pattern can be seen to follow the direction of displacement of the laser. This move is in agreement with equation (2), and for small difference $\Delta\theta_i$ in incident angle the change in angle of incidence is equal to the change in angle of scattering observation to maximize the correlation function. This results in a memory line [10, 15]. When increasing $\Delta\theta_i$, a maximum of correlation can be found following angular condition (2). To test the general validity of condition (2), we set the scattering system in an arbitrary configuration, i.e. $\theta_{i1} = 5°$, $\theta_{s1} = 15°$, $\theta_{i2} = 10°$. Equation (2) provides a peak of correlation for $\theta_{s2} = 20.2°$. Table 1 summarizes the results for the maximal value of the correlation lobe as a function of sample and polarization.

Table 1 Maximal correlation value for $\theta_{s2} = 20.2°$ as a function of sample and polarization.

	Sample A	Sample B	Sample C
s-s	0.60	Below 0.1	Below 0.1
p-p	0.65	Below 0.1	0.40

Some values were below 0.1 and could not be distinguished from background fluctuation contributions. From the results of table 1, it can be noticed that depending on the sample investigated, condition (1) may not lead to a distinguishable correlation peak, and for sample C, this value is highly dependent on the polarization configuration. Equation (2) being linked to the memory effect, the absence of a correlation peak when Eq. (2) is satisfied may indicate a decrease in the memory effect when increasing $\Delta\theta_i$. Table 2 represents the maximal value of the correlation peak observed for sample B as a function of θ_{i2}, for $\theta_{i1} = 5°$, $\theta_{s1} = 15°$, and θ_{s2} satisfying Eq. (2). It is noticeable that contrarily to what happens in section 3.1, the decrease in correlation is not due to a change in the average scale of the speckle pattern.

Table 2 Correlation value for sample B, s-s, when increasing $\Delta\theta_i$, but keeping θ_{i2} satisfying Eq. (2).

	$\theta_{i2} = 6°$	$\theta_{i2} = 7°$	$\theta_{i2} = 8°$	$\theta_{i2} = 9°$
Correlation	0.9	0.7	0.3	Below 0.1

3.3 Presence of fringes The angle-resolved scattering curve for samples B and C displayed diffused fringes, with the positions of maxima of intensity changing with the angle of incidence in agreement with [1]. Figure 3 represents the correlation and intensity curves obtained in the conditions of section 3.2 for sample C in p-p polarization. It can be noticed that the angular intensity oscillations result in induced correlation oscillations, although the involved speckle patterns do not show signs of similitude, but present similar global intensity distribution. On the contrary, the correlation peak present for $\theta_{s2} = 20.2°$, does come from similar speckle patterns. This problem may rise interest in the application of other correlation calculators, rather than direct blind point-to-point intensity product between the two images.

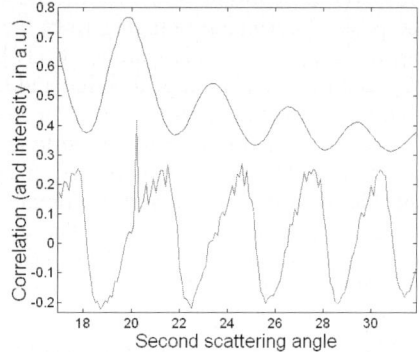

Fig. 3 Correlation (down) and intensity (up) for sample C as a function of θ_{s2}, for p-p polarization.

3.4 $C^{(10)}$ correlation condition In [7], a second angle condition for intensity correlation maximum is provided:

$$\sin\theta_{i1} + \sin\theta_{i2} = \sin\theta_{s1} + \sin\theta_{s2} \qquad (3)$$

corresponding to correlations denoted $C^{(10)}$, that should be of the same order of magnitude as correlations obeying condition (2). In [10], this type of correlation was reported taking $\theta_{i1} = \theta_{i2}$, and looking at scattering positions symmetric around the specular direction. A vertical symmetry of the speckle patterns was also noticed for images obtained symmetrically around the specular direction. Due to the presence of the specular beam, and to comply with the geometrical parameters of our set-up, it was not possible to record speckle pattern closer than 4° with respect to specular direction. No symmetry was noted in the two patterns obtained symmetrically with respect to the specular direction.

To try to observe intensity correlation according to Eq. (3), we tested the following angular configurations: (a) $\theta_{i1} = 10°$, $\theta_{s1} = 0°$, $\theta_{i2} = 0°$, and scanning around $\theta_{s2} = 10°$, (b) $\theta_{i1} = 5°$, $\theta_{s1} = 15°$, $\theta_{i2} = 10°$, and scanning around $\theta_{s2} = 0.11°$, (c) $\theta_{i1} = 10°$, $\theta_{s1} = 0°$, $\theta_{i2} = -10°$, and scanning around $\theta_{s2} = 0°$. None of these configurations coupled with any combination of sample and polarization could provide a set of correlation values that could be extracted from background fluctuation contributions.

The absence of peak corresponding to condition (3) is explained in [7] by the different physical origin giving rise to $C^{(10)}$ and $C^{(1)}$. Particularly, $C^{(10)}$ is not enhanced by phase coherent processes related to guided waves that propagate within the dielectric film. Condition (3) being intimately linked to symmetry around the specular [10] or illuminating directions (c), the absence of symmetry in the speckle patterns taken symmetrically around them might come from the too large minimal angles (4°) of imaging imposed presently by our set-up.

4 Conclusion   We presented our first experimental results for angular speckle correlation of light scattered by silica-on-silicon wafers. $C^{(1)}$ correlation was observed for all samples, but found to decrease differently depending on sample and polarization when $\Delta\theta i$ is increased. On the contrary, no correlation corresponding to $C^{(10)}$ was observed, nor the symmetry of the speckle pattern around specular direction.

Acknowledgement   A.B. wishes to thank the Alcatel Research Center in Photonics of Stuttgart for financial support of his Ph.D. work.

References

[1] J. Q. Lu, J. A. Sánchez-Gil, E. R. Méndez, Z.-H. Gu, A. A. Maradudin. J. Opt. Soc. Am. A **15** (1), 185 (1998).
[2] Y. S. Kaganovskii et al., Opt. Lett. **23**, 316 (1998).
[3] D. Léger, E. Mathieu, J.C. Perrin, Appl. Opt. **14**, 872 (1975).
[4] S. Feng, C. Kane, P.A. Lee, A. D. Stone, Phys. Rev. Lett. **61**, 834 (1988).
[5] I. Freund et al., Phys. Rev. Lett. **61**, 2328 (1988); R. Berkovits et al., Phys. Rev. B **41**, 2635 (1990).
[6] A. R. McGurn et al., Phys. Rev. B **39**, 13160 (1989); M. F. Knotts et al., J. Opt. Soc. Am. A **9**, 1822 (1992).
[7] A. R. McGurn and A. A. Maradudin, Phys. Rev. B **58**, 5022 (1998).
[8] J. A. Sánchez Gil, Phys. Rev. B **55**, 15928 (1997).
[9] J. Q. Lu and Z.-H. Gu, Appl. Opt. **36** (19), 4562 (1997);
[10] Z.-H. Gu and A. R. McGurn, Proc. SPIE **3784**, 285 (1999).
[11] M. Kawashi, Opt. Quant. Electron. **22**, 391 (1990).
[12] Y. S. Kaganovskii et al., J. Opt. Soc. Am. A **16** (2), 331 (1999).
[13] J. W. Goodman, in: Laser Speckle and Related Phenomena, edited by J. C. Dainty (Springer-Verlag, Berlin, 1984), pp. 9-75.
[14] J. D. Briers, in: Speckle Metrology, edited by R. S. Sirohi, chapter 8, pp. 373-426.
[15] Z. Q. Lin and Z.-H. Gu, Waves Random Media **7**, 435 (1997).

phys. stat. sol. (c) **0**, No. 8, 3032–3036 (2003) / **DOI** 10.1002/pssc.200303839

Amplitude and gradient scattering in waveguides with corrugated surfaces

F. M. Izrailev[*,1], **G. A. Luna-Acosta**[1], **J. A. Méndez-Bermúdez**[1], and **M. Rendón**[2]

[1] Instituto de Física, Universidad Autónoma de Puebla, Apartado Postal J-48, Puebla, Pue., 72570, México

[2] Facultad de Ciencias de la Electrónica, Universidad Autónoma de Puebla, Puebla, Pue., 72570, México

Received 30 May 2003, revised 4 August 2003, accepted 11 August 2003
Published online 10 November 2003

PACS 05.45.Mt, 41.20.Jb, 42.25.Dd, 71.23.An

We study chaotic properties of eigenstates for periodic quasi-1D waveguides with "regular" and "random" surfaces. Main attention is paid to the role of the so-called "gradient scattering" which is due to large gradients in the scattering walls. We demonstrate numerically and explain theoretically that the gradient scattering can be quite strong even if the amplitude of scattering profiles is very small in comparison with the width of waveguides.

During last decade much attention has been paid to the theory of quasi-1D disordered solids with the so-called *bulk scattering*. By this term one describes the situation where the whole volume of a scattering region contains scatters whose density determines the mean free path λ for propagation of electrons. According to the theory, apart from λ, transport properties in finite samples are described by two other characteristic lengths: the size L of a sample and the *localization length* l_∞. The latter is determined by the degree of decrease of the amplitude of eigenstates along infinite samples with the same scattering characteristics. The core of the modern theory of transport for such quasi-1D systems is the so-called *single-parameter* scaling. It was shown that when the mean free path is much less than both L and l_∞, *all* statistical characteristics of the transport are fully described by only one scaling parameter which is the ratio of the localization length to the size of a sample (see, e.g. [1] and references therein).

Another kind of quasi-1D systems that has attracted much attention in the past few years, is the many-mode waveguide with rough surfaces. In this case the scattering is entirely related to the statistical characteristics of the scattering walls, therefore, one can speak about *surface scattering*. For some time it was believed that surface scattering can be analytically described by modified methods thoroughly developed for bulk scattering. However, recent numerical studies of such systems [2, 3] have revealed a principal difference between surface and bulk scattering (see discussion and references in [4]). Specifically, it was found that the transport through quasi-1D waveguides with rough surfaces essentially depends on many characteristic lengths, not on one length as in the case of the bulk scattering. This fact is due to a non-isotropic character of scattering in the channel space. In particular, the transmission coefficient smoothly decreases with an increase of the angle of incoming waves, since characterictic lengths for backscattering are different for different channels [3, 5].

---

[*] Corresponding author: e-mail: izrailev@sirio.ifuap.buap.mx, Phone: +52 222 229 56 10, Fax: +52 222 229 56 11

The latter subject of surface scattering has a direct link to the problem of *quantum chaos*. The point is that waveguides with rough walls can be treated from the viewpoint of classical and quantum mechanics that describe a particle moving inside billiards and having multiple reflections from the walls. One of the problems of quantum chaos is the quantum-classical correspondence for the situation when, in the classical limit, global properties of the motion of a particle are strongly chaotic. More specifically, it is of great interest to find what is the fingerprint of classical chaos in quantum eigenstates of closed/periodic billiards, as well as the relation of statistical properties of the transport through open billiards to the underlying classical chaos.

In this paper we investigate the properties of quantum eigenstates of billiards with regular and rough walls, with the application to the wave scattering through quasi-1D waveguides with surface scattering. To be specific, we consider quasi-1D waveguides that are periodic in the x−direction with period 2π. The upper and lower walls are given by the functions, $f_1(x) = f_1(x + 2\pi) = d + a_1 \xi_1(x)$ and $f_2(x) = f_2(x + 2\pi) = a_2 \xi_2(x)$ where d is the average width of the waveguides and $a_{1,2} \ll d$ stand for the amplitude of the scattering walls. Our interest is in the structure of eigenstates of the corresponding Hamiltonian with zero boundary conditions on the two walls.

For our purpose it is convenient to pass to the variables $u = x$, $v = \dfrac{f_2(x) - y}{f_2(x) - f_1(x)}$ in which the new Hamiltonian $\hat{H} = \hat{H}^0 + \hat{V}$ describes a particle moving inside a waveguide with flat boundaries in the new u, v coordinates (see details in [6]). Here $\hat{H}^0 = \dfrac{1}{2m_e}(\hat{P}_u^2 + \hat{P}_v^2)$ and the effective potential $\hat{V}(u, v, \hat{P}_u, \hat{P}_v)$ depends on the functions f_1 and f_2, with $\hat{P}_u, \hat{P}_v$ as the canonical momenta. The solution of the Schrödinger equation for $\hat{H}$ can be written in the form $\psi_E(u, v) = \exp(iu)\,\psi_\chi(u, v)$. Since statistical properties of eigenstates of $\hat{H}$ do not depend on specific value of the Bloch index χ inside the first Brillouin band, all numerical data were obtained for a specific value of χ. By expanding $\psi_E(u, v)$ in the basis of $\hat{H}_0$, one can find the matrix representaion of $\hat{H}$ in the unperturbed basis specified by the two indexes n and m [7].

The Hamiltonian matrix elements are given by

$$
\begin{aligned}
H^k_{mnm'n'} = \frac{\hbar^2}{2\pi} \Bigg\{ &\pi(n+k)^2\,\delta_{nn'}\delta_{mm'} + \left[\frac{m^2\pi^2}{2}\left(J^2_{n'n} + J^3_{n'n} + J^6_{n'n}\right) + \left(\frac{1}{8} + \frac{m^2\pi^2}{6}\right)J^4_{n'n} \right]\delta_{mm'} \\
&+ i(n+n'+2k)\frac{mm'}{m^2 - m'^2}\left[J^1_{n'n} - (-1)^{m+m'}\left(J^1_{n'n} + J^5_{n'n}\right)\right] \\
&+ \frac{2mm'(m^2 + m'^2)}{(m^2 - m'^2)^2}\left[-J^6_{n'n} + (-1)^{m+m'}\left(J^6_{n'n} + J^4_{n'n}\right)\right] \Bigg\}
\end{aligned}
\tag{1}
$$

where

$$
J^1_{n'n} = \int_0^{2\pi} \frac{f_2'}{f_1 - f_2}\,e^{i(n'-n)u}\,du; \qquad J^2_{n'n} = \int_0^{2\pi} \frac{(f_2')^2}{(f_1 - f_2)^2}\,e^{i(n'-n)u}\,du,
\tag{2}
$$

$$
J^3_{n'n} = \int_0^{2\pi} \frac{1}{(f_1 - f_2)^2}\,e^{i(n'-n)u}\,du; \qquad J^4_{n'n} = \int_0^{2\pi} \frac{(f_1' - f_2')^2}{(f_1 - f_2)^2}\,e^{i(n'-n)u}\,du,
\tag{3}
$$

$$
J^5_{n'n} = \int_0^{2\pi} \frac{f_1' - f_2'}{f_1 - f_2}\,e^{i(n'-n)u}\,du; \qquad J^6_{n'n} = \int_0^{2\pi} \frac{f_2'(f_1' - f_2')}{(f_1 - f_2)^2}\,e^{i(n'-n)u}\,du.
\tag{4}
$$

with $f_{1,2}' \equiv \partial f_{1,2}(u)/\partial u$.

One should note that the matrix elements in the new variables u, v depends both on $f_{1,2}$ and on their derivatives $f_{1,2}'$. This very fact demonstrates a highly non-trivial role of the scattering walls since it is a problem to separate the influence of the *amplitude scattering* from the *gradient scattering*. It should be stressed that the above analytical expressions for the Hamiltonian matrix elements are obtained explicitly, without any approximation for the strength of perturbation. This is in contrast with existing theories that are based on perturbative approaches (see, for example, Ref. [8], where the smallness of the amplitudes a_1, a_2 and derivatives $f'_{1,2}$ is essential). In what follows, we perform a numerical study of the model by using the above expressions, by paying main attention to the influence of the roughness of the profiles that directly depends on the derivatives $f'_{1,2}$. Specifically, we compare the case of a "regular" upper profile $\xi_1(u) = \cos(u)$ that has been studied in detail in [6, 7] and [9], with the "random" one $\xi_1(u) = \sum_N^{N_T} A_N \cos(Nu)$ where the amplitudes A_N are chosen at random and $N_T = 100$. In both cases we assume $\xi_2(u) = 0$ and we refer to this case as to the "symmetric case".

In contrast to our previous studies [7], we address the question about the role of the gradient scattering, that is due to the derivatives $f'_{1,2}$, in comparison with the amplitude scattering that is determined by the amplitudes a_1, a_2 of the profiles. For this we consider the "asymmetric case" with $f_1 = d + a\xi(u)$ and $f_2 = a\xi(u)$ with $a = a_1 = a_2$, $\xi(u) = \xi_1(u)$. As one can see from the expressions for the matrix elements, in this case the scattering is only due to the "gradient terms" since $f_1 - f_2 = d = $ const. Specifically, we have $J^4_{n'n} = J^5_{n'n} = J^6_{n'n} = 0$, and the rest of the integrals are reduced to

$$J^1_{n'n} = \epsilon \int_0^{2\pi} \xi_u \, e^{i(n'-n)u} \, du \, ; \qquad J^2_{n'n} = \epsilon^2 \int_0^{2\pi} \xi_u^2 \, e^{i(n'-n)u} \, du \, ; \qquad J^3_{n'n} = \frac{1}{d^2} \int_0^{2\pi} e^{i(n'-n)u} \, du \, . \qquad (5)$$

where $\xi_u \equiv \partial \xi(x)/\partial u$, and $\epsilon \equiv a/d$.

One natural representation of the Hamiltonian matrix $H_{l,l'}(\chi) = \langle l| \hat{H} |l'\rangle_\chi$ is the "channel representation" for which one fixes the values of n starting from the lowest one, $n = -N_{max}$, and running over all values of m. In numerical simulations we have to make a cutoff for the values of m and n in the Hamiltonian matrix. Our data refer to the ranges, $1 \leq m \leq M_{max}$ and $|n| \leq N_{max}$ with $N_{max} = 32$, $M_{max} = 62$, for which the total size of the Hamiltonian matrix is $L = (2N_{max} + 1) M_{max} = 4030$.

From the structure of the Hamiltonian matrices shown in Fig. 1 one can make some important conclusions. First, by passing from "regular" waveguedes with $N_T = 1$ to "random" ones with $N_T = 100$ in both symmetric and asymmetric cases, the Hamiltonian matrices tend to be fully filled by off-diagonal elements. However, the matrices correponding to the random waveguides keep the block structure, thus indicating some regularity in spite of a completely random character of the surface scattering. This fact was shown to result in a kind of "non-ergodicity" for the eigenstates of the total Hamiltonian $\hat{H}$ for any high energy, see details in Ref. [10]. It is clear that standard theoretical approaches based on completely random matrices can not adequately discribe the structure of eigenstates. As was shown in Ref. [5], for open waveguides with rough surfaces and large number of channels, there are many characteristic lengths in contrast to the bulk scattering where the so-called single-parameter scaling holds.

First, Figure 1 clearly demonstrates that the eigenstates of $\hat{H}$ are expected to be much more extended in the unperturbed basis for rough profiles, than for one-cosine profiles. Second, it is instructive to make a comparison between the symmetric and asymmetric cases (compare (a) with (b) and (c) with (d) in Fig. 1). The data show that the matrices for the assymetric cases have smaller elements than those for the symmetric ones. This is in correspondence with the fact that large part of the terms in the expressions for the matrix elements of $\hat{H}$ vanishes due to the absence of amplitude scattering. Therefore, one can expect that the eigenstates are less random for the asymmetric case.

In order to analyze the structure of eigenstates of $\hat{H}$ in detail, we have diagonalized the Hamiltonian matrices shown in Fig. 1 and constructed the "state matrices" $|C^\alpha_l|^2$. Here C^α_l are the amplitudes of the eigenstates in the basis representaion given by the index l. Namely, the index l refers to unperturbed basis states that correspond to the unperturbed Hamiltonian $\hat{H}^0$. The index α refers to a specific exact eigenstate. All eigenstes are reordered in increasing energy, with $\alpha = 1$ the ground state. Therefore, to understand how strongly localized/extended are the exact eigenstates in the unperturbed basis, one should fix the value of α and explore the dependence of $|C^\alpha_l|^2$ on l.

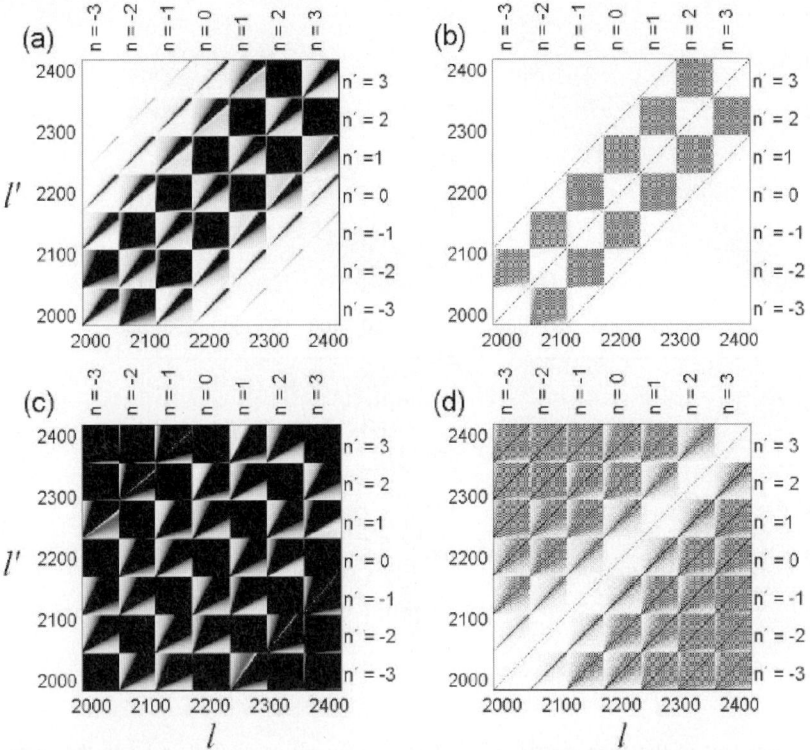

Fig. 1   Central part of the Hamiltonian matrix. The 62×62 blocks corresponding to $n, n' = [-3, 3]$ are shown. The larger the amplitudes of the matrix elements the more black the corresponding regions are. Four cases refer to: (a) one-cosine symmetric waveguide, (b) one-cosine asymmetric waveguide, (c) symmetric waveguide with a rough wall, $N_T = 100$, (d) asymmetric waveguide with rough walls, $N_T = 100$.

Let us now compare the symmetric cases with the asymmetric ones, see Fig. 2. The most important conclusion that can be deduced from the data is that the eigenstates are, in general, more extended in the asymmetric cases. Specifically, for the same small values of α there are more components C_l^α with large values of l, in comparison with the symmetric case, compare (c) with (d). At a first glance, this looks strange since for asymmetric profiles the Hamiltonian matrices are obviously less "random" than for the symmetric ones, as is mentioned above. Close inspection of the data in Fig. 2 shows that there is an additional effect that is also important in connection with the structure of eigenstates. Namely, the eigenstates for asymmetrical cases (b) and (d) turn out to be more *sparse* in comparison with the cases (a) and (c). This is manifested by a large number of "holes" along each line in (b) and (d) for fixed values of α, in comparison with the cases (a) and (c). Thus, the eigenstates for waveguides with only gradient scattering (asymmetric case) are more extended in the unperturbed basis, and, at the same time, more sparse than for the wavegudies with both gradient and amplitude scattering (symmetric case).

This phenomenon is important in view of the scattering properties through waveguides of *finite* size with profiles such as those considered here. As is known, chaotic structure of eigenstates of closed (or periodic) waveguides/billiards is directly related to the scattering properties of open systems with the same profiles. For example, the degree of localization of eigenstates of closed systems determines the degree of localization of scattering states, and correspondingly, the value of transmission through open systems.

In conclusion, we have studied the structure of eigenstates of quasi-1D periodic waveguides with regular and random walls. Main attention was paid to the role of gradient scattering in comparison with the amplitude scattering. It was shown that for the case when the scattering walls have a large number of harmonics, the gradient scattering is very strong. This is demonstrated by the data obtained

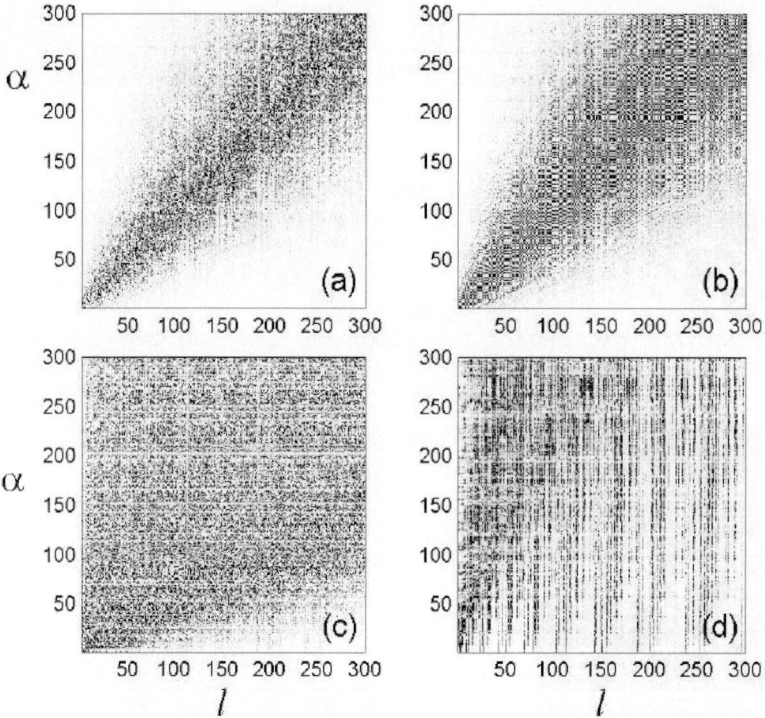

Fig. 2 Lower part of the state matrix $|C_l^\alpha|^2$. The data for all cases (a-d) refer to the same profiles for which the structure of Hamiltonian matrices is shown in Fig.1. Here darker regions correspond to larger values of $|C_l^\alpha|^2$.

for waveguides with asymmetric walls, for which the amplitude scattering is absent. It was also revealed that the role of the gradient scattering is highly non-trivial. Specifically, the gradient scattering turns out to be relatively strong in the absence of amplitude scattering.

Acknowledgements We are grateful to N. Makarov for fruitful discussions. This work was supported by the CONACyT (Mexico) Grant No. 34668-E, and IIG3G02, VIEP, BUAP.

References

[1] Y.V. Fyodorov and A.D. Mirlin, Int. J. Mod. Phys. B **8**, 3795 (1994); K.B. Efetov, *Supersymmetry in Disorder and Chaos* (Cambridge University Press, Cambridge, England, 1997); A.D. Mirlin, Phys. Rep. **326**, 259 (2000).

[2] A. García-Martín, J.A. Torres, J.J. Sáenz, and M. Nieto-Vesperinas, Appl. Phys. Lett. **71**, 1912 (1997); A. García-Martín, J. J. Sáenz, and M. Nieto-Vesperinas, Phys. Rev. Lett. **84**, 3578 (2000).

[3] J.A. Sánchez-Gil, V. Freilikher, I. Yurkevich, and A.A. Maradudin, Phys. Rev. B **59**, 5915 (1999).

[4] M. Leadbeater, V.I. Falko, and C.J. Lambert, Phys. Rev. Lett. **81**, 1274 (1998).

[5] F.M. Izrailev and N.M. Makarov, Phys. Rev. B **67**, 113402 (2003).

[6] G.A. Luna-Acosta, K. Na, L.E. Reichl, and A. Krokhin, Phys. Rev. E **53**, 3271 (1996).

[7] G.A. Luna-Acosta, J.A. Méndez-Bermúrdez, and F.M. Izrailev, Phys. Lett. A **274**, 192 (2000); G.A. Luna-Acosta, J.A. Méndez-Bermúrdez, and F.M. Izrailev, Physica E **12**, 267 (2002); Phys. Rev. E **65**, 046605 (2002).

[8] N.M. Makarov and Yu.V. Tarasov, J. Phys.: Condens. Matter, **10**, 1523 (1998); N.M. Makarov and Yu.V. Tarasov, Phys. Rev. B, **64**, 235306 (2001).

[9] G.A. Luna-Acosta, A.A. Krokhin, M.A. Rodriguez, and P.H. Hernandez-Tejeda, Phys. Rev. B **54**, 11410 (1996); G.A. Luna-Acosta, M.A. Rodriguez, A.A. Krokhin, K. Na, and R.A. Méndez, Rev. Mex. Fis. **44**, S3 7 (1998); G.B. Akguc and L.E. Reichl, J. Stat. Phys. **98**, 813 (2000); B. Huckestein, R. Ketzmerick, and C.H. Lewenkopf, Phys. Rev. Lett. **84**, 5504 (2000); W. Li, L.E. Reichl, and B. Wu, Phys. Rev. E **65**, 056220 (2002).

[10] J.A. Méndez-Bermúrdez, G.A. Luna-Acosta, and F.M. Izrailev, to be published.

phys. stat. sol. (c) **0**, No. 8, 3037–3041 (2003) / **DOI** 10.1002/pssc.200303840

Controlled transparency of many-mode waveguides with rough surface

F. M. Izrailev[1] and **N. M. Makarov**[*, 2]

[1] Instituto de Física, Universidad Autónoma de Puebla, Apartado Postal J-48, Puebla, Pue. 72570, México
[2] Instituto de Ciencias, Universidad Autónoma de Puebla, Priv. 17 Norte No 3417, Puebla, Pue. 72050, México

Received 30 May 2003, revised 4 August 2003, accepted 11 August 2003
Published online 10 November 2003

PACS 42.79.Gn, 72.10.–d, 72.15.Rn, 73.20.Fz, 73.20.Jc, 73.23.–b

In a unified approach we consider transport properties of 1D and quasi-1D waveguides with rough surfaces. Main attention is paid to the possibility of perfect transmission of waves due to specific long-range correlations in the surface profiles. First, we show how to construct random profiles that lead to a complete transparency of waveguides with one open channel. Then, we present analytical results for many-mode waveguides. It was revealed that by a proper choice of correlations in surface profiles the transmission through such quasi-1D waveguides is described by a coset of non-interacting 1D channels with a perfect transmission along each one. The number of these conducting modes is governed by the control parameter, and can be equal to the total number of channels. Therefore, the waveguides can be completely transparent in some region of frequency of incoming waves. This unexpected phenomenon is discussed in connection with the violation of the single-parameter scaling for surface scattering.

1 1D waveguiding systems

During last few years much attention has been paid to studying transport properties of one-dimensional (1D) systems with correlated disorder (see, e.g., Ref. [1] and references therein). The interest is due to intriguing results that may have potential applications both to electromagnetic and electron/optic devices. In particular, it was shown [2] that any desired combination of transparent and nontransparent frequency windows can be achieved by a proper construction of random potentials with specific long-range correlations. Experimental realization [3] of such potentials with delta-like scatters has confirmed the theoretical predictions for single-mode waveguides.

The subject of wave propagation through surface-corrugated waveguides is important for application in optics fibers, remote sensing, shallow water waves, etc., as well as in view of electronic transport in mesoscopic conducting structures. For this reason, we have recently analyzed anomalous properties of surface-governed transmission in waveguides with one open channel [4]. As a result, a theoretical method was proposed to construct disordered rough surfaces with specific long-range correlations along the waveguide. It was shown analytically and by direct numerical simulations as well, that single-mode waveguides with desired selective transparency can be fabricated with this method.

The idea of the controlled transparency in single-mode surface-corrugated waveguides is based on two principal points. First, the transport through such systems, as through any 1D disordered structure, is completely described by the ratio L/L_{loc} of the waveguide length L to the quantity L_{loc} known as the localization length [5, 6]. This concept is called *one-parameter scaling*. The localization length is determined by the parameters of a waveguide, and by statistical characteristics of its rough surface. For large localization length, $L/L_{loc} \ll 1$, the waveguide is practically transparent and its transmittance T is almost equal to unity. Otherwise, when $L_{loc}/L \ll 1$, the transmittance is exponentially small because of

---

[*] Corresponding author: e-mail: makarov@siu.buap.mx, Phone: +52 222 242 1072, Fax: +52 222 242 1072

strong wave localization. Second, the inverse localization length is proportional to the lengthwise Fourier transform $W(k_x)$ of the binary correlator of random surface profile, $L_{loc}^{-1} \propto W(2k_1)$ where k_1 is the longitudinal wave number of single propagating waveguide mode. Therefore, if the roughness power spectrum $W(2k_1)$ abruptly vanishes within some interval of wave number k_1, then the localization length L_{loc} diverges and the waveguide is fully transparent. Consequently, with the use of the methods developed in the theory of surface profile generation [7], we can construct such random surfaces that result in a complete transparency of waveguides within a predefined part of the allowed wave-number region [4]. A sharp transition between localized and ballistic transport at a given point implies the roughness power spectrum $W(k_x)$ having a discontinuity at this point. This means that the corrugated profile $\xi(x)$ of waveguide surface should have particular long-range correlations along the waveguide (along the coordinate x). Note that surfaces with the properties that give rise to rectangular roughness spectrum are not exotic. They have been recently fabricated in experimental studies of an enhanced backscattering [8].

2 Quasi-1D waveguides with correlated surface disorder   Now, let us discuss transport properties of *many-mode plane* (quasi-1D) waveguides with correlated surface disorder. Below we show that in this case the role of long-range correlations is much more sophisticated in comparison with that discussed above for single-mode waveguides. The reason is that unlike the single-mode situation, the concept of one-parameter scaling is no more true for the transport through surface-corrugated many-mode waveguides. There are two points that should be stressed in this respect. First, the correlations discussed above result in suppression of the interaction between different propagating modes. This non-trivial fact turns out to be crucial for the reduction of a system of mixed channels with quasi-1D transport, to the coset of independent waveguide modes with purely 1D transport. Second, the same correlations lead to a complete transparency of each independent channel, similar to what happens in strictly single-mode geometry.

To start with, we should note that in the spirit of the Landauer's concept [9], in our case of many-mode waveguides the total average transmittance $<T>$ can be expressed as a sum of partial transmittances T_n for every nth propagating normal mode (conducting channels),

$$<T> = \sum_{n=1}^{N_d} T_n \, . \tag{1}$$

Here $N_d = [kd/\pi]$ is the total number of propagating modes determined by the integer part [...] of the ratio kd/π, and d is the average waveguide width. The total wave number k is equal to ω/c for a classical wave of frequency ω, and to the Fermi wave number for electrons.

From the general theory of surface scattering [10] it follows that transmission properties of any nth conducting channel are determined by two attenuation lengths, the length $L_n^{(f)}$ of forward scattering and the backscattering length $L_n^{(b)}$. These scattering lengths are expressed by

$$\frac{1}{L_n^{(f)}} = \sigma^2 \frac{(\pi n/d)^2}{k_n d} \sum_{n'=1}^{N_d} \frac{(\pi n'/d)^2}{k_{n'} d} W(k_n - k_{n'}) \, , \tag{2}$$

$$\frac{1}{L_n^{(b)}} = \sigma^2 \frac{(\pi n/d)^2}{k_n d} \sum_{n'=1}^{N_d} \frac{(\pi n'/d)^2}{k_{n'} d} W(k_n + k_{n'}) \, , \tag{3}$$

with σ being the root-mean-square roughness height and $k_n = \sqrt{k^2 - (\pi n/d)^2}$ the lengthwise wave number of nth propagating mode. These results can be obtained by the diagrammatic Green's function approach [10] as well as by the technique developed in Ref. [11]. Note that in a single-mode waveguide with $n = N_d = 1$ the sum over n' contains only one term with $n' = 1$. In this case Eq. (3) for the back-scattering length $L_1^{(b)}$ gives the discussed above value of the localization length L_{loc} that is four times larger than $L_1^{(b)}$.

One can see from Eqs. (2) and (3) that, in general, both attenuation lengths are contributed by scattering of a given nth propagating mode into all other modes. This is the case when, for example, surface profiles $\xi(x)$ are either of white-noise type with constant power spectrum $W(k_x)$, or the random function with widely used fast decreasing Gaussian correlator (or, the same, with a slow decrease of its Fourier transform). Besides, these expressions manifest rather strong dependence on the mode index n. Specifically, the larger the mode number n is, the smaller are the corresponding attenuation lengths and as a consequence, the stronger is the scattering of this mode into the others. As was shown in Ref. [12], even in the absence of correlations in $\xi(x)$ a very interesting phenomenon of the coexistence of ballistic, diffusive, and localized transport arises, which seems to be generic for propagation through many-mode waveguides with disordered surfaces. Namely, while lowest modes can be in the ballistic regime, the intermediate and highest propagating modes exhibit the diffusive and localized behavior, respectively. As a result, we come to the concept of the *hierarchy of mode attenuation lengths* instead of the one-parameter scaling.

Now let us demonstrate that the situation fundamentally changes when the surface roughness has specific long-range correlations. As an example, we consider the random surface profile $\xi(x)$ with the simplest power spectrum in the form of a "window function",

$$W(k_x) = (\pi / k_c)\Theta(k_c - |k_x|) . \tag{4}$$

Here $\Theta(z)$ stands for the unit-step function and the characteristic wave number $k_c > 0$ is the controlling parameter to be specified below.

It is evidently that in the case under consideration the number of modes into which a given nth mode is scattered, i.e. the actual number of the summands of Eqs. (2) and (3), is entirely determined by the width k_c of the rectangular spectrum (4). Moreover, if the distance $|k_n - k_{n\pm1}|$ between neighboring wave numbers is larger than the controlling width k_c,

$$|k_n - k_{n\pm1}| > k_c , \tag{5}$$

then the transitions between all modes are forbidden. In this case the sum over n' in Eq. (2) for the forward scattering length $L_n^{(f)}$ contains only one term with $n' = n$ which describes intra-mode scattering only. At the same time, each term in the sum (3) is equal to zero so that the backscattering length is infinite, $L_n^{(b)} = \infty$.

From the above consideration a remarkable effect can be revealed. Namely, all the propagating modes with index n for which the condition (5) holds, are fully independent of other waveguide modes, in spite of their interaction with a rough surface. In other words, they represent a coset of 1D non-interacting channels. As is well known from the standard theory of 1D localization (see, e.g., Refs. [5,6]), the transmission through any 1D disordered structure is determined by the backscattering length only and does not depend on the forward scattering. Since the former is infinite for every such channel, its partial transmittance is equal to unity, $T_n = 1$. As for other propagating modes with index n that is in contradiction with the condition (5), they remain to be mixed by the surface scattering because the roughness power spectrum (4) for them is non-zero, $W(k_n - k_{n'}) = \pi / k_c$. Since these *mixed modes* have finite attenuation lengths, for large enough waveguide length L they do not contribute to the total transmittance (1) and the latter is equal to the total number of *independent transparent modes*.

The further analytical treatment is allowed for large number of conducting channels

$$N_d = [kd / \pi] \approx kd / \pi \gg 1. \tag{6}$$

In this case the condition (5) is equivalent to the requirement $|\partial k_n / \partial n| > k_c$, which can be written in the explicit form

$$n > N_m \equiv \left[(kd / \pi)\{1 + (k_c d / \pi)^{-2}\}^{-1/2}\right], \tag{7}$$

where, as above, the square brackets stand for the integer part of the inner expression. Thus, all propagating modes with $n > N_m$ are independent and fully transparent, otherwise, they are mixed and characterized by finite attenuation lengths. Therefore, the integer N_m should be regarded as the total number of mixed non-transparent modes while the total number of independent transparent channels is given by the difference $N_t = N_d - N_m$.

One can see from Eq. (7) that the numbers N_m and N_t of mixed non-transparent and independent transparent modes are determined by two parameters: the mode parameter $\alpha = kd / \pi$ and the dimensionless *correlation parameter* (CP) $\alpha_c = k_c d / \pi$. In the case of "weak" correlations when $\alpha_c \gg 1$, the number of mixed modes $N_m \approx [\alpha(1 - \alpha_c^{-2} / 2)]$ is of the order of $N_d = [\alpha]$. Consequently, in this case the number of independent transparent modes $N_t = N_d - N_m$ is small, or there are no such modes at all. If the CP α_c tends to infinity ($\alpha_c \to \infty$) the rough surface profile becomes white-noise-like and, naturally, $N_m \to N_d$. The most appropriate case is when surface roughness is strongly correlated so that the CP is small, $\alpha_c \ll 1$. Then the number of mixed non-transparent modes $N_m \approx [\alpha_c \alpha]$ is much less than the total number of propagating modes N_d and the number of transparent modes N_t is large. When α_c decreases and becomes anomalously small ($\alpha_c < \alpha^{-1} \ll 1$), the number N_m vanishes and all modes turn out to be independent and fully transparent. In this case the correlated disorder results in a perfect transmission of waves, in spite of their scattering from a rough surface.

Finally, let us briefly discuss the expression for the transmittance of many-mode waveguides with correlated surface roughness of the above kind,

$$<T> = [\alpha] - [\alpha(1 + \alpha_c^{-2})^{-1/2}]. \tag{8}$$

It is clear that the transmittance (8) reveals a step-wise dependence on the value of the mode parameter $\alpha = kd / \pi$. The analogous effect is known to occur for the conductance of quasi-1D ballistic (non-disordered) structures (see, e.g., Ref. [13]). However, in our surface-disordered model the step-wise dependence consists not only of the usual (ballistic) *steps up* from the first term in Eq. (8) but also of the *steps down*. The latter belong to the second term and are formed by the correlated surface scattering. The steps up arise for integer values of $\alpha = kd / \pi$ only. In contrast to this, the positions of the steps down are determined by the correlation parameter $\alpha_c = k_c d / \pi$ and can occur for non-integer values of kd / π. Evidently, within some intervals the steps of such different types cancel each other. Therefore, the experimental picture of the discussed dependence is expected to be very interesting and sophisticated.

3 Summary
We have studied the role of long-range correlations in surface profiles for the transport through quasi-1D waveguides. It was found that the correlations that result in a complete transmission of waves in the case of one-mode waveguides, lead to a quite unexpected phenomenon when the number of modes is large. Specifically, we show that these long-range correlations give rise to the appearance of non-interacting 1D channels, the number of which is controlled by the correlation parameter. The remarkable point is that these channels turn out to be completely transparent in some range of frequency of incoming waves. As a result, the total transmission of waveguides can be significantly enhanced in com-

parison with uncorrelated surface profiles (or profiles with Gaussian correlations of finite length). Moreover, the number of independent transparent channels can be as large as the total number of modes. In this case, the waveguides are fully transparent. This effect is directly related to the fact that for many-mode surface scattering there are many characteristic lengths that determine the total transport. In other words, the famous single-parameter scaling known to be held for bulk scattering, in our case is not valid. As a result, surface scattering transport through different channels can be separated by a proper choice of long-range correlations along surface profiles. Our study may find practical applications, for example, when fabricating of waveguides and superlattices with selective transport in a given frequency range of incoming waves. Although our consideration is based on the first order approximation (2-3), the higher correction terms can not change drastically the whole picture for the mobility edge transition. For this reason it is naturally to expect that global properties of the transport are correctly described by our theory. Specifically, transitions from metallic to localized regimes discussed above for the correlated disorder, are expected to be sharp enough in order to observe them experimentally.

Acknowledgements This work was partially supported by the Consejo Nacional de Ciencia y Tecnología (CONA-CYT, México) under Grants No. 34668-E, 36047-E and by the Universidad Autónoma de Puebla (BUAP, México) under Grant II-104G02.

References

[1] P. Carpena, P. Bernaola-Galván, P. Ch. Ivanov, and H. E. Stanley, Nature (London) **418**, 955 (2002).
[2] F. M. Izrailev and A. Krokhin, Phys. Rev. Lett. **82**, 4062 (1999); A. A. Krokhin and F. M. Izrailev, Ann. Phys. (Leipzig) **8**, 153 (1999); F. M. Izrailev, A. A. Krokhin, and S. E. Ulloa, Phys. Rev. B **63**, 041102(R) (2001).
[3] U. Kuhl, F. M. Izrailev, A. A. Krokhin, and H.-J. Stöckmann, Appl. Phys. Lett. **77**, 633 (2000).
[4] F. M. Izrailev and N. M. Makarov, Opt. Lett. **26**, 1604 (2001).
[5] I. M. Lifshits, S. A. Gredeskul, and L. A. Pastur, Introduction to the Theory of Disordered Systems (Wiley, New York, 1988).
[6] N. M. Makarov and I. V. Yurkevich, Zh. Eksp. Teor. Fiz **96**, 1106 (1989) [Sov. Phys. JETP **69**, 628 (1989)]; V. Freilikher, N. M. Makarov, and I. V. Yurkevich, Phys. Rev. B **41**, 8033 (1990).
[7] S. O. Rice, in: Selected Papers on Noise and Stochastic Processes, ed. by N. Wax (Dover, New York, 1954), p. 180.
[8] C. S. West and K. A. O'Donnell, J. Opt. Soc. Am. A **12**, 390 (1995).
[9] R. Landauer, Physica Scripta **T42**, 110 (1992).
[10] F. G. Bass and I. M. Fuks, Wave Scattering from Statistically Rough Surfaces (Pergamon, New York, 1979).
[11] A. R. McGurn and A. A. Maradudin, Phys. Rev. B **30**, 3136 (1984).
[12] J. A. Sánchez-Gil, V. Freilikher, I. V. Yurkevich, and A. A. Maradudin, Phys. Rev. Lett. **80**, 948 (1998); J. A. Sánchez-Gil, V. Freilikher, A. A. Maradudin, and I. V. Yurkevich, Phys. Rev. B **59**, 5915 (1999).
[13] B. J. van Wees, H. van Houten, C. W. J. Beenakker et al., Phys. Rev. Lett. **60**, 848 (1988).

phys. stat. sol. (c) **0**, No. 8, 3042–3045 (2003) / **DOI** 10.1002/pssc.200303863

Magnetic field effects on the optical response of corrugated films: Voigt geometry

J. H. Jacobo-Escobar[*,1] and **Gregorio H. Cocoletzi**[2]

[1] FCQB de la Universidad Autónoma de Sinaloa, Culiacán, Sinaloa, México
[2] Instituto de Física, Universidad Autónoma de Puebla, Apartado Postal J-48, Puebla 72570, México

Received 30 May 2003, revised 4 August 2003, accepted 11 August 2003
Published online 27 November 2003

PACS 78.68.+m

Using the Rayleigh–Fano modal theory it is investigated the interaction of p-polarized light with the corrugated surface of a metallic film. Calculations of the magnetoreflectance in the presence of an external magnetic field in the Voigt configuration are carried out. Studies are performed on the nonreciprocity property and the coupling of the incident light with the surface magnetoplasmons. It is found that the calculated dispersion relation of the surface magnetoplasmons predicts the possible experimental excitation of such modes.

1 Introduction Effects of external magnetic fields on the optical properties of magnetoplasmons in thin films and superlattices have received considerable attention in recent works. Different configurations are considered, with the optical properties showing a dependence on the direction and strength of the applied field. According to the direction of the external field, the following configurations may be defined; perpendicular, Faraday or Voigt geometries. Magnetoplasmons in corrugated films and corrugated films in contact with semiinfinite superlattices, taking into account an external magnetic field in the perpendicular configuration, have been recently investgated. In such studies the surface was modeled by $\xi(x) = \xi_0 \cos\left(\dfrac{2\pi}{a} x\right)$. On the other hand, device miniaturization demands the optical characterization of surfaces. Therefore, it is important to investigate surface modes propagation. Studies of surface modes excitations may be performed using the surface roughness or the attenuated total reflectivity (ATR) techniques. In this work we investigate theoretically the optical excitation of surface magnetoplasmons in a film with a corrugated surface, taking into account an external magnetic field in the Voigt configuration (VG). In the VG geometry, the external field is applied parallel to the film surface and perpendicular to the corrugated surface profile. The model dielectric response tensor, we use in the formalism, is obtained from the classical equation of motion of electrons in the presence of applied electric and magnetic fields. Therefore, the approach is valid within the regime of validity of the dielectric tensor. We focus our attention on the conditions of light coupling with the surface magnetoplasmons, mainly on the corrugated surface. Using the Rayleigh–Fano modal theory, the amplitudes of the scattered fields are calculated to zero and first order on the corrugation height. The minima of the specular reflection and the first order resonances are interpreted as the coupling of the incident light with the surface magnetoplasmons. In addition, since the Voigt geometry preseves the polarization of light and shows the nonreciprocity property [5, 6], this issue is also explored.

---

* Corresponding author: e-mail: ddtee@uas.uasnet.mx

2 Theory Let us consider a metallic film of dielectric tensor ϵ with a coordinate system in such a way that the flat surface is at $z = 0$ and the corrugated deterministic surface is at $z = L + \xi(x)$, where $\xi(x)$ is the sinusoidal profile along the x-axis. According to this coordinate system, the surface modes propagate along the x-axis and the Voigt geometry is achieved when the external magnetic field $\boldsymbol{B}_0$ is applied in the y-direction. The studies consider p-polarized light incident onto the corrugated surface at an angle θ respect to the z-axis, with a wave vector $\boldsymbol{k} = (k_0, 0, \alpha_{0v})$. To investigate the optical response of the deterministic corrugated film we use the Maxwell wave equation that has the form

$$\nabla \times (\nabla \times \boldsymbol{E}) - q_0^2 \epsilon \boldsymbol{E} = 0, \tag{1}$$

where $q_0 = \omega/c$. Using the plane wave solutions one may obtain

$$k^2 \boldsymbol{E} - (\boldsymbol{k}\boldsymbol{E})\,\boldsymbol{k} - q_0^2 \epsilon \boldsymbol{E} = 0, \tag{2}$$

with $\boldsymbol{k} = (k_p, 0, \beta_p)$. The non-zero elements of the dielectric tensor ϵ, as obtained from the equation of motion of electrons subject to external electric and magnetic fields, are of the form

$$\epsilon_{xx} = \epsilon_{zz} = \epsilon_\infty \left[1 - \frac{\omega_p^2(\omega + i\tau)}{\omega[(\omega + i\tau)^2 - \omega_c^2]} \right], \tag{3}$$

$$\epsilon_{yy} = \epsilon_\infty \left[1 - \frac{\omega_p^2}{\omega(\omega + i\tau)]} \right] \tag{4}$$

$$\epsilon_{xz} = -\epsilon_{zx} = \epsilon_\infty \left[i\,\frac{\omega_c}{\omega} \left[\frac{\omega_p^2}{(\omega + i\tau)^2 - \omega_c^2} \right] \right] \tag{5}$$

where ω_c, and ω_p are the cyclotron and plasma frequencies, respectively. τ is the damping term and ϵ_∞ is the background dielectric constant. The substitution of the dielectric tensor elements in (2) allows one to write the secular equation from where we obtain the dispersion relation

$$\beta_p^2 + k_p^2 = q_0^2 \epsilon_v, \epsilon_v = \frac{\epsilon_{xz}^2 + \epsilon_{zz}^2}{\epsilon_{zz}}. \tag{6}$$

Applying the boundary conditions [3, 4], and after some algebra [2, 4], it is possible to obtain the reduced *Rayleigh equations*,

$$\sum_p \begin{pmatrix} M_{lp}^{(11)} & M_{lp}^{(12)} \\ M_{lp}^{(12)} & M_{lp}^{(22)} \end{pmatrix} (R_p p) = \begin{pmatrix} V_l \\ W_l \end{pmatrix}, \tag{7}$$

where R_p and F_p are the amplitudes of the scattered fields in the vacuum side and in the film, respectively. And the matrix elements have the form

$$M_{lp}^{(11)} = (i)^{l-p}\, e^{i\alpha_{pv}L}\, J_{l-p}(\alpha_{pv}\xi_0), \tag{8}$$

$$M_{lp}^{(12)} = -q_0[\delta_{p1}S_p(i)^{l-p}\, e^{i\beta_p L} + \delta_{p2}(-i)^{l-p}\, e^{-i\beta_p L}]\, J_{l-p}(\beta_p \xi_0), \tag{9}$$

$$M_{lp}^{(21)} = (i)^{l-p} \left(\frac{\alpha_{pv}^2 - k_p k_l}{q_0 \alpha_{pv}} \right) e^{i\alpha_{pv}L}\, J_{l-p}(\alpha_{pv}\xi_0), \tag{10}$$

$$M_{lp}^{(22)} = -(S_p\, e^{i\beta_p L}\, I_{p1} + e^{-i\beta_p L}\, I_{p2}), \tag{11}$$

$$V_l = -(-i)^l\, e^{-i\alpha_{0v}L}\, J_l(\alpha_{0v}\xi_0), \tag{12}$$

$$W_l = (-i)^l \left(\frac{q_0^2 - k_0 k_l}{q_0 \alpha_{0v}} \right) e^{-i\alpha_{0v}L}\, J_l(\alpha_{0v}\xi_0). \tag{13}$$

In these equations, $J_\nu(\gamma)$ are the Bessel functions of the first class and order ν. The coefficients S_p are obtained using the boundary conditions at the two surfaces and I_{pi} are defined by the integrals

$$I_{pi} = \int_0^a dx \left[\frac{\epsilon_{xz} \pm \beta_p \delta_{pi} \xi(x)' + (\epsilon_{xz} - k_p \delta_{pi}) \xi(x)'^2}{\epsilon_{xz} + \xi(x)' \epsilon_{zz}} \right] e^{-i\beta_p \xi(x)} e^{i\Delta k x}, \qquad \Delta k = k_p - k_l. \qquad (14)$$

3 Results and discussion In this section we present the numerical studies. We consider deterministic corrugated surfaces and use the modal theory developed above. The system under study is a Mg layer of thickness $L + \xi(x)$ with a frequency and magnetic field dependent dielectric tensor. The parameters are $\hbar\omega_p = 10.5$ eV for the plasma frequency, $\nu = 0.06\omega_p$ for the damping factor and $\epsilon_\infty = 1$ for the background dielectric constant. The film is surrounded by vacuum, then $\epsilon_{vac} = 1$. Calculations are performed for $|R_0|^2$ when light of p-polarization is incident onto the corrugated surface at an angle θ. The wave vectors are $\mathbf{k}_0 = (k_0, 0, \alpha_0)$ of the incident light and $\mathbf{k}_p = (k_p, 0, \beta_p)$ of the back scattered light. These wave vectors obey the dispersion relations.

Using the frequency at the minima of $|R_0|^2$ in $k_p = q_0 \sin\theta + (2\pi/a)p$ we construct the dispersion relation of the surface magnetoplasmons. An example is presented in Fig. 1 for a film of $L = 600$ Å, $a = 4000$ Å, $\xi_0/a = 0.06$ and $\omega_c/\omega_p = 1$. The straight line is the light line and the other curves are the calculated dispersion realtions. ω_{smp}^-/ω_p and ω_{smp}^+/ω_p are the limits of the surface magnetoplasmons as indicated in the figure. These are obtained using the asymptotic limit [7, 8]

$$\frac{\omega_{smp}^\pm}{\omega_p} = \frac{1}{2} \left(\sqrt{2 + \Omega_c^2} \mp \Omega_c \right), \qquad (15)$$

where $\Omega_c = \dfrac{\omega_c}{\omega_p}$. Therefore, this figure shows the possible coupling of the incident light with the surface magnetoplasmons.

An interesting feature of the magnetic field effects on the optical response of magnetoplasmons in the Voigt geometry is the nonreciprocity property. This phenomenon is considered in Fig. 2 for a film of $L = 600$ Å, $a = 4000$ Å, $\xi_0/a = 0.06$, an angle of incidence of $\theta = 30^0$ and $\omega_c/\omega_p = 1$. Both

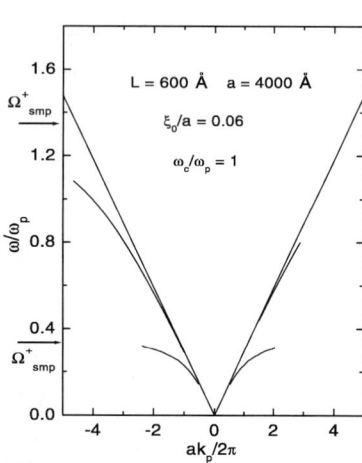

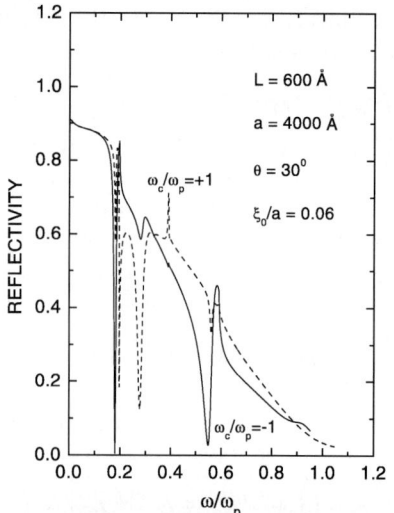

Fig. 1 It is shown the dispersion relation $\omega = \omega(k_p)$ of the surface magnetoplasmons obtained using the frequencies at the minima of $|R_0|^2$. The parameters are indicated.

Fig. 2 In this figure we present $|R_0|^2$ considering two different cyclotron frequencies $\omega_c/\omega_p = \pm 1$. These frequencies correspond to two external magnetic fields applied in opposite directions.

phys. stat. sol. (c) **0**, No. 8 (2003) / www.physica-status-solidi.com

curves were calculated using the same absolute value of the cyclotron frequency but the fields in opposite directions. The minima are somewhat shifted, and have different depth and broad. Results indicate that $\omega(k_p) \neq \omega(-k_p)$, which is a consequence of the fact that the magnetic field $\boldsymbol{B_0}$ is an axial vector, the change of k_p for $-k_p$ is equivalent to the change of $\boldsymbol{B}_0$ for $-\boldsymbol{B}_0$, that is, this result is a symmetry property [6].

4 Conclusions   We have presented a study of the optical response of magnetoplasmons in a metallic deterministic corrugated film taking into account a magnetic field in the Voigt geometry. The Rayleigh–Fano modal theory has been applied to calculate the specular and the first order scattered fields when light with p-polarization is incident onto the corrugated surface. The specular reflectivity show two minima due to the light coupling with the antisymmetric and symmetric surface modes. Each of these two minima split into two minima when an external field is applied in the Voigt geometry. The gap of the splitting increases with the magnetic field strength. On the other hand, studies of the perpendicular configuration show that the equivalent minima gap closes with the increase of the magnetic field strength. The use of the frequency at the minima of the reflectivity allows one to calculate the dispersion relation of surface magnetoplasmons and show the possible experimental excitation and detection of such modes.

Acknowledgements   The work of GHC was supported by SEP-CONACYT, México Project # 2003-01-21-001-051.

References

[1] E. D. Palik et al., Phys. Rev. B **13**, No. 16, 2497 (1976).
[2] J. H. Jacobo-Escobar, X. I. Saldãna, and G. H. Cocoletzi, J. Phys.: Condens. Matter **10**, 5807 (1998).
[3] J. H. Jacobo-Escobar and G. H. Cocoletzi, J. Phys.: Condens. Matter **11**, 1961 (1999).
[4] G. Martinez, J. H. Jacobo-Escobar, P. H. Hernández, and G. H. Cocoletzi, Phys. Rev. B **59**, No. 19 (1998).
[5] M. S. Kushwaha and P. Halevi, Solid State Commun. **64**, No. 11 (1987).
[6] R. E. Camley, Surface Science Reports **7** (1987).
[7] K. W. Chiu and J. J. Quinn, Phys. Rev. B **5**, No. 12 (1972).
[8] G. Borstel and H. J. Falge, Appl. Phys. **16** (1978).

phys. stat. sol. (c) **0**, No. 8, 3046–3049 (2003) / **DOI** 10.1002/pssc.200303831

Phenomenology of magnetic second harmonic generation from low symmetry surfaces and interfaces

L. Carroll and **J. F. McGilp***

Department of Physics, Trinity College, Dublin 2, Ireland

Received 30 May 2003, revised 4 August 2003, accepted 11 August 2003
Published online 10 November 2003

PACS 42.65.Ky, 75.70.Rf, 78.20.Ls

Low dimensional magnetic structures show interesting and novel phenomena such as oscillatory magnetic coupling and giant magnetoresistance. Magnetic second harmonic generation (MSHG) can provide unique information on magnetic surfaces and interfaces because, within the dipole approximation, broken space-inversion symmetry at the surface or interface of centrosymmetric media, and broken time-reversal symmetry arising from the magnetization, are both required in order to observe a magnetic-field-dependent second harmonic response. However, the additional reduction in symmetry arising from the magetization produces many non-zero susceptibility tensor components, particularly in the case of vicinal, stepped surfaces of $1m$ symmetry, and care is needed in designing experiments that will produce readily interpretable results. Phenomenological expressions for the MSHG response from systems of $1m$ symmetry are presented, where combinations of input and output polarizations and magnetic field orientations allow the essential physics of these systems to be explored, particularly in relation to distinguishing terrace and step contributions to the magnetization from vicinal surfaces and interfaces.

1 Introduction

Low dimensional magnetic systems show interesting new physical phenomenon, some of which have already led to technological innovation (for a recent review see [1]). The most famous example is Giant Magneto-Resistance (GMR), which is a phenomenon in layered magnetic structures associated with enhanced sensitivity of the electrical resistivity to external magnetic fields [2]. GMR is already exploited in high performance read heads for high density magnetic storage hard disc systems. Linear magneto-optic techniques, particularly the magneto-optic Kerr effect (MOKE), have proved very useful in probing the magnetization of thin films and are now quite widely used [3]. More recently, nonlinear magneto-optics, and specifically magnetic second harmonic generation (MSHG) and the nonlinear Kerr effect (NOMOKE), have begun to be used to probe surface and interface magnetism (for a recent review see [4]). In contrast to linear magneto-optics, nonlinear magneto-optic probes are able to distinguish surface and interface magnetization from bulk magnetization in thin films of centrosymmetric media. An elegant example is provided by the growth of Co films on Cu(001), where oscillations with 1 ML period in the MSHG signal was observed, *in situ*, providing strong evidence of periodic oscillations in the *surface* magnetization only [5]. Some evidence of an increase in the magnetic moment of Co atoms at step sites was inferred from the MSHG data, in agreement with *ab initio* calculations which indicate that step atoms may have a larger magnetic moment than terrace atoms [6]. Recently, substrates with regular arrays of steps have been used to induce new magnetic anisotropies and spin reorientations as a function of coverage and temperature [1]. However, the role of lateral confinement in the presence of steps needs to be distinguished from the role of the steps themselves in contributing to the magnetic anisotropy. For example, studies of Fe deposition on Pd(001) infer a step-induced magnetic moment in the Pd atoms at

* Corresponding author: e-mail: jmcgilp@tcd.ie, Phone: +353 160 817 33, Fax: +353 167 711 59

the step edges [7]. In this paper it is shown that, for some experimental geometries, it is possible to use MSHG to distinguish terrace and step contributions to the magnetic anisotropy of surfaces and interfaces.

2 Phenomenology

With centrosymmetric magnetic material, the presence of a magnetization, M, will not effect the inversion symmetry (M is an axial vector), but will introduce extra surface tensor elements, some of which are odd in the magnetization and change sign when the direction of M is reversed [4, 8]. For centrosymmetric material, the magnetic second harmonic polarization becomes

$$P_i(2\omega,\pm M) = \varepsilon_0 \left\{ \left(\chi_{ijk}^{s,+} \pm \chi_{ijk}^{s,-} \right) E_j(\omega) E_k(\omega) + \text{higher order terms} \right\} \tag{1}$$

where $\chi_{ijk}^{s,\pm}(-M) = \pm\chi_{ijk}^{s,\pm}(M)$ is the second-order susceptibility tensor component reflecting the structure and symmetry properties of the magnetic surface or interface: $\chi_{ijk}^{s,-}(M)$ is an odd component, changing sign when the magnetic field is reversed.

Tables of even and odd tensor components for the (001), (110) and (111) surfaces of fcc crystals [8], and for isotropic surfaces under various excitation geometries [9], have been published previously. Here the analysis is extended to stepped surfaces and interfaces of overall $1m$ symmetry, comprising terraces of $mm2$ symmetry and steps of $1m$ symmetry. There will be a region in the vicinity of the step where atoms experience a potential of $1m$ symmetry and, away from this region, atoms will experience the $mm2$ potential of the terrace [10]. The extent of this step region is expected to be dependent on both the material and the physical property being probed. These symmetries are important because they describe vicinal W(110) and Mo(110) surfaces possessing ordered arrays of monatomic steps. The growth of Fe on vicinal W(110) has shown very interesting magnetic behaviour, which changes dramatically with Fe coverage [11, 12], making this a useful model system against which to test the phenomenology of the nonlinear magneto-optic response from stepped surfaces and interfaces.

The xz-plane is taken as the single mirror plane of such a surface or interface, with step edges running in the y-direction. The magnetization vector of the material in the presence of an applied magnetic field can be expressed, equivalently, in terms circular electric currents and the application of symmetry elements to these currents allows odd and even components to be identified. For example, if the magnetization vector lies along the z-axis, then this is equivalent to a circular current in the xy-plane. A mirror plane operation in either the xz- or yz-plane will reverse the handedness of the circular current and thus the direction of the magnetization vector. Table 1, below, lists the independent nonvanishing MSHG components for $mm2$ symmetry for magnetization in the x- and z-directions. This table applies to the terraces of a vicinal body-centred cubic (110) surface, with the magnetization direction either in the surface plane perpendicular to the step edges, or normal to the surface. If the xz-plane is the plane of incidence of the optical probe, then this corresponds to longitudinal and polar geometries, respectively.

Table 1 Independent nonvanishing MSHG components of $\chi_{ijk}^{s,\pm}$ for $mm2$ symmetry, with surface magnetization in the x- and z-directions.

Direction of M	Even parity	Odd parity
x	$xzx=xxz, yzy=yyz,$ zxx, zyy, zzz	$xyx=xxy, yxx, yyy,$ $yzz, zyz=zzy$
z	$xzx=xxz, yzy=yyz,$ zxx, zyy, zzz	$xzy=xyz, yzx=yxz,$ $zxy=zyx$

In Table 2 the results are extended to surfaces of $1m$ symmetry, where the same set of tensor components are associated with magnetization in the x- and z-directions.

Table 2   Independent nonvanishing components of $\chi_{ijk}^{s,\pm}$ for $1m$ symmetry, with surface magnetization in the x- and z-directions.

Direction of M	Even parity	Odd parity
x, z	xxx, xyy, xzz, $xzx=xxz$, $yxy=yyx$, $yzy=yyz$, zxx, zyy, zzz, $zxz=zzx$	$xyx=xxy$, $xzy=xyz$, yxx, yyy, yzz, $yzx=yxz$, $zxy=zyx$, $zyz=zzy$

3   Discussion

The results in Tables 1 and 2 show that magnetization effectively lowers the symmetry of the system, increasing both the number of nonvanishing tensor components and the difficulty of interpreting data in the absence of full theoretical calculations, which remain a daunting task. However, certain geometries do simplify the interpretation of experimental results. The use of normal incidence eliminates all z-dependent tensor components: for normally-incident radiation linearly polarized at an angle, φ, to the x-axis, the MSHG intensity polarized along the x- and y-axes is given by

$$I_x^{2\omega}(\varphi) \sim \left| \chi_{xxx}^{s,\pm} \cos^2 \varphi + \chi_{xyy}^{s,\pm} \sin^2 \varphi + \chi_{xyx}^{s,\pm} \sin 2\varphi \right|^2, \tag{2}$$

$$I_y^{2\omega}(\varphi) \sim \left| \chi_{yxx}^{s,\pm} \cos^2 \varphi + \chi_{yyy}^{s,\pm} \sin^2 \varphi + \chi_{yxy}^{s,\pm} \sin 2\varphi \right|^2. \tag{3}$$

Table 3 shows the in-surface-plane components for the terraces and steps using this geometry. It can be seen that $1m$ step and $mm2$ terrace (and also higher symmetry terrace) contributions can, in principle, be distinguished by using the polar and longitudinal response. The polar MSHG response originates only at the steps and may be identified by reversing the magnetic field direction. The experimental geometry of normal incidence combined with a polar field will typically require either a split pole piece or a small optical access hole. The longitudinal response originates at both the steps and terraces, with the same components being excited. Quantitative separation of the step and terrace contribution in the most general case will thus require determining the relative phase of these various contributions, which can be accomplished by interfering the sample signal with a reference signal [13].

Table 3   Independent MSHG components of $\chi_{ijk}^{s,\pm}$ for $mm2$ and $1m$ symmetry, with surface magnetization in the x- and z-directions, for SH light polarized in the x- and y-directions in normal incidence experiments.

Symmetry	Direction of M	Even parity		Odd parity	
		I_x	I_y	I_x	I_y
$mm2$	x	–	–	$xyx=xxy$	yxx, yyy
$mm2$	z	–	–	–	–
$1m$	x, z	xxx, xyy	$yxy=yyx$	$xyx=xxy$	yxx, yyy

MSHG provides a simple and direct way of identifying step magnetization and distinguishing it from terrace magnetization, because the different symmetry of the two types of structure is reflected in the components of the third rank nonlinear magneto-optic tensor. In contrast, the linear magneto-optic response is governed by a second rank tensor with fewer independent components and, for example, does not distinguish between $mm2$ and $1m$ surface symmetries.

Turning briefly to off-normal geometries, many z-dependent components may be accessed, as shown in Table 2. Separating the step contribution from the terrace contribution becomes more difficult, but is explored here because in-surface-plane components may be small (previous work indicates, however, that this may not be a problem for Fe surfaces [9]). A simple off-normal geometry that can distinguish step and terrace contributions exploits p-input and s-polarized second harmonic output, with longitudinal magnetization. For an xz optical plane of incidence, only odd terms are excited, $\chi_{yxx}^{s,-}$ and $\chi_{yzz}^{s,-}$ being common to both symmetries, and $\chi_{yzx}^{s,-} = \chi_{yxz}^{s,-}$ originating only at the steps. However, rotating both the sample and the magnetic field direction by $90°$ to produce a yz plane of incidence with longitudinal magnetization, results in no signal from the terraces, and $\chi_{xyy}^{s,+}$, $\chi_{xzz}^{s,+}$ and $\chi_{xzy}^{s,-} = \chi_{xyz}^{s,-}$ contributing from the steps. Reversing the magnetic field direction then allows a step contribution to the magnetization to be identified. MSHG measurements of the W(110)/Fe model system using these experimental geometries are underway [14].

4 Conclusion

It has been shown that MSHG provides a simple and direct way of identifying step magnetization and distinguishing it from terrace magnetization of $mm2$ (and higher) symmetry, because the different symmetry of the two types of structure is reflected in the components of the third rank nonlinear magneto-optic tensor. Substrates with regular arrays of steps are currently being used to induce new magnetic anisotropies and spin reorientations [1] and MSHG offers the interesting prospect of clarifying the roles of terrace and step crystallographic structure and spin in determining the magnetic behaviour of these new materials.

Acknowledgement This work was supported by Enterprise Ireland Grant No. SC/2001/109/.

References

[1] S. D. Bader, Surf. Sci. **500**, 172 (2002).
[2] M. N. Baibich, J. M. Broto, A. Fert, F. Nguyen Van Dau, F. Petroff, P. Etienne, G. Creuzet, A. Friederich, and J. Chazelas, Phys. Rev. Lett. **61**, 2472 (1988).
[3] S. D. Bader, J. Magn. Magn. Mater. **100**, 440 (1991).
[4] T. Rasing, J. Magn. Magn. Mater. **175**, 35 (1997).
[5] Q. Y. Jin, H. Regensburger, R. Vollmer, and J. Kirschner, Phys. Rev. Lett. **80**, 4056 (1998).
[6] A. V. Smirnov and A. M. Bratkovsky, Phys. Rev. B **54**, R17371 (1996).
[7] H. J. Choi, R. K. Kawakami, E. J. Escorcia-Aparicio, Z. Q. Qiu, J. Pearson, J. S. Jiang, L. Dongqi, and S. D. Bader, Phys. Rev. Lett. **82**, 1947 (1999).
[8] P. Ru-Pin, H. D. Wei, and Y. R. Shen, Phys. Rev. B **39**, 1229 (1989).
[9] B. Koopmans, M. G. Koerkamp, T. Rasing, and H. van den Berg, Phys. Rev. Lett. **74**, 3692 (1995).
[10] J. R. Power, J. D. Omahony, S. Chandola, and J. F. McGilp, Phys. Rev. Lett. **75**, 1138 (1995).
[11] J. Hauschild, H. J. Elmers, and U. Gradmann, Phys. Rev. B **57**, R677 (1998).
[12] H. J. Elmers, J. Hauschild, and U. Gradmann, Phys. Rev. B **59**, 3688 (1999).
[13] R. K. Chang, J. Ducuing, and N. Bloembergen, Phys. Rev. Lett. **15**, 6 (1965).
[14] L. Carroll and J. F. McGilp, *to be published*.

phys. stat. sol. (c) **0**, No. 8, 3050–3054 (2003) / **DOI** 10.1002/pssc.200303832

Anisotropic second harmonic generation from Si(111)-4x1-In

S. Chandola, L. Carroll, and **J. F. McGilp**[*]

Department of Physics, Trinity College, Dublin 2, Ireland

Received 30 May 2003, revised 4 August 2003, accepted 11 August 2003
Published online 10 November 2003

PACS 42.65.Ky, 78.67.Lt, 78.68.+m

Previous optical second harmonic generation (SHG) studies of single domain Si(111)-4x1-In revealed a large in-surface-plane optical anisotropy, with an enhanced hyperpolarisability aligned along the In chains. This single wavelength study indicated that SHG might provide a non-contact, non-destructive method of characterising low dimensional metallic behaviour. However, it was not clear whether the observed enhancement is due to the low dimensional metallicity, or to the presence of surface and interface states at either the fundamental or SH energy. Spectroscopic SHG studies from single domain Si(111)-4x1-In are reported. A very large anisotropy in the SH response is observed using excitation energies in the 1.36-1.65 eV region, with the dominant hyperpolarizability orthogonal to the chain direction. The results help to clarify the approach likely to be necessary to use SHG as a probe of low dimensional metallic behaviour on planar surfaces.

1 Introduction

Self-assembly during growth at Si surfaces possessing ordered arrays of monatomic and diatomic steps appears particularly promising in the drive to produce low dimensional nanoscale structures for information technology applications. The directional nature of the Si bonding favours the formation of highly regular arrays of steps with very low kink densities on vicinal Si single crystal surfaces, with periodicities which can range from 2 to 80 nm [1]. These stepped surfaces can serve as model templates for the growth, by self-assembly, of low dimensional, aligned, nanoscale structures of high perfection. While simple nanostructures grown by self-assembly and characterised under ultra-high vacuum (UHV) conditions are of considerable interest, structures stable to the ambient are the clear technological goal. This will often involve the growth of a layer or cap that protects the structure of interest from degradation under ambient conditions. Conventional surface characterisation cannot probe these buried structures and it has been argued recently that surface and interface optical techniques are particularly suited to the characterisation of capped material grown on stepped structures [2].

Much remains to be understood, however, about the origins of the linear and nonlinear optical response from such low dimensional, aligned nanostructures. Si(111)-4x1-In is a particularly interesting model system to study, as it currently appears to be the best candidate for a *quasi*-one-dimensional (*quasi*-1D) metallic system on a 2D solid surface. The In atoms form two parallel zig-zag chains which are separated by a zig-zag Si chain [3] and there is strong evidence from both experiment [4, 5], and theory [6-8], that this structure forms a *quasi*-1D metallic phase. The system also forms a low temperature phase where the period along the chain is doubled, possibly indicating a Peierls-like transition [9].

Only a few surface optics studies have been published so far. Reflection anisotropy spectroscopy (RAS) probes the in-surface-plane linear optical anisotropy, and studies show the dominant polarizability to be across the chains at ~2 eV [10], and along the chains below ~1 eV [11, 12]. It is possible that this

[*] Corresponding author: e-mail: jmcgilp@tcd.ie, Phone: +353 160 817 33, Fax: +353 167 711 59

lower energy response is related to the low dimensional metallicity of the chains. Optical second harmonic generation (SHG) studies, using 1.17 eV excitation, reveals a large in-surface-plane nonlinear optical anisotropy, with an enhanced hyperpolarisability aligned along the In chains [13, 14]. This single wavelength study indicated that SHG might provide a non-contact, non-destructive method of characterising low dimensional metallic behaviour. However, it was not clear whether the observed enhancement is due to the low dimensional metallicity, or to the presence of localised interface states at either the fundamental or SH energy. Spectroscopic SHG studies from single domain Si(111)-4x1-In are reported which indicate the approach likely to be necessary to use SHG as a probe of low dimensional metallic behaviour.

2 Experiment

All measurements were carried out in UHV with a base pressure of $\sim 1 \times 10^{-10}$ mbar. Vicinal n-type Si(111) samples, cut 3° towards the $[\bar{1}\,\bar{1}\,2]$, x-direction, produced a regular array of ordered single-height steps after thermal treatment, as confirmed by low energy electron diffraction (LEED). Approximately 1 monolayer (ML) of In was deposited onto the clean substrate held at 400°C and, on cooling to room temperature, a well-ordered, single domain, (4x1)-In overlayer was formed. Visual inspection of the LEED pattern showed no traces of other domains or the original Si(111)-7x7 structure. The SHG experiments used $\sim$ 130 fs pulses from a diode-pumped, mode-locked Ti:sapphire laser, at a repetition rate of 76 MHz and with an average laser power in the range 1.0-1.4 W, depending on wavelength. The beam was focused to a 200μm spot on the sample at an incident angle of 67.5° to the sample normal. The variation of s-polarized SH intensity with input polarization angle, α, was measured with the plane of incidence orthogonal (x-axis), and parallel (y-axis), to the direction of the step edges. A SH signal obtained by reflection from z-cut quartz wedge was used to normalise the SH response at the different wavelengths.

The Si(111)-4x1-In structure is known to be sensitive to the details of surface preparation, particularly in relation to the low temperature phase [9, 12, 15]. Once prepared, however, the room temperature phase remains stable for many days under UHV conditions [11, 12]. The SH response of the initially clean Si(111)-4x1-In structure, using 800 nm excitation, showed no change over a ten day period in the UHV chamber. The SH response, using 750, 800 and 910 nm excitation wavelengths, was recorded on the same surface over a period of days, during which time both the SH response and the 4x1 LEED pattern remained unchanged. This approach has the advantage that the response at different wavelengths from the same surface is being compared, but has the disadvantage that there will be hydrogen adsorbed on the Si(111)-4x1-In structure. Flashing the sample at the end of the measurements produced the characteristic double pressure pulse associated with hydrogen adsorption on Si(111).

For excitation at frequency, ω, in the xz or yz plane of incidence (where z is normal to the surface), the variation of s-polarized SH intensity, with input polarization angle, α, is of the form

$$I_{2\omega}(\alpha) = C\left| \{F\cos^2\alpha + G\sin^2\alpha + H\sin 2\alpha\}I_\omega \right|^2 \tag{1}$$

where α is measured from the plane of incidence, and C is a collection of constants [16]. For the yz plane of incidence, tensor components of the general form χ_{yjk}^s are probed (where χ_{ijk}^s refers to the i-coordinate SH field for j, k excitation), because the s-polarized output has an SH field vector perpendicular to the plane of incidence. Single domain Si(111)-4x1-In has $1m$ symmetry [3], with an xz mirror plane orthogonal to the step edges [13]. F then depends on χ_{yxx}^s and χ_{yzz}^s, G on χ_{yyy}^s, and H on χ_{xzx}^s, as well as Fresnel and local-field factors. For the xz plane of incidence, $F = G = 0$ and H depends on χ_{yxy}^s and χ_{yzy}^s, giving a $\sin^2 2\alpha$ variation of intensity with polarization angle, α.

© 2003 WILEY-VCH Verlag GmbH & Co. KGaA, Weinheim

3   Results and discussion

Figure 1 shows the variation with input polarisation angle, α, of the s-polarised SH intensity from Si(111)-4x1-In, using 750, 800 and 910 nm excitation. The SH s-polarization along the chains (xz optical plane of incidence) is shown in Fig. 1(a), and the SH s-polarization across the chains is shown in Fig. 1(b). The scale in Fig. 1(a) has been expanded by a factor of 10 to allow the degree of anisotropy of the nonlinear response to be seen. Figure 1 may also be compared with results obtained using 1064nm excitation, published previously [13]. The very weak xz optical plane results for 800 nm can be fitted to a $\sin^2 2\alpha$ dependence, as discussed above, while the results for 750 nm and 910 nm are too noisy.

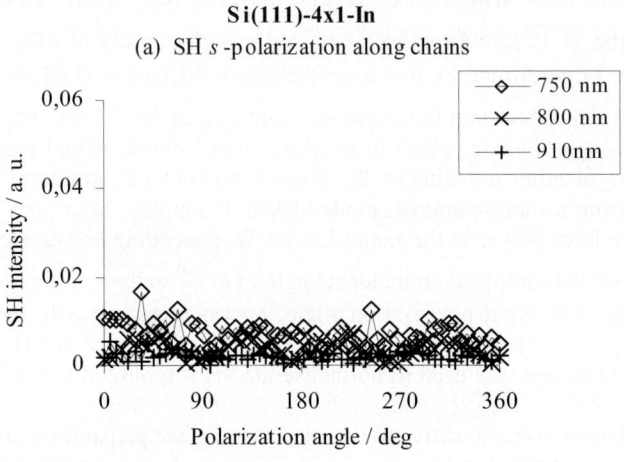

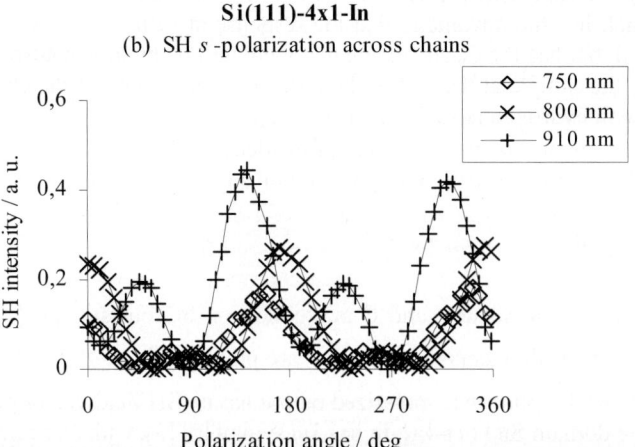

Fig. 1   Variation with input polarisation angle, α, of the s-polarised SH intensity, using 750, 800 and 910 nm excitation (a) SH s-polarization along chains (scale expanded by x10) (b) SH s-polarization across chains. The data points are connected by solid lines to provide a guide to the eye.

The anisotropy of the in-surface-plane intensity in Fig. 1 is ~10 for 750 nm (limited by noise), ~35 for 800 nm and a very large ~75 for 910nm. However, the dominant hyperpolarizability is *across* the chains, in contrast to the results for 1064nm excitation, where the dominant hyperpolarizability is *along* the chains [13, 14]. The p-polarized SH response was measured and also showed a dominant hyperpolarizability *across* the chains at all three wavelengths. The interesting change of peak shape and position in Fig. 1(b) arises from the variation of Fresnel factors and tensor components with wavelength, as various optical transitions are accessed in this spectral region.

The excitation wavelengths correspond to energies of 1.36, 1.55 and 1.65 eV, and SH energies of 2.72, 3.10 and 3.31 eV, all these energies, with the exception of the last, lying below the direct optical gap of Si. Comparison with recent RAS results allows progress to be made in interpreting these results and emphasises the complementary nature of linear and nonlinear optical spectroscopies [2]. Figure 2 shows the excitation and SH energies superimposed on the RAS signal from Si(111)-4x1-In [11]. The dominant in-surface-plane linear polarizability is orthogonal to the chains in the energy regions accessed using 750, 800 and 910 nm excitation. The previous results using 1064 nm have an excitation energy of 1.17 eV and a SH energy of 2.33 eV, also marked in Fig. 2. The excitation energy is in a region where the dominant in-surface-plane linear polarizability is along the chains, while the SH energy is again in the across-chain region.

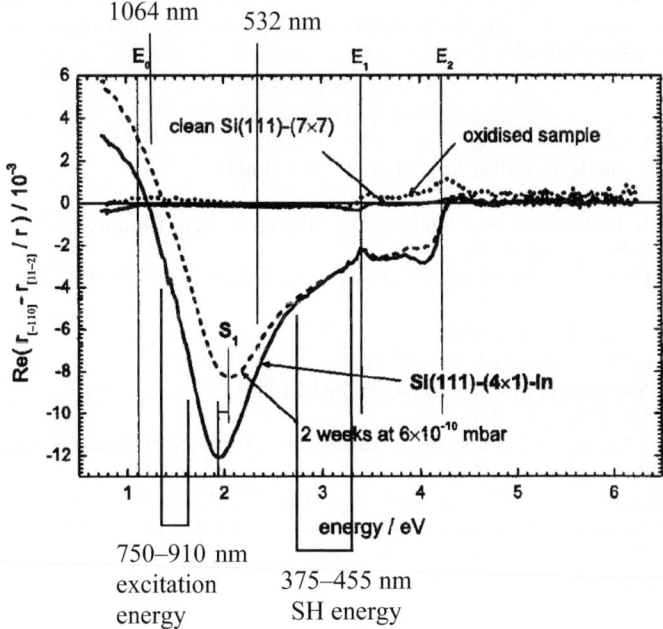

Fig. 2 Excitation and SH energies superimposed on the RAS signal from Si(111)-4x1-In (after [11]).

The results show that the presence of surface and interface electronic transitions in the spectral region being probed can overwhelm any general increase in the SH intensity arising from low dimensional metallicity, and that surface and interface electronic transitions may well be the origin of the 1064 nm results also. These conclusions are independent of the presence of adsorbed hydrogen. The results do show that the nonlinear anisotropy associated with low dimensional structures, whether metallic or not, can be extremely large (~6000 bigger than the RAS anisotropy) and is clearly worth exploring further. However, the most promising approach for yielding unambiguous, easily interpreted results concerning metallicity appears to be to use significantly lower excitation energies, in the 0.5 eV region and below, where stronger optical coupling of light polarized in the chain direction would be expected. The recent linear optics RAS results from Si(111)-4x1-In [11, 12], and anisotropic reflection difference results at 0.6 eV from Pb nanowires on Si(335) [17], provide support for this view. It should be noted that RAS and SHG provide complementary information, and both techniques are very promising as non-contact, non-destructive methods for characterising low dimensional structures [2].

4　Conclusion

Spectroscopic SHG studies from single domain Si(111)-4x1-In using excitation energies in the 1.35-1.65 eV region (SH energies 2.7-3.3 eV) reveal an anisotropic response which can be extremely large, but the dominant hyperpolarizability in this spectral region is found to be orthogonal to the chain direction. Comparison with recent RAS results indicates that surface and interface state optical transitions may dominate the SH response of Si(111)-4x1-In in this energy region, making the detection of low dimensional metallicity *via* an enhanced hyperpolarizability along the chains difficult. This is likely to be a general problem and it is suggested that the anisotropic in-surface-plane optical response at lower excitation energies, ~0.5 eV and below, may provide an unambiguous signature of low dimensional metallicity on planar surfaces.

Acknowledgements　　This work was supported by Enterprise Ireland Grant No. SC/2001/109/. The Authors thank J. D. O'Mahony for useful discussions and a critical reading of the manuscript. Instrumental contributions by F. Pedreschi and J. D. O'Mahony are gratefully acknowledged.

References

[1] F. J. Himpsel, J. E. Ortega, G. J. Mankey, and R. F. Willis, Adv. Phys. **47**, 511 (1998).
[2] J. F. McGilp, phys. stat. sol. (a) **188**, 1361 (2001).
[3] O. Bunk, G. Falkenberg, J. H. Zeysing, L. Lottermoser, R. L. Johnson, M. Nielsen, F. Berg-Rasmussen, J. Baker, and R. Feidenhans'l, Phys. Rev. B **59**, 12228 (1999).
[4] T. Abukawa, M. Sasaki, F. Hisamatsu, T. Goto, T. Kinoshita, A. Kakizaki, and S. Kono, Surf. Sci. **325**, 33 (1995).
[5] I. G. Hill and A. B. McLean, Phys. Rev. Lett. **82**, 2155 (1999).
[6] J. Nakamura, S. Watanabe, and M. Aono, Phys. Rev. B **63**, 193307/1 (2001).
[7] J.-H. Cho, D.-H. Oh, K. S. Kim, and L. Kleinman, Phys. Rev. B **64**, 235302/1 (2001).
[8] R. H. Miwa and G. P. Srivastava, Surf. Sci. **473**, 123 (2001).
[9] H. W. Yeom, S. Takeda, E. Rotenberg, I. Matsuda, K. Horikoshi, J. Schaefer, C. M. Lee, S. D. Kevan, T. Ohta, T. Nagao, and S. Hasegawa, Phys. Rev. Lett. **82**, 4898 (1999).
[10] F. Pedreschi, J. D. O'Mahony, P. Weightman, and J. R. Power, Appl. Phys. Lett. **73**, 2152 (1998).
[11] K. Fleischer, S. Chandola, N. Esser, W. Richter, and J. F. McGilp, phys. stat. sol. (a) **188**, 1411 (2001).
[12] K. Fleischer, S. Chandola, N. Esser, W. Richter, and J. F. McGilp, Phys. Rev. B, in press.
[13] S. Chandola and J. F. McGilp, phys. stat. sol. (a) **175**, 189 (1999).
[14] S. Chandola and J. F. McGilp, phys. stat. sol. (a) **184**, 111 (2001).
[15] C. Kumpf, O. Bunk, J. H. Zeysing, Y. Su, M. Nielsen, R. L. Johnson, R. Feidenhans'l, and K. Bechgaard, Phys. Rev. Lett. **85**, 4916 (2000).
[16] J. R. Power, J. D. O'Mahony, S. Chandola, and J. F. McGilp, Phys. Rev. Lett. **75**, 1138 (1995).
[17] M. Jalochowski, M. Strozak, and R. Zdyb, Appl. Surf. Sci., in press.

phys. stat. sol. (c) **0**, No. 8, 3055–3059 (2003) / **DOI** 10.1002/pssc.200303841

Reflectance-difference and second-harmonic generation: a meeting of two surface spectroscopies

Jinhee Kwon[1] and **M. C. Downer**[*, 1]

[1] Texas Materials Institute, University of Texas at Austin, Texas 78712, USA

Received 30 May, 2003, revised 4 August 2003, accepted 11 August 2003
Published online 10 November 2003

PACS 42.65.–Ky, 78.40.Fy, 78.68.+m

We quantitatively compare reflectance-difference spectra (RDS) and second harmonic generation (SHG) spectra of native-oxidized vicinal $Si(001)/SiO_2$ interfaces with off-cut angles $\alpha = 0°$, $4°$, $6°$, $8°$, and $10°$ from (001) toward [110] direction. The RD spectrum and the spectrum of the first-order Fourier component of the SHG azimuthal anisotropy reveal similar derivative-like E_1 resonance lineshapes, suggesting a common origin in the step-edges and the possibility of formulating a common microscopic model of the RD and SHG responses.

1   Introduction

Reflectance-difference spectroscopy (RDS) and second-harmonic generation (SHG) are currently the two dominant optical methods for non-invasive analysis of surfaces and interfaces. RDS exploits the isotropy of the bulk contribution to near-normal incidence reflectance of cubic materials, so that observed anisotropies can be attributed to lower symmetry surfaces or interfaces [1]. SHG exploits the dipole-forbidden character of the second-order bulk nonlinear susceptibility $\chi^{(2)}$ of centrosymmetric substrates, so that observed SH signals can be attributed to centrosymmetry breaking surfaces or interfaces [2]. The native-oxidized vicinal Si(001) interfaces studied in this work exemplify a large class of surfaces and interfaces that fall within the range of application of both methods. Although RDS and SHG have been used previously as complementary probes in such cases [3], SHG has usually been restricted to a single wavelength and to phenomenological analysis.

Recently microscopic calculations of RDS and SHG surface spectra from a common theoretical foundation have begun to be formulated [4,5]. To test such calculations, and to forge fundamental connections between these two spectroscopies, SHG and RD must both be acquired *spectroscopically* from common samples. Here we present a systematic set of such data for vicinal $Si(001)/SiO_2$. Preliminary phenomenological analysis reveals similar derivative-like E_1 resonances in the RD spectra and the spectra of the first-order Fourier component of the SHG azimuthal anisotropy, and motivates the search for a common microscopic origin in the step edges.

2   Experimental procedure

The Cz-grown, lightly p-doped (1 Ω-cm) Si(001) wafers were cut vicinally at $\alpha = 0°$, $4°$, $6°$, $8°$, and $10°$ $\pm 0.5°$ from (001) toward the [110] direction, producing a regular step structure with step density increasing with α (Fig. 1a, lower inset) [6]. All samples had native oxides about 15 Å thick.

[*] Corresponding author: e-mail: downer@physics.utexas.edu, Phone: +01 512 471 6054, Fax: +01 512 471 9637

RDS data were acquired over a spectral range $1.5 < h\nu < 4.6\ eV$. White light from a Xe arc lamp, polarized linearly at $45°$ from the step edges, reflected from the sample at near-normal incidence (Fig. 1a, upper left inset). This reflected light was phase-modulated photoelastically at $\omega_{mod} = 50\ KHz$, polarization-analyzed, then spectrally analyzed by a monochromator. The ω_{mod} and $2\omega_{mod}$ components of the detected intensity are related to the imaginary and real part, respectively, of the complex anisotropy $\Delta r/r = 2(r_\| - r_\perp)/(r_\| + r_\perp)$, where $r_\perp$ and $r_\|$ are the complex reflectance for light polarized along the $\perp = [110]$ and $\| = [1\text{-}10]$ principal axes, respectively [7].

SH signals were generated by unamplified ~150 fs Ti:sapphire laser pulses, focused unto the sample at incidence angle $\theta = 42°$ with s or p polarization. Reflected p-polarized SHG intensity was measured as a function of the sample azimuthal angle ϕ between the plane of incidence and the [110] direction at each fundamental wavelength. Azimuthal scans were repeated in 10 nm increments as the laser tuned from 900 to 720 nm ($2.8 < 2h\nu < 3.5\ eV$). All SHG data were normalized to reference SHG from a z-cut quartz wedge to correct for variations in intensity and pulse structure of incident laser during sample rotation or tuning. Accumulation of laser-induced charge on the oxide surface during data acquisition can introduce spurious SHG signals [8]. Such time-dependent artifacts were eliminated by performing the measurements in UHV, to suppress the catalytic role of ambient oxygen in the charging process [8], and by displacing the focused laser spot slightly from the sample rotation axis, so that its movement prevented charge build-up over time.

3   Results, analysis and discussion

Figure 1 shows the (a) real and (b) imaginary components of a portion of the RD spectra within a photon energy range $2.9\ eV < h\nu < 3.6\ eV$ near the E_1 critical point that corresponds to the range of $2h\nu$ in the SHG measurements. All RD spectra were featureless at $h\nu < 3.0$ eV, and an α-dependent E_2 feature (not shown) was observed at $h\nu \sim 4.2$ eV.

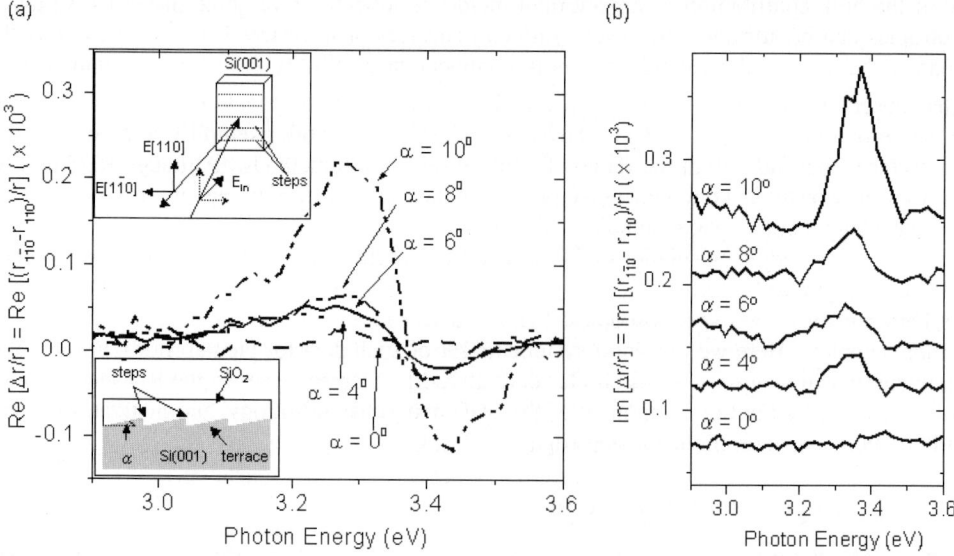

Fig. 1   Normalized (a) real and (b) imaginary parts of the RD spectra of native-oxidized vicinal Si(001) at five off-cut angles for $2.9\ eV < h\nu < 3.6\ eV$. The $Re(\Delta r/r)$ line shape at $\alpha = 4°$ is consistent with data reported by Yasuda et al. [9]. Upper inset in (a): relative orientations of incident polarization, step edges and [110] and $[1\bar{1}0]$ crystalline directions. Lower inset in (a): cross-section of interface in plane perpendicular to $[1\bar{1}0]$ direction. Baselines in (b) are offset for clearer presentation.

The data in Fig. 1 show clearly that the main features in $Re(\Delta r/r)$ and $Im(\Delta r/r)$ grow monotonically with α, and vanish for $\alpha = 0^\circ$, confirming that they originate from the step edges. The $Re(\Delta r/r)$ line shapes near E_1 is proportional to the derivative dR/dE of the reflection spectrum $R(E)$ of silicon with respect to the energy. This shape is consistent with a blue- (red) shift of the E_1 gap along the [110] ([1$\bar{1}$0]) direction - i.e. $\perp$ and $\parallel$ to the step edges. Likewise $Im(\Delta r/r)$ resembles the difference between two slightly shifted spectra of the reflective phase across the E_1 gap. As expected from this picture, the peak of $Im(\Delta r/r)$ coincides with the zero-crossing of $Re(\Delta r/r)$.

Linear reflectance is proportional to an effective interface susceptibility $\chi_{ij}^{(1)}$ which, as a 2nd rank tensor, allows up to 2-fold anisotropy. Thus the reflected field can be represented phenomenologically by a Fourier sum $E_R^\omega = \sum_{n=0}^{2} a_n^{(\omega)} \cos n\phi$, where, for RDS, ϕ represents the angle between the polarization and the [110] direction. Thus $\Delta r/r$ in Fig. 1 can involve $a_1^{(\omega)}$ and $a_2^{(\omega)}$ Fourier coefficients.

Figure 2 shows representative normalized SHG intensity data $I^{2\omega}(\phi, 2h\nu)$ for $\alpha = 10^\circ$. SHG data of other samples was recorded similarly.

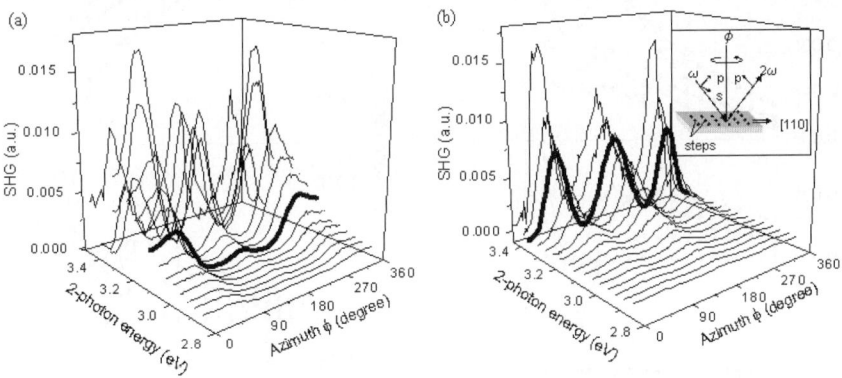

Fig. 2 Typical SHG data (thin curves) of oxidized vicinal Si(001) for (a) p-in/p-out and (b) s-in/p-out polarization configurations for the with $\alpha = 10^\circ$ sample. Thick curves show typical fits based on Eq. (2) in text. Inset of (b) shows geometry of SHG measurements.

The second harmonic polarization density can be written [2]

$$P_i^{(2\omega)} = \varepsilon_o \chi_{ijk}^{(2)} E_j^{(\omega)} E_k^{(\omega)} + \varepsilon \chi_{ijkl}^{q} E_j^{(\omega)} \nabla_k E_l^{(\omega)}, \tag{1}$$

where $\chi_{ijk}^{(2)}$ is the dominant interface dipole susceptibility caused by breaking of inversion symmetry at the Si/SiO$_2$ interface, and χ_{ijkl}^{q} is a background bulk electric quadrupole susceptibility. For the oriented Si(001) surface ($\alpha = 0^\circ$), $\chi_{ijk}^{(2)}$ is isotropic and χ_{ijkl}^{q} contains 4-fold anisotropy as well as an isotropic component with respect to ϕ rotation [2]. Step-edges at vicinal Si(001) induce additional anisotropies into $\chi_{ijk}^{(2)}$ up to the maximum 3-fold allowed for a 3rd rank susceptibility tensor. Thus for each incident polarization g = s or p, the SH field can be represented phenomenologically by the Fourier sum [9,10]

$$E_{g,p}^{(2\omega)} = \sum_{n=0}^{4} a_n^{g,p}(\omega) \cos(n\phi) \tag{2}$$

where $a_n^{g,p}(\omega)$ is a linear combination of products of $\chi_{ijk}^{(2)}$ or χ_{ijkl}^{q} tensor components and Fresnel factors. The isotropic ($n = 0$) component remains strong for vicinal samples, so intensity anisotropies appear mainly through cross terms of the form $a_n a_o$. Step-induced 1- and 3-fold anisotropic components are immediately evident in Fig. 2. The 4-fold component from χ_{ijkl}^{q}, though clearly evident for $\alpha = 0^\circ$, is masked by stronger 1- and 2-fold anisotropies for the vicinal samples.

Since SHG and RDS can share 1- and 2-fold anisotropies, underlying connections between them are most likely to be found in the $a_1(\omega)$ and $a_2(\omega)$ coefficients in Eq. (2). We therefore fitted all measured SHG intensity data $|E_{g,p}(2\omega, \phi)|^2$ using Eq. (2). In general, after arbitrarily setting the phase of one of the complex a_n coefficients to zero, this fit would involve 9 parameters. However, we found that an excellent fit to all details of the data was obtained with only the 5 parameters a_0 thru a_4 with *relative* phase at each 2ω held fixed at either 0 or π (*i.e.* sign difference only). Typical fitted intensity anisotropies are shown by thick curves in Fig. 2. Moreover, when a_2 and/or a_4 were set to zero, the quality of the fit was only slightly affected and the fitted a_0, a_1 and a_3 were not altered significantly. We concluded that the fitted a_0, a_1 and a_3 are robust features of the fit, while remaining parameters are not well determined by our set of data. Note that fixing the relative phases of a_0, a_1 and a_3 at each frequency still allows their overall phase factor $e^{i\Phi(\omega)}$ to vary with frequency, as expected near a resonance. However, since we fitted intensity anisotropy $I^{2\omega}(\phi)$ separately at each frequency, $\Phi(\omega)$ is not determined.

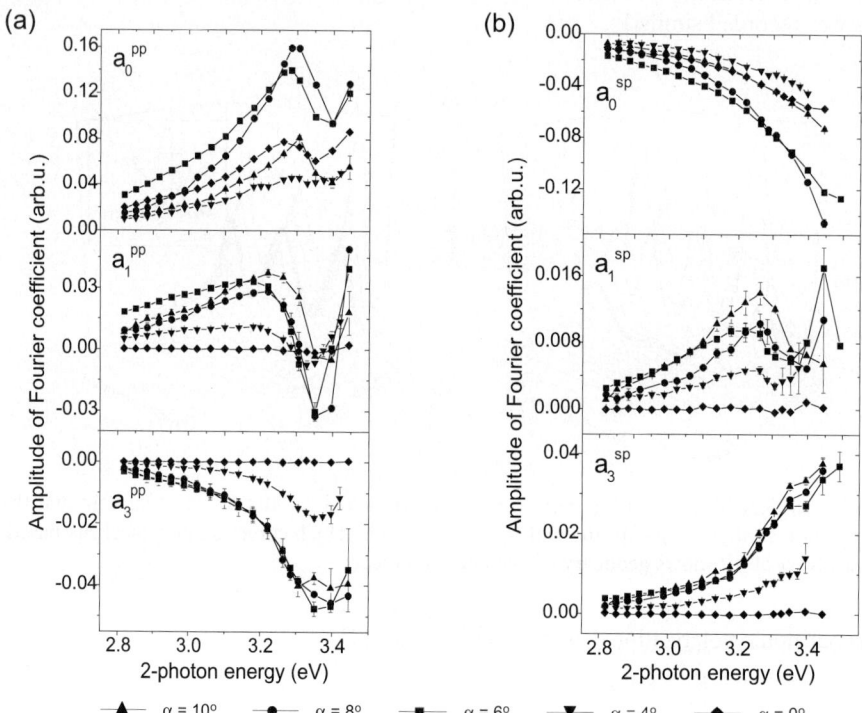

Fig. 3 Spectroscopic Fourier coefficients a_n (n = 0, 1, 3) of five different vicinal Si/SiO$_2$ interface as a result of fitting the data [p-in/p-out] (a) and [s-in/p-out] (b).

Figure 3 shows the spectroscopic Fourier coefficients $a_{0,1,3}^{g,p}$ derived from fitting of all SHG data for (a) p-in/p-out and (b) s-in/p-out configurations. The error bars indicate the range of variation that occurs in fits that include different sets of parameters. The isotropic $a_0^{p,p}$ (Fig. 3a, top), the largest parameter, shows the familiar red-shifted (~3.3 eV) E$_1$ resonance peak that has been widely documented in SHG spectroscopy of this interface [2]. For $\alpha = 0°$, $a_0^{p,p}$ is proportional to a linear combination of all three independent, non-zero components $\chi_{zzz}^{(2)}$, $\chi_{xzx}^{(2)}$ and $\chi_{zxx}^{(2)}$ of the dipole susceptibility of the Si/SiO$_2$ interface. For $\alpha > 0°$, $a_0^{p,p}$ can be attributed essentially to the same source from the terraces, accounting for the preservation of its basic spectral shape, although it is perturbed by additional contributions resulting from lowering of the interface symmetry. $a_0^{s,p}$ is smaller in amplitude than $a_0^{p,p}$, reflecting weaker s-in/p-out SHG signal levels. For $\alpha = 0°$, $a_0^{s,p}$ is proportional to $\chi_{zxx}^{(2)}$, so its rather different spectral structure can be assigned to the $\chi_{zxx}^{(2)}$ component.

The most important result in Fig. 3 is the striking resemblance in spectral shape between the step-induced $a_1^{p,p}(\omega)$ (Fig. 1a, middle panel) and the RD spectra in Fig. 1a. Both show a derivative-like line shapes which cross zero near the E_1 critical point, and increase in amplitude with the vicinal angle α. This similarity between the $Re(\Delta r/r)$ and $a_1^{p,p}(\omega)$ suggests the existence of common microscopic origins for RDS and SHG spectra.

The resemblance between $a_1^{s,p}(\omega)$ and $Re(\Delta r/r)$ is weaker -- in particular, $a_1^{s,p}(\omega)$ lacks a clear derivative-like spectral shape. Thus the strongest connections between SHG and RDS may reside in step-induced perturbations to the $\chi_{xzx}^{(2)}$ and $\chi_{zzz}^{(2)}$ tensor components. The step-induced a_3 components have no counterpart in RDS. In fact, apart from dependence on α and a resonant enhancement near the E_1 critical point, they bear no obvious resemblance to the RD spectra. Explanation of these phenomenological trends is the challenge for a common microscopic theory of the RD and SHG spectra.

4 Conclusion

We have obtained the linear RD and nonlinear SH spectra of vicinal Si/SiO_2 for five different vicinal angles. The main result is that the first-order step-induced Fourier coefficient $a_1^{p,p}(\omega)$ shows a derivative-like, α-dependent E_1 feature that closely resembles the RD spectra. The other significant step-induced coefficient $a_3(\omega)$ has no RDS counterpart, and bears little resemblance to the RD spectrum. The RDS-SHG linkage presented here is phenomenological and suggestive. Nevertheless it should motivate further microscopic calculations and experiments. Regarding calculations, we are currently analyzing the spectra presented here using a simplified bond-hyperpolarizability model [12], as a first step toward developing a common microscopic model of SHG and RDS. This result will be presented elsewhere. Regarding experiments, the spectroscopic RD/SHG techniques used in the present work will be extended to clean, reconstructed surfaces in UHV, which have well-defined structures that are more amenable than oxidized surfaces to first-principles calculations.

Acknowledgements This work was supported by the U. S. National Science Foundation (Grants DMR-0207295 and PHY-0114336) and the Robert Welch Foundation (Grant F-1038).

References

[1] D. E. Aspnes and A. A. Studna, Phys. Rev. Lett. **54**, 1956 (1985).
[2] G. Lüpke, Surf. Sci. Rep. **35**, 75 (1999) and references therein.
[3] J. F. McGilp, D. Weaire, and C. Patterson, eds., Epioptics: Linear and Nonlinear Optical Spectroscopy of Surfaces and Interfaces (Springer Verlag, Berlin, 1995).
[4] V. I. Gavrilenko et al., Phys. Rev. B **63**, 165325 (2001).
[5] W. G. Schmidt and F. Bechstedt, Phys. Rev. B **63**, 045322 (2001).
[6] O. L. Alerhand et al., Phys. Rev. Lett. 64, 2406 (1990).
[7] D. E. Aspnes et al., J. Vac. Soc. Am. A **6**, 1327 (1988).
[8] H. M. Van Driel et al., Phys. Rev. B **59**, 2164 (1999).
[9] T. Yasuda, D. E. Aspnes et al., J. Vac. Sci. Technol. A **12**, 1152 (1994).
[10] G. Lüpke, D. J. Bottomley, and H. M. van Driel, Phys. Rev. B **47**, 10839 (1993).
[11] G. Lüpke, D. J. Bottomley, and H. M. van Driel. J. Opt. Soc. Am. B **11**, 33 (1994).
[12] G. D. Powell, J.-F. Wang, and D. E. Aspnes, Phys. Rev. B **65**, 205320 (2002).

phys. stat. sol. (c) **0**, No. 8, 3060–3064 (2003) / **DOI** 10.1002/pssc.200303845

Bond hyperpolarizabilities – SHG simplified?

J. F. McGilp[*] and **L. Carroll**

Department of Physics, Trinity College, Dublin 2, Ireland

Received 30 May 2003, revised 4 August 2003, accepted 11 August 2003
Published online 10 November 2003

PACS 42.65.An, 42.65.Ky, 78.68.+m

It has been suggested that bond-hyperpolarizability models of optical second harmonic generation (SHG) may be simplified by assuming that the SH radiation originates from the anharmonic motion of bond charges strictly along bond directions [G. D. Powell, J. F. Wang, and D. E. Aspnes, Phys. Rev. B **65**, 205320/1 (2002)]. This assumption allows more physical insight by connecting the SH response to the response of the bonds in a simple and direct manner. However, previous theoretical calculations of bond hyperpolarizabilities indicate that transverse components may be large [2]. The simplified model is modified here to account directly for the variation of the number of step bonds with vicinal angle. The results are promising but further work is required before general conclusions can be drawn about the role of the transverse components.

1 Introduction

A current limitation of optical second harmonic generation (SHG) as a probe of surface and interface structure is the difficulty of calculating the nonlinear optical response from first principles and thus interpreting the results at a fundamental level. Interpretation is rarely taken beyond a discussion of the contribution of the various second order susceptibility tensor components, determined from the standard phenomenological analysis. Powell et al. [1], and others [3, 4], have suggested assuming interfacial bonds have one (complex) hyperpolarizability coefficient directed along the bond axis. This assumption of a single axial component simplifies the problem and allows interpretation of rotational SHG plots in terms of the hyperpolarizability contribution from the different bonds at the interface. The approach is particularly promising for lower symmetry interfaces, such as those associated with stepped surfaces, where the retention of, at the most, only a single mirror plane symmetry, leads to so many non-zero nonlinear susceptibility components that standard phenomenological analysis is very difficult [5].

Powell et al. [1] tested their model on vicinal Si(111)-SiO$_2$ and Si(001)-SiO$_2$ interfaces, using the experimental data of Lupke et al. [5]. The results are reasonably consistent for these $1m$ symmetry interfaces for the different polarization combinations, but the vicinal angle of 1.12° for the nominally 5°-offcut Si(111)-SiO$_2$, obtained by fitting to the experimental data, is clearly incorrect, because the data of Lupke et al. [5] agree very well with the earlier results of van Hasselt et al. [6], also for 5°-offcut Si(111)-SiO$_2$. The agreement with experiment, apart from the vicinal angle, is surprising for three reasons. Firstly, theoretical calculations of bond hyperpolarizabilities for Si-X bonds, X = Ga, As, at Si(111) surfaces indicate that transverse components may be comparable in size to the axial components [2], and it appears likely that the Si-O bond has appreciable transverse components. Secondly, it is well established that the bulk quadrupolar contribution, neglected in this model, is comparable in size to the dipolar interface contribution for Si-SiO$_2$ interfaces [5]. Thirdly, the model does not distinguish between step bonds and terrace back bonds with the same orientation (Fig. 1), the fit producing a combined value.

---

[*] Corresponding author: e-mail: jmcgilp@tcd.ie, Phone: +353 160 817 33, Fax: +353 167 711 59

However, the effect of varying the tilt angle, although altering the projection of these bonds in the same way, alters the number of these bonds in a different way. For small tilt angle, α, the number of *step* bonds varies as $\tan\alpha$, while the number of terrace *down* (*back*) bonds with the same orientation varies as $(1 - \tan\alpha)$. In this paper the third point is addressed to show that it is possible to obtain a reasonable fit using the experimental tilt angle of 5°. Although promising, this result should be regarded as tentative, as the bulk quadrupolar contribution remains to be included for this system when 765 nm excitation is used. It is clear that more evaluation is required before general conclusions can be drawn about the usefulness of the interesting approach of Powell et al. [1].

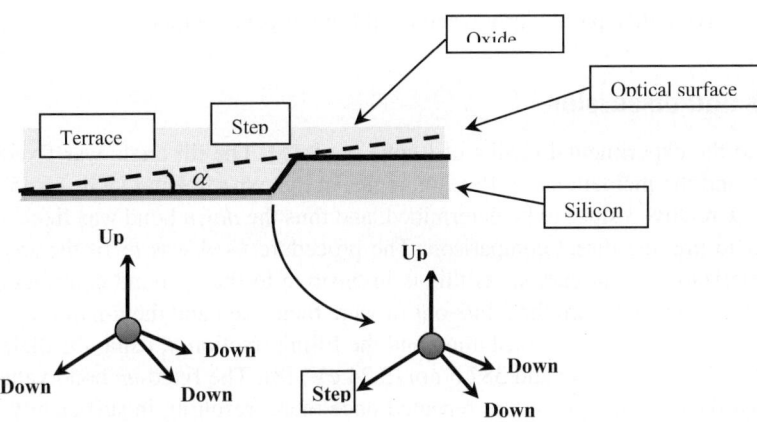

Fig. 1 Schematic of the interface geometry of Si(111), for vicinal angle α, showing the bonds associated with a terrace and a step Si atom.

2 The Hyperpolarizability Model

The model is a refinement of the simplified bond hyperpolarizability model (SBHM) of Powell et al. [1] and the same notation is used, except that β is used for bond hyperpolarizabilities. In outline, the SH field in the far-field region is given by

$$\vec{E}_{ff} = k^2 \frac{e^{ikr}}{4\pi\varepsilon_0 r} \left[\sum_j \vec{p}_j - \hat{k}\left(\hat{k} \cdot \sum_j \vec{p}_j \right) \right] \tag{1}$$

where the sum is over the different bond types, j, and $\vec{k} = k\hat{k}$ is the SH wavevector in the direction of the detector. The induced second order polarization is given by

$$\sum_j \vec{p}_j = \sum_j \beta_j (\hat{b}_j \cdot \vec{E})^2 \hat{b}_j \tag{2}$$

where β_j is the second order bond polarizability (the lowest order hyperpolarizability), $\hat{b}_j$ are unit vectors along the bond axes, and $\vec{E}$ is the electric field at the fundamental wavelength. Using the SBHM assumptions, the model duplicated the results of Powell *et al* successfully. Figure 1 shows that there are three distinct bond types for vicinal Si(111), the two *down* (*back*) bonds being equivalent by symmetry. The bond angles are assumed to be those of bulk Si. The vicinal interface against which the model is tested adds some complexity to the problem. The wavevectors of the incoming and outgoing radiation relate to the optical plane, requiring appropriate projection of the bond directions. In addition, as discussed above, the number of *step* bonds varies as $\tan\alpha$, for small tilt angle α, while the number of ter-

© 2003 WILEY-VCH Verlag GmbH & Co. KGaA, Weinheim

race *down* bonds with the same orientation varies as $(1 - \tan\alpha)$. The statistical weights for the three distinct bond contributions scale as $up : down : step = 1 : (3 - \tan\alpha) : \tan\alpha$. In order to maintain the simplicity of the model, *up* and *down* bonds at steps are assumed to have the same hyperpolarizability as those on the terrace. Finally, the fields experienced by the interface bonds need to be modelled. Local fields are neglected and the Bruggeman effective medium model is used [7]. The effective dielectric function of the interface layer lies between the limiting values of the oxide layer and the bulk Si. This is another simplifying assumption, but has the advantage of reducing the adjustment of the fundamental and SH fields to a single fitted parameter, *f*, the filling fraction of oxide. Of course, it is quite possible that the field gradient across the interface is sufficiently large that, for example, the *up* bond and *down* bonds experience different effective fields, particularly when local field effects are included.

3   Results and discussion

The model was fitted to the experimental results of Lupke et al. [5]. The tilt angle was fixed at the experimental value of 5°, and the incident and reflection angles at the experimental value of 45°. The data of Lupke *et al* only allow relative values to be determined, and thus the *down* bond was fixed at the Powell *et al* value in order to provide direct comparison. The procedure used was to fit the *step* β value using the *s*-in/*s*-out polarization combination, as this is insensitive to the *up* bond contribution and the filling fraction. The values obtained from the *s*-in/*s*-out fit were then fixed and the *p*-in/*p*-out polarization combination was used to fit the *up* bond contribution and the filling fraction, *f*, using the dielectric functions of Si and SiO$_2$ at 765 nm (1.62 eV) and 382.5 nm (3.24 eV) [8]. The fixed *up* bond value was then inserted in the *s*-in/*s*-out data and the procedure repeated once more, resulting in sufficiently converged values. Finally, the *p*-in/*s*-out and *s*-in/*p*-out response were generated using the parameters determined from the *s*-in/*s*-out and *p*-in/*p*-out fits. Figure 2 shows that this significantly more constrained model produces fits that are comparable to, but not quite as good as, those obtained by Powell *et al*.

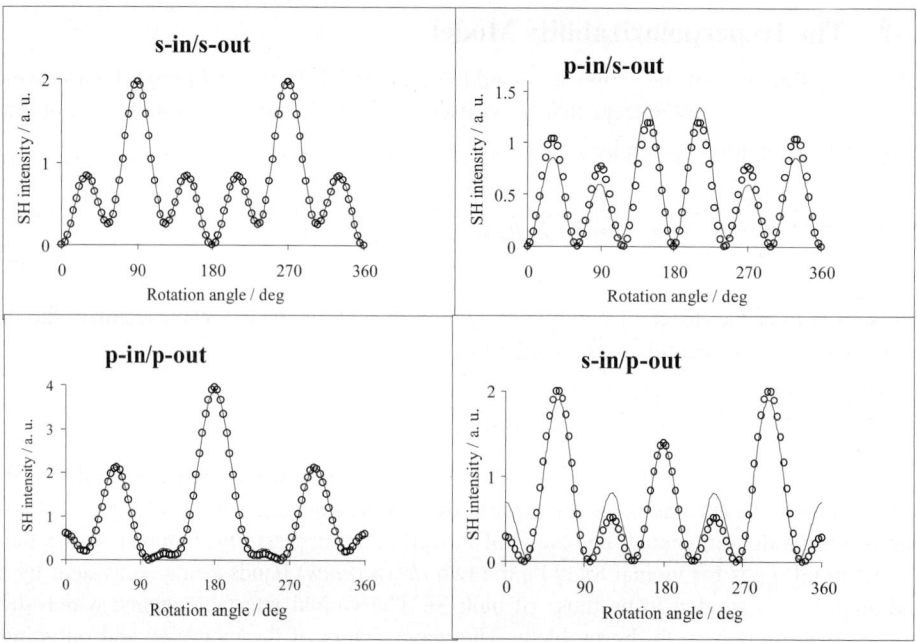

Fig. 2   Fits of the hyperpolarizability model to the SHG data (open circles) of Lupke et al. [5], for various polarization combinations.

Table 1 Comparison of fixed and fitted parameters used in Fig. 2, with those of Powell et al. [1], for 5° vicinal Si(111)-SiO$_2$.

Parameter	SBHM [1]	This work
tilt angle	1.12°	5° (fixed)
filling fraction	1.0 (implicit)	0.69
down bond	2.19	2.19 (fixed)
step bond	2.76 + i1.43	11.6 + i20.3
up bond	1. 77 + i0.17	10.1 + i3.08

The parameters are compared with the SBHM results in Table 1. It can be seen that, by including the number of *step* bonds explicitly, the much larger β value of the *step* bond is revealed: the SBHM "step" values include both *step* and *down* bond contributions. The large imaginary component is consistent with a step-induced resonance at 3.26 eV identified previously (SH energy 3.24 eV) [9].

It is also interesting that the *up* bond has a large real component. The *up* bond is a Si-O bond at the interface and might be expected to have a larger β value than the *down* Si-Si bonds, within the constraints of the model. However, in contrast to the one-fold-symmetric *step* contribution, the isotropic interface contribution is centred at 3.3 eV [9], which may explain the lower value of the imaginary part. Finally, the SBHM fitted values for the tilt (1.12°) and the effective dielectric function experienced by the interface bonds (that of SiO$_2$) are replaced by more intuitively appealing values. The tilt angle is the experimental value of 5°, and the effective dielectric function lies between the values of SiO$_2$ and Si.

The importance of the SBHM approach suggested by Powell *et al* can be seen, for example, in Fig. 3, where various contributions to the overall SH intensity are explored. In Fig. 3 (a) and (b), the tilt is removed to leave a singular interface with no *step* bonds. This has a large effect on the 90° and 270° peaks in the *s*-in/*s*-out plot, and a smaller effect on the 180° peak of the *p*-in/*p*-out plot. When the tilt is restored and $\beta_{step} \equiv \beta_{down}$, the large change shows the effect of the higher β value of the *step* bonds in these peaks (Fig. 3 (c) and (d)). Finally, with β_{step} restored and $\beta_{up} \equiv \beta_{down}$, Fig. 3 (e) shows that the *s*-in/*s*-out plot is not sensitive to the *up* bonds. Fig. 3 (f) shows that, even for the *p*-in/*p*-out combination, the *up* bond has only a relatively small effect on the profile, leading to the conclusion that the β_{up} value has the largest uncertainty associated with it, which may help explain the large variation in the *up* values in Table 1.

The above discussion should be regarded as tentative, however, because standard phenomenological analysis has revealed that, for the Si(111)-SiO$_2$ system using 765nm excitation, the bulk quadrupolar terms make a contribution comparable in size to that of the dipolar interface terms [5]. It is clear that, in order to progress further, the bulk contribution must be included. The simplest approach for this particular system is probably to use the phenomenological values obtained by Lupke et al. [5] to generate a bulk SH polarizability that can be added coherently to the dipolar term. The addition of the bulk term should allow conclusions to be drawn about the various approximations employed in the model, particularly the assumption that the SH response is dominated by the axial component of the bond hyperpolarizability.

4 Conclusion

The SBHM of Powell et al., modified here to account for *step* bond density, may simplify the interpretation of the SH response from complex systems. However, the bulk quadrupolar response remains to be included in the SH response of the vicinal Si(111)-SiO$_2$ system using 765 nm excitation before conclusions can be drawn from the analysis of this complex system. Field gradients at the interface, and transverse hyperpolarizabilities, may turn out to limit the usefulness of the approach, but the SBHM is certainly worth pursuing at this stage because of the physical insights it promises.

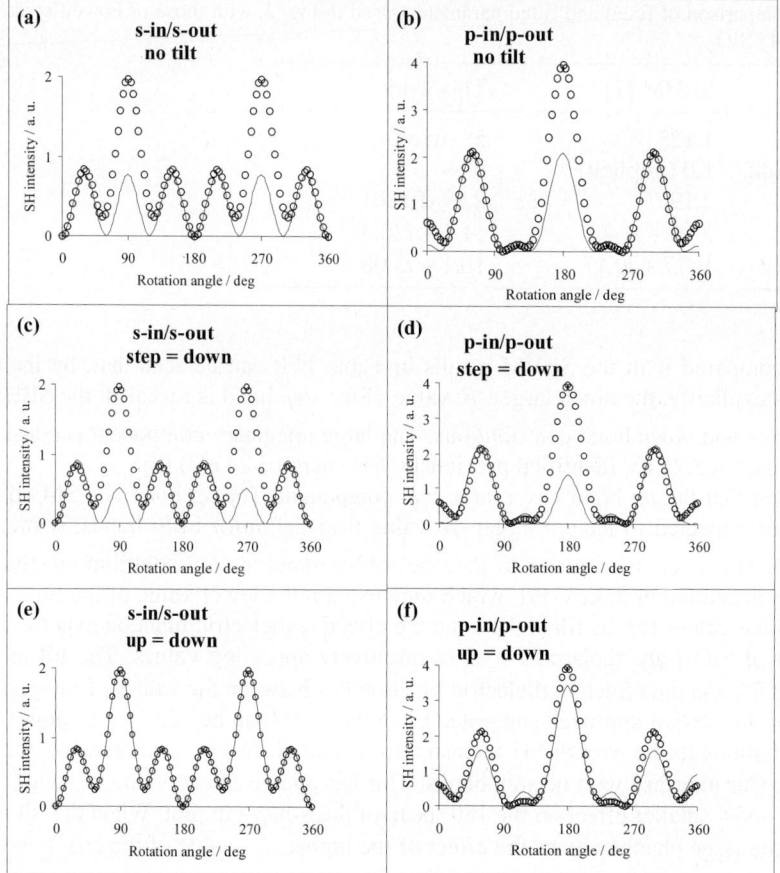

Fig. 3   Contributions to the SH intensity from various components: (a), (b) no tilt (singular surface) and thus also no step bond contribution; (c), (d) 5° tilt, with $\beta_{step} = \beta_{down}$; (e), (f) 5° tilt, with $\beta_{up} = \beta_{down}$.

Acknowledgements   Useful discussions are acknowledged with Dave Aspnes, Mike Downer and Tony Heinz. This work was supported by Enterprise Ireland Grant No. SC/2001/109/.

References

[1] G. D. Powell, J. F. Wang, and D. E. Aspnes, Phys. Rev. B **65**, 205320/1 (2002).

[2] C. H. Patterson, D. Weaire, and J. F. McGilp, J. Phys.: Condens. Matter **4**, 4017 (1992).

[3] B. S. Mendoza and W. L. Mochan, Phys. Rev. B **55**, 2489 (1997).

[4] N. Arzate and B. S. Mendoza, Phys. Rev. B **63**, 113303/1 (2001).

[5] G. Lupke, D. J. Bottomley, and H. M. van Driel, J. Opt. Soc. Am. B **11**, 33 (1994).

[6] C. W. van Hasselt, M. A. Verheijen, and T. Rasing, Phys. Rev. B **42**, 9263 (1990).

[7] J. F. McGilp, Prog. Surf. Sci. **49**, 1 (1995).

[8] SOPRA, nk database, http://www.sopra-sa.com/.

[9] C. Meyer, G. Lupke, U. Emmerichs, F. Wolter, H. Kurz, C. H. Bjorkman, and G. Lucovsky, Phys. Rev. Lett. **74**, 3001 (1995).

phys. stat. sol. (c) **0**, No. 8, 3065–3069 (2003) / **DOI** 10.1002/pssc.200303849

Second-harmonic generation spectroscopy
on reconstructed Si(111) surfaces

K. Pedersen[*, 1] and **P. Morgen**[2]

[1] Institute of Physics, Aalborg University, Pontoppidanstræde 103, DK 9220 Aalborg Øst, Denmark
[2] Fysisk Institut, University of Southern Denmark, Odense University, Odense, Denmark

Received 30 May 2003, revised 4 August 2003, accepted 11 August 2003
Published online 10 November 2003

PACS 42.65.An, 73.20.At

Optical second harmonic generation spectra obtained from the clean 7×7 surface are compared to those of surfaces with Au and Ag induced ($\sqrt{3} \times \sqrt{3}$)-reconstructions. All three reconstructions lead to broad spectra in the region below the bulk direct interband transition energy due to excitation of surface states. Variations in the spectra with sample temperature and oxygen exposure have been used to identify resonances in the spectra. In this way several transitions consistent with energy separations between surface states known from photoemission have been identified.

1 Introduction

The surface state spectrum of Si surfaces depends critically on the surface atomic arrangements [1]. Optical second harmonic generation (SHG) from Si surfaces has been shown to be totally dominated by excitations of these surface states when the SH photon energy is below the onset of direct bulk interband transitions [2-7]. In particular the rotationally anisotropic part of the SH signal is useful for studies of surface state transitions since this signal requires ordered surface structures. Previous SHG studies of the 7×7 surface have identified at least two excitation mechanisms in a rather structure-less SHG spectrum below the direct interband transition energy at 3.4 eV. Various techniques have been applied in order to decompose this broad spectrum into different contributions. From other surface probes such as photoemission [8] and scanning tunnelling spectroscopy [9] it is known that states related to adatoms, rest atom, and backbond states respond differently to variations in temperature and adsorption of foreign atoms. Furthermore, sum frequency generation spectroscopy can help to distinguish between excitations at the fundamental and SH frequencies. Using these techniques Suzuki et al. [5] concluded that a resonance in the anisotropic response appearing at a fundamental photon energy of 1.2 eV is due to a two-photon transition from backbond to adatom states while a resonance at 1.4 eV is caused by a one-photon transition from rest atom to adatom states.

In order to extend the basis for studies of SH excitations on Si(111) surfaces the present work compares SHG spectra from the 7×7 surface to those of Au and Ag induced ($\sqrt{3} \times \sqrt{3}$)-reconstructed surfaces. The metal-Si bonds of these ordered structures lead to new surface states, both below and above the Fermi level. As for the 7×7 surface these surface states are well studied by photoemission and inverse photoemission.

---

[*] Corresponding author: e-mail: i13kp@physics.auc.dk

2 Experimental details

The experiments were performed on samples mounted in a vacuum system equipped with low energy electron diffraction (LEED) and a manipulator allowing liquid nitrogen cooling (to 170K) and direct resistive heating of the samples. Sample cleanliness was verified by Auger electron spectroscopy in combination with LEED. The samples were cut from a 1-mm thick *n*-type wafer with a resistivity of 5 Ω-cm. Noble metals were evaporated from a thermal source consisting of a tungsten coil wrapped around a boron nitride crucible. A quartz crystal oscillator was used to calibrate the deposition rate. The SHG experiments were performed with an optical parametric oscillator pumped by the third harmonic of 6-ns pulses from a Q-switched Nd:YAG laser. All experiments have been performed at 60° angle of incidence. The SH signals were detected by a photo multiplier tube mounted on the exit of a monochromator and connected to a box-car integrator. The SH signals have been normalized against the SH signal generated in reflection from a wedge shaped quartz crystal.

3 Experimental results and discussion

Figure 1 shows spectra for the anisotropic contribution to SHG from Si(111)7×7 recorded at room temperature and at 550 K. The response at room temperature covers most of the spectral range. It is slowly decreasing with the photon energy, but a sharp drop in the signal is seen at the high-energy end. The effect of the increased temperature is small at the low-energy end while a pronounced change in the spectra is seen at higher energies. This clearly demonstrates that different transitions are involved in the broad room-temperature spectrum. A peak appears at the energy of the E'_0 direct interband transitions as the surface state response decays with temperature. The sharp drop in the room-temperature response is thus due to destructive interference between a surface state part and a contribution from E'_0 interband transitions [2, 3, 11].

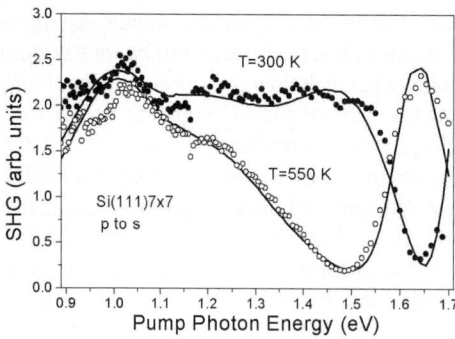
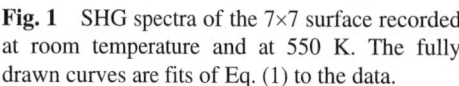
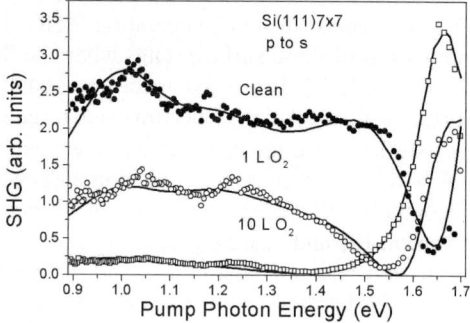

Fig. 1 SHG spectra of the 7×7 surface recorded at room temperature and at 550 K. The fully drawn curves are fits of Eq. (1) to the data.

Fig. 2 Effect of oxygen exposure on SHG spectra of the 7×7 surface. The fully drawn curves are fits of Eq. (1) to the data.

Oxygen exposure is known to quench the surface states of the 7×7 structure. Fig. 2 shows the decay of the surface state response for growing exposure to oxygen molecules. As expected the response in the whole region below the bulk interband transition energy disappears upon oxygen exposure. However, like it was seen for the temperature dependence, the features around 1.45 eV are the most sensitive to the perturbation of the surface. In fact, the sharp edge in the response at 1.5 eV was found to be a very good indication of the cleanliness of the surface. Apart from the faster decay around 1.45 eV the rest of the spectrum was found to decay at the same rate. The oxygen exposure reveals the E'_0 interband transition resonance peak that for the clean surface is hidden by the interference with the surface state resonance.

A Si(111)$\sqrt{3}\times\sqrt{3}$-reconstructed surface was obtained by depositing 2 atomic layers of Ag on the 7×7 surface followed by annealing at 800 K for a few minutes. Fig. 3 shows the anisotropic nonlinear response of this surface at 300 and 550 K. Like the 7×7 surface the $\sqrt{3}$-Ag surface has a broad spectrum below the direct E_0' interband transitions and the part at the highest energy is clearly more sensitive to the sample temperature than the low-energy part. It is thus clear that more than one resonance contributes to the surface state transitions. Compared to the spectra of the 7×7 surface the decay of the signal around 1.5 eV is less sharp for the $\sqrt{3}$-Ag surface, indicating a different position of the surface state resonance. However, from the two spectra it is clear that the surface state resonance still interferes destructively with the E_0' resonance.

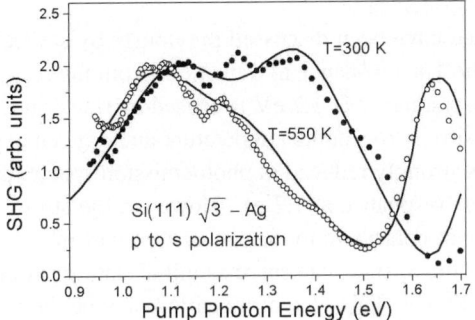

Fig. 3 SHG spectra of the $\sqrt{3}$-Ag surface recorded at room temperature and at 550 K. The fully drawn curves are fits of Eq. (1) to the data.

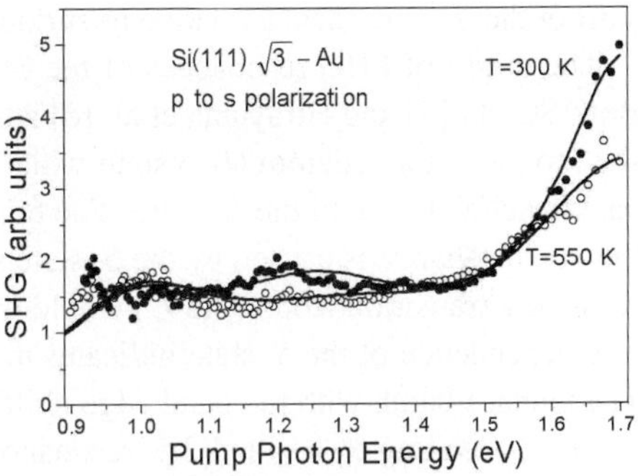

Fig. 4 SHG spectra of the $\sqrt{3}$-Au surface recorded at room temperature and at 550 K. The fully drawn curves are fits of Eq. (1) to the data.

Room-temperature deposition of Au on the 7×7 surface followed by annealing at 800 K leads to formation of a $\sqrt{3}$-Au surface structure. Figure 4 shows spectra recorded at 300 and 550 K. Unlike for the 7×7 and the $\sqrt{3}$-Ag reconstructed surfaces the spectra of the $\sqrt{3}$-Au surface show no significant effect of the sample temperature on the surface state the part of the spectrum. Another difference is the absence of the destructive interference in the region just below the E_0' transition.

The three different surface reconstructions all lead to broad SHG spectra from surface state transitions. However, from the dependence on temperature and oxygen exposure it is clear that different transitions contribute to the broad spectra. In order to decompose the spectra into a number of resonances the nonlinear surface response has been modelled by a coherent superposition of 1ω and 2ω resonances with excitonic line shapes as described by Erley, Butz, and Daum [10] and by Suzuki [5]:

$$\chi_s^{(2)}(2\omega) = \sum_{n,m}\left\{\frac{f_n\exp(i\phi_n)}{(\omega-\omega_n+\frac{1}{2}i\gamma_n)} + \frac{f_m\exp(i\phi_m)}{(2\omega-\omega_m+\frac{1}{2}i\gamma_m)}\right\} \tag{1}$$

Here ω_n is the resonance frequency, γ_n is a damping constant representing the width of the resonance, and f_n and ϕ_n are the amplitudes and phases, respectively of the resonances. In order to fit the experimental spectra Eq. (1) has been used together with the theory of Sipe [11] and experimental linear properties from Palik [12]. The linear properties only give a weak monotonous contribution to the variations in the spectra.

From previous works it is well known that the presence of the surface gives rise to a 2ω resonance at the energy of direct bulk interband transition [2-7]. The position of the resonance is affected by surface conditions and sample temperature. Due to the interference with surface state transitions the position of this resonance is not well determined from the room-temperature data in Fig. 1. However, as a starting point, we take the position (3.37 eV) and width determined from the oxidised surface (Fig. 2). In a previous work on SHG from the 7×7 surface Suzuki [5] used a model with two surface state resonances at 1.4 and 1.2 eV in addition to the resonance at E_0'. In the present data a 2ω resonance at 1.2 eV would not be sufficient to describe the low-energy part of the spectra. An additional resonance has thus been introduced at 1.0 eV. Based on sum-frequency spectra Suzuki [5] concluded that the resonance at 1.4 eV is a 1ω-resonance while the structure at 1.2 eV is caused by a 2ω-resonance. It is assumed here that the resonance behind the structure at 1.0 eV is a 2ω-resonance. Furthermore, the 1ω-resonance appears at a a slightly higher energy, namely 1.5 eV in the present data. These four resonances, having temperature and oxygen exposure dependent positions, amplitudes, and widths, have been used to obtain the fits given by the fully drawn curves on Figs. 1 and 2. The destructive interference is modelled by a 90° phase shift of the E_0'-resonance relative to surface state resonances.

The origin of SHG resonances of the 7×7 surface have been discussed previously by several authors. Both Suzuki [5] and Hirayama et al. [6] attribute the 1ω resonance to transitions from the rest atom (S_2) state to the empty adatom (U_1) state while the 2ω resonance at 1.2 eV is caused by transitions from the backbond (S_3) state to the U_1 state. The higher sensitivity to sample temperature and oxygen exposure of the 1ω resonance is caused by the S_2 state which is strongly reduced in photoemission spectra above 500 K. Also a transition from S_2 to U_2 would give a 2ω-resonance at 1.2 eV. However, the strong temperature dependence of the S_2 state indicates that it cannot contribute to the low-energy part of the spectrum. The valence band, with the band edge a little above the S_2 state, can serve as initial state for transitions to U_1 and U_2 leading to 1ω and 2ω resonances near 1.0 eV. This type of transition may be the responsible for the part of the spectrum appearing below 1.2 eV.

The spectra of the $\sqrt{3}$-Ag surface have been fitted with two surface state resonances, giving features at 1.05 and 1.35 eV, in addition to the E_0'-resonance. Since the major structure in inverse photoemission spectra appear around 2.0 eV (U') above the Fermi level [13] both resonances are assumed to be at the SH frequency. The introduction of metal atoms in the surface removes surface states associated with Si-Si bonds and new states from Si-metal bonds are formed. Occupied states are found at –0.2 eV (S_1') and ~ –1.0 eV (S_2') relative to the Fermi level [14]. The S_1' state is known to have p_x and p_y character can thus contribute to the anisotropic polarization. The number of resonances agree with the results of Hirayama et al. [6] who found peaks at 1.15 and 1.45 eV in the normal incidence spectra of the $\sqrt{3}$-Ag surface. They attributed the 1.15-eV peak to a two-photon transition from the S_1' to the U' state and suggested that the other peak (at 1.45 eV) is caused by excitation of a multipole plasmon. In a SHG study of the dynamics of the $\sqrt{3}$-Ag structure Venkataraghavan et al. [15] suggested that the 1.45-eV peak is due to a two-photon transition from the S_2' to the U' state. Hirayama et al.[6] explained the excitation of a plasmon by normal incidence light by the coupling of electronic motion parallel and perpendicular to the surface. If such a plasmon exists it should show up as a clear structure in SHG spectra recorded with p-to-p polarization. However, we did not find any structure in p-to-p spectra near 1.45 eV. We thus ascribe the resonance at 1.35 eV to a two-photon transition from the S_2' to the U' state. According to this interpretation of the resonances our results show that the S_2' state decays quickly with increasing temperature.

The $\sqrt{3}$-Au structure has an empty surface state ~1.7 eV (U'') above the Fermi level [13] and occupied surface states [16] at ~ -0.2 eV (S_1'' and S_2'') and ~ -1.2 eV (S_3''). The resonances reached with the photon energies of the present experiments are thus all at the SH frequency. It is difficult to identify resonances in the broad, almost structure-less, spectrum below the E_0' transition, in particular since the temperature dependence of the spectrum does not help to reveal different transitions. Fig. 4 shows fit to the data with 2ω resonances at 1.0 eV, 1.25 eV, and 1.69 eV. The Resonance at 1.0 eV corresponds to transitions from S_1'' to U'' while the broad resonance at 1.25 eV seems to appear at too low an energy to correspond to a S_3'' to U'' transition. A more detailed investigation of $\sqrt{3}$-Au spectra, for instance along with spectra of the Au-induced 6×6 reconstructions, is thus necessary before the surface state transitions can be identified.

4 Conclusion

The SH spectra of various Si(111) surface reconstructions all show surface state resonances in the region below the direct interband absorption energy. For the 7×7 and the $\sqrt{3}$-Ag surfaces a number of resonances have been identified and possible initial and final states for the transitions have been suggested. The identification of resonances is more difficult for the $\sqrt{3}$-Au system since it is less sensitive to adsorption and temperature variations. For further progress in interpretation of the spectra extension of the spectral region compared to the present range would be useful. Presently, resonances from 1 eV and down are not fully resolved. Furthermore, the attribution of structures in the 1.0 - 1.2 eV region to two-photon resonances can be tested by recording spectra with the double photon energy. Interpretation of such spectra will however be complicated by interference with resonances associated with critical points above E_0'.

Acknowledgement   The authors acknowledge the financial support from Carlsbergfondet.

References

[1] T. F. Heinz, M. M. T. Loy, and W. A. Thompson, Phys. Rev. Lett. **54**, 63 (1985).
[2] U. Höfer, Appl. Phys..A **63**, 533 (1996).
[3] K. Pedersen and P. Morgen, Phys. Rev. B **52**, R2277 (1995); **53**, 9544 (1996).
[4] K. Pedersen and P. Morgen, Surf. Sci. **377-379**, 393 (1997).
[5] T. Suzuki, Phys. Rev. B **61**, R5117 (2000).
[6] H. Hirayama, T. Komizo, T. Kawata, and K. Takayanagi, Phys. Rev. B **63**, 155413 (2001).
[7] Bernardo S. Mendoza, Maurizia Palummo, Giovanni Onida, and Rodolfo Del Sole, Phys. Rev. B **63**, 205406 (2001).
[8] R. I. G. Uhrberg, G. V. Hansson, J. M. Nicholls, P. E. S. Persson, and S. A. Flodstrum, Phys. Rev. B **31**, 3805 (1985).
[9] R. J. Hammers, R. M. Tromp, and J. E. Demuth, Phys. Rev. Lett. **56**, 1972 (1987).
[10] G. Erley, R. Butz, W. Daum, Phys. Rev. B **59**, 2915 (1999).
[11] J. E. Sipe, D. J. Moss, and H. M. van Driel, Phys. Rev. B **35**, 1129 (1987).
[12] E. D. Palik (Ed.), Handbook of Optical Properties of Solids (Academic Press, Orlando, 1985).
[13] J. M. Nicholls, F. Salvan, and B. Reihl, Phys. Rev. B **34**, 2945 (1986).
[14] L. S. O. Johansson, E. Landemark, C. J. Karlsson, and R. I. G. Uhrberg, Phys. Rev. Lett. **63**, 2092 (1989).
[15] R. Venkataraghavan, M. Aono, and T. Suzuki, Surf. Sci. **517**, 65 (2002).
[16] C. J. Karlsson, E. Landemark, L. S. O. Johansson, and R. I. G. Uhrberg, Phys. Rev. B **42**, 9546 (1990).

phys. stat. sol. (c) **0**, No. 8, 3070–3074 (2003) / **DOI** 10.1002/pssc.200303834

Second-harmonic far-field microscopy
of random nanostructured gold surfaces

Victor Coello[*, 1], **Jonas Beermann**[2], and **Sergey Bozhevolnyi**[2]

[1] CICESE Monterrey, P. de Alba S/N Edificio de Posgrado FCFM-UANL,
C. P. 66450 S. N. de los Garza N. L., Mexico.
[2] Institute of Physics, Aalborg University, Pontoppidanstræde 103, DK-9220 Aalborg Øst, Denmark.

Received 30 May 2003, revised 4 August 2003, accepted 11 August 2003
Published online 10 November 2003

PACS 42.65.Ky, 73.20.Mf, 78.67.Bf, 81.07.-b

We use a scanning far-field optical second-harmonic microscope in reflection mode to image a gold film surface covered with randomly distributed scatterers. We recorded simultaneously images of the first (FH) and second harmonic (SH) signal in the wavelength range of 750-830 nm for different densities of scatterers. The SH images showed localized enhancement (~10^3 the background) in the form of small (~0.7μm) round bright spots that exhibited both wavelength and polarization dependences in their brightness. We concluded that the overall behaviour of the observed phenomenon is related to the overlap of FH and SH eigenmodes.

1 Introduction

Spatially (on nanometer scale) localized optical field enhancement (up to several orders of magnitude times the background) remains as one of the most exciting effects in light scattering by *metal* nanostructures [1,2]. Such effect plays a major role in surface enhanced phenomena, e.g., Raman scattering and second harmonic (SH) generation, and it is entirely originated due to coherent multiple scattering and interference in a random media. Some examples of practical applications of these studies are high-density optical data storage, contrast enhancement in local spectroscopy, and SH generation. The advent of the scanning near-field optical microscopy (SNOM) techniques [3] brought important achievements in *local* probe of optical field enhancements. Systematic studies have been proposed in order to compare different samples and scattering configurations establishing information criteria beyond a qualitative observation of round bright spots [4]. SH field localization is one of the phenomenon more exhaustive studied in this context and traditionally the investigations have been carried out by means of angular resolved measurements of the (far-field) SH radiation. Local SH generation from rough metal surfaces has already been measured using a SNOM [5,6]. Tip-surface distance dependences of the signal were used in order to demonstrate the presence of evanescent SH field components. However wavelength dependences of the optical signal were not available making the reported SH field localization nature only partly elucidated [5]. Second harmonic scanning optical microscopy (SHSOM), which uses spatially resolved detection of SH radiation opened up an alternative and novel technique for local probing of enhanced SH generation at rough metal surfaces. SHSOM has been successfully used for imaging of periodically poled ferroelectric domains [7,8], domains in polycristalline metals [9], poled silica waveguides [10], semiconductor quantum dots [11], and individual microstructures [12]. Using a SHSOM, here we report the *direct* observations of strong and spatially localized SH enhancement in random metal nanostructures. The signal enhancement appeared to be sensitive to wavelength and polarization of the light which is consistent with SH localized enhancement.

[*] Corresponding author: e-mail: vcoello@cicese.mx, Phone: +52 81 8478 0507, Fax: +52 81 8478 0508

2 Experimental techniques

The experimental setup is shown in Fig. 1. It consists of a scanning optical microscope in reflection geometry built on the base of a commercial microscope and a computer controlled two-dimensional (*XY*) translation stage. The *XY* stage is able of moving in steps down to 50 nm with the accuracy of 4 nm within a scanning area of 25x25 mm^2. We used a mode-locked pulsed Ti-Sapphire laser tunable in the range of 730-920 nm with pulse duration of ~200 fs, 80 MHz of repetition rate and an average output power of ~300 mW. The linearly polarized light beam from the laser is used as a source of sample illumination at the fundamental harmonic (FH) frequency. The laser beam pass through a wavelength selective splitter and it is focused at normal incidence on the sample surface (spot size ~1μm) with a Mitutoyo infinity-corrected long working distance x 100 objective. In order to avoid thermal damage on the sample, the average incident power was kept at the level of ~20 mW (intensity at the surface ~2 × 10^6 W/cm^2). The sample reflects the FH and the generated SH radiation back to the same objective. The FH and SH signals are separated with the beam splitter and, after passing the appropriated filters and polarizers, detected with a photodiode and a photomultiplier tube respectively. Finally both signals are recorded as a function of the scanning coordinate obtaining simultaneously FH and SH images of the sample surface. The resolution of the SHSOM has been evaluated to be ~0.7μm using domain walls of an electric field poled KTiOPO$_4$ quasi-phase matching crystal [13]. The value is consistent with the aforementioned FH spot size, determined from the SH images of test structures, since the SH intensity scales quadratically with the FH.

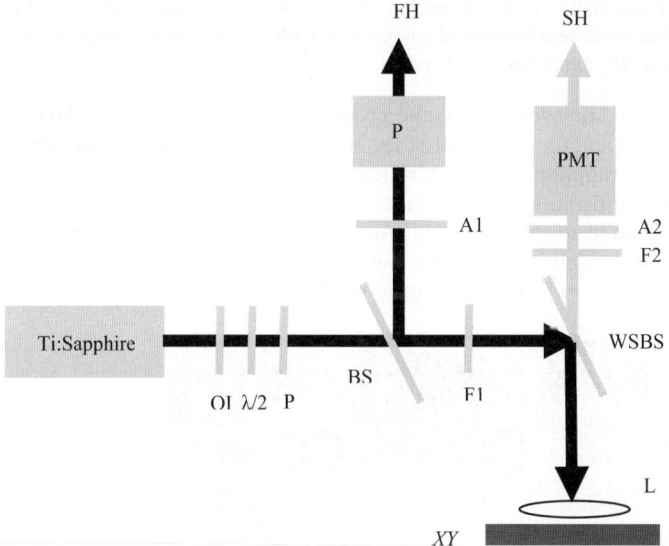

Fig. 1 Schematic of the experimental setup. OI: optical isolator, λ/2: halfwave-plate, P: polarizer, BS: beam splitter, F1 and F2: filters, WSBS: wavelength selective beam splitter, L: objective, S: sample, XY: stage, A1 and A2: analysers, PMT photomultiplier tube, and PD: photodiode.

3 Results and discussion

The sample under investigation has been fabricated by using thermal evaporation and electron beam lithography techniques (more information about the procedure is found elsewhere [14]). The final sample structure consists of high density scattering regions composed of gold bumps (~ 70 nm high) randomly distributed over a thin (~50 nm) gold film. The scattering regions density is approximately 50 scatterers per 1μm^2 and contains 2μm-wide channels free from scattereres that were used for investigations of guiding of surface plasmon polaritons (SPPs) [14].

We recorded simultaneously FH and SH images of a dense scattering region in the wavelength range of 750-830 nm. The overall behaviour of images of the same particular signal was very similar (Fig. 2(a-j)). In general one can appreciate bright and dark regions which are a collection of small and round bright spots similar to those reported as evidence of localized SPPs [4]. Concerning FH images, they exhibited very low optical contrast with the signal varying in the range of (0.9-1.2) S_0, with S_0 being the average signal measured at the flat gold film surface.

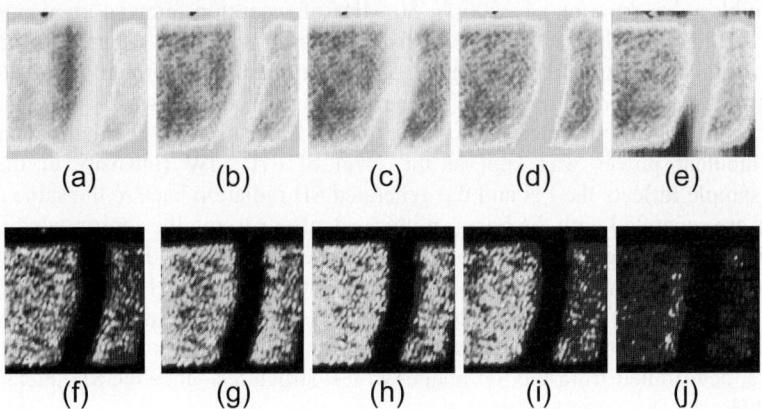

Fig. 2 Gray scale FH (a-e) and corresponding SH (f-j) images of $23 \times 23 \mu m^2$ obtained for 750 (a,f), 770 (b,g), 790 (c,h), 810 (d,i), and 830 (e,j) nm of the FH wavelength. The polarization of the FH (incident) and the SH (detected) signal were kept horizontal and vertical with respect to the presented images respectively. The maximum of the SH signal was 2300 cps (f).

However, such images show interesting features as the presence of, both, bright and dark round spots whose contrast appeared to be wavelength dependent (Fig. 3(a-d)). The presence of the observed spots suggests that the total detected FH radiation can be considered as a superposition of the FH beam reflected from the flat gold surface and the FH field scattered by strongly interacting gold bumps. The latter contribution is in general related to the regime of multiple scattering of the light which exhibits strong polarization and frequency dependence [4] and can be expected to occur, for example, due to localization of resonant dipolar excitations at nanostructured surfaces.

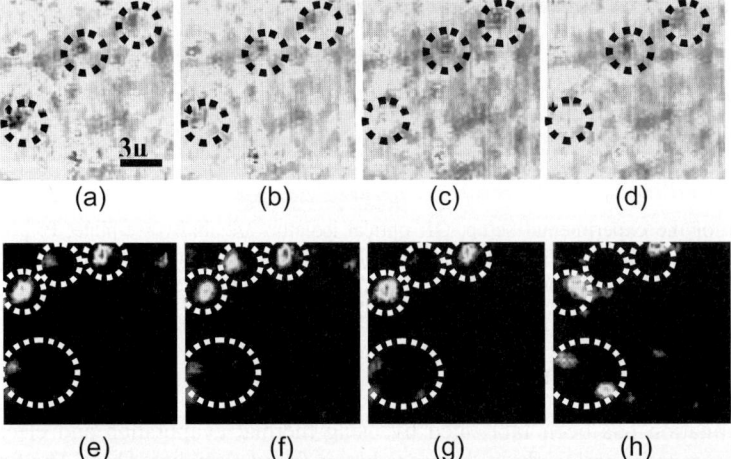

Fig. 3 Gray scale FH (a-d) and corresponding SH (e-h) images of $10 \times 10 \mu m^2$ obtained for 770 (a,e), 780 (b,f), 790 (c,g), and 800 (d,h) nm of the FH wavelength. The polarization of the FH (incident) and the SH (detected) signal were kept vertical and horizontal with respect to the presented images respectively. The maximum of the SH signal was 1000 cps (e). Representative dark (a-d) and bright (e-h) spots were enclosed in circles for better visualization.

In general, the process involved in the multiple scattering regime of the light are rather complicated and difficult to predict. However, one should expect the most efficient excitation for well-localized modes with one strong field maximum. Light scattering via excitation of such a mode (arising from the nano particles) should be similar to the dipole scattering resulting in the excitation of SP polariton modes (being in turn scattered in the surface plane and into the substrate as well as absorbed due to the internal damping). These processes contribute to the decrease of the total flux in the direction of reflection and, thereby, formation of dark spots (see spots enclosed in circles in Fig. 3(a-d)). SH images showed a better contrast than the correspondent FH and the presence of bright round spots of ~0.7 μm of radius (Fig. 3(e,h) and Fig. 4(d-f)). The images showed a noticeable re-distribution of the intensity. The bright spots (enclosed in circles) seen clearly in Fig. 3(e) are practically re-distributed in Fig. 3(h). Therefore one can claim that the bright spots are not correlated with surface defects since they change position along with illumination wavelength. During the course of experiments, we noticed that very often dark FH spots coincide with bright SH ones (see enclosed spots in Fig. 4(a-f)). These features suggest the following explanation, excitation of an FH eigenmode (leading to the local FH enhancement) results in a strong SH signal only if the SH field, which is associated with the generated nonlinear polarization, is further enhanced due to excitation of the corresponding SH eigenmode. This means that the FH and SH eigenmodes should overlap in the surface plane. However one should remark that in general such a correspondence is very difficult to observe because of the relative low contrast of FH images (for example is not obvious from Fig. 3). We observed that the change in the polarization of the detected SH incident resulted in the modification of SH images (Fig 4(d-f)), which is consistent with the suggested explanation. Here one should bear in mind that the polarization is not preserved in the multiple scattering regime of ligth.

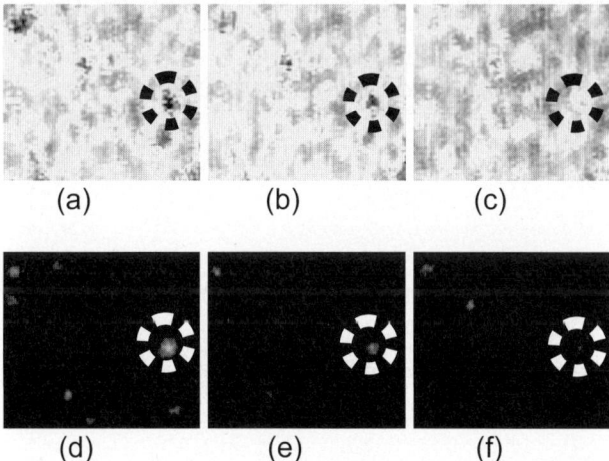

(a) (b) (c)

(d) (e) (f)

Fig. 4 Gray scale FH (a-c) and corresponding SH (d-f) images of 8x8μm² obtained for 770 (a,d), 780 (b,e), and 790 (c,f) nm of the FH wavelength. The polarization of the FH (incident) and the SH (detected) signal were both kept vertical with respect to the presented images respectively. The maximum of the SH signal was 1200 cps (d). Representative dark (a-c) and bright (d-f) spots were enclosed in circles for better visualization

In fact, the overlap (in location and polarization) condition for localized modes can be perceived as an analog to the phase matching condition for propagating waves. Indeed, the co-existence of the FH and SH eigenmodes means that the SH field driven by the FH eigenmode is in phase with multiply scattered SH fields (that actually form the SH eigenmode).

4   Conclusions

We have imaged a 50 nm gold nanostructured surface with a SHSOM in the wavelength range of 750-830 nm and for different polarization configurations of the light. FH and SH images showed wavelength and polarization dependences in their optical intensity distributions which is in agreement with the expected for multiple scattering phenomena. Round and bright (~10^3 the background signals) spots were observed in the SH images. The spots were attributed to the overlap in location and polarization of FH and SH eigenmodes in the surface plane. The overall behaviour of the SH images is related as a *direct* evidence of enhanced nano-optical fields (localized SPPs) generated via non-linear interaction in metal nanostructures. A local control of these fields and consequent maximized efficiency of them should be further elucidated. Promising applications in this context can be found in non-linear optical probing of molecules, nano-modification, and in general in nanoscience and nanotechnology.

References

[1] V. M. Markel and T. F. George (eds.) Optics of Nanostructured Materials (Wiley, New York, 2001).
[2] V. M. Shalaev (ed.) Optical Properties of Nanostructured Random Media (Springer-Verlag, Berlin, 2002).
[3] D. Courjon, Near Field Microscopy and Near Field Optics, Imperial College Press (London, 2003).
[4] S. I. Bozhevolnyi, in: Optics of Nanostructured Materials, Ref. [1], pp. 330-358, and references therein.
[5] I. Smolyaninov, A. V. Zayats, and C. C. Davis, Phys. Rev. B **56**, 9290 (1997).
[6] A. V. Zayats, T. Kalkbrenner, V. Sandoghdar, and J. Mlynek, Phys. Rev. B **61**, 4545 (2000).
[7] Y. Uesu, S. Kurimura, and Y. Yamamoto, Appl. Phys. Lett. **66**, 2165 (1995).
[8] S. Kurimura and Y. Uesu, J. Appl. Phys. **81**, 369 (1997).
[9] S. I. Bozhevolnyi, K. Pedersen, T. Sketrup, X. Zhang, and M. Belmonte, Opt. Commun. **152**, 221 (1998).
[10] K. Pedersen and S. I. Bozhevolnyi, phys. stat. sol. (a) **175**, 201 (1999).
[11] K. Pedersen, S. I. Bozhevolnyi, J. Arentoft, M. Kristensen, and C. Laurent-Lund, J. Appl. Phys. **88**, 3872 (2000).
[12] B. Vohnsen, S. I. Bozhevolnyi, K. Pedersen, J. Erland, J. R. Jensen, and J. M. Hvam, Opt. Commun. **189**, 305 (2001).
[13] B. Vohnsen and S. I. Bozhevolnyi, J. Microscopy **202**, 244 (2001).
[14] S. I. Bozhevolnyi, V. S. Volkov, and K. Leosson, Phys. Rev. Lett. **89**, 186801 (2002).

phys. stat. sol. (c) **0**, No. 8, 3075–3080 (2003) / **DOI** 10.1002/pssc.200303836

IR-SNOM on lithium fluoride films with regular arrays based on colour centres

A. Cricenti[*, 1], **G. Longo**[1], **V. Mussi**[1], **R. Generosi**[1], **M. Luce**[1], **P. Perfetti**[1], **D. Vobornik**[2], **G. Margaritondo**[2], **P. Thielen**[3], **J. S. Sanghera**[3], **I. D. Aggarwal**[3], **N. H. Tolk**[4], **G. Baldacchini**[5], **F. Bonfigli**[5], **F. Flora**[5], **T. Marolo**[5], **R. M. Montereali**[5], **A. Faenov**[6], **T. Pikuz**[6], **F. Somma**[7], and **D. W. Piston**[8]

[1] Istituto di Stuttura della Materia, CNR, via Fosso del Cavaliere 100, 00133 Roma, Italy
[2] Institut de Physique Appliquée, Ecole Polytecnique Fédérale, CH-1015 Lausanne, Switzerland
[3] Optical Sciences Division, U.S. Naval Research Laboratory, Washington, DC 20375, USA
[4] Department of Physics and Astronomy, Vanderbilt University, Nashville, TN, 31235, USA
[5] ENEA, UTS Tecnologie Fisiche Avanzate, C.R. Frascati, Via E. Fermi 45, 00044 Frascati, Roma, Italy
[6] MISDC of VNIIFTRI Mendeleevo, Moscow region, 141570, Russia
[7] Physics Department, University Roma Tre, V. della Vasca Navale 84, 00146 Roma, Italy
[8] Department of Molecular Physiology and Biophysics, Vanderbilt University, Nashville, TN 37232, USA

Received 30 May 2003, revised 4 August 2003, accepted 11 August 2003
Published online 10 November 2003

PACS 42.82.Cr, 61.82.Ms, 68.37.Uv, 68.55.Jk, 78.55.Fv

LiF films have been grown on silicon substrate, irradiated with soft x-rays to create fluorescent regular micrometric-spaced arrays based on colour centres, and studied by Scanning Near-field Optical Microscope (SNOM). Strong variations in the local reflectivity have been observed in the infrared region between 6.1 and 9.2 μm and tentatively ascribed to a modulated variation of the refractive index of the coloured zone with respect to that of the uncoloured LiF matrix.

1 Introduction

The properties of colour centres (CCs) in Lithium Fluoride (LiF) crystals and thin films became recently object of renewed interest [1], mainly related to the fact that the stable formation of active, visible emitting electronic defects, induced by many kinds of ionising radiation, could be exploited for the realisation of miniaturised innovative optical devices [2–4].

Quite recently [5] an innovative irradiation technique, using EUV radiation or soft X-rays by a laser plasma source, has been shown to be an efficient tool to create optically active CCs in LiF crystals. In such crystals, a coloured pattern with high spatial resolution on large areas was produced and detected through a strong visible photoluminescence emitted at room temperature. Moreover, when the colouring is performed by using low penetrating radiation, it is possible to prepare thin layers with high concentration of lattice defects located at the surface of the crystalline material. Focussed particle beams can also be used to generate fluorescent colour-centres patterns in the insulating LiF matrix [6]. A significant improvement in high lateral resolution characterization of these patterned LiF samples can come from the Scanning Near-field Optical Microscope (SNOM) [7–10]. In fact, SNOM has proven to be a powerful tool in this context, thanks to its ability to analyse local optical properties and scattering phenomena at the sample surface [11].

---

[*] Corresponding author: e-mail: cricenti@ism.cnr.it

Here we present some results obtained with the SNOM, coupled with an infrared radiation (IR) source, on LiF films grown on a silicon substrate containing primary and aggregate point defects. Infrared spectroscopy with a Free Electron Laser (FEL) is an excellent probe of material science samples, since the IR radiation has a deep penetration and can analyse otherwise buried structures [12]. By combining the tuneable Vanderbilt FEL IR source with high quality IR fiber tips, we were able to investigate for the first time the IR signatures from LiF CCs regularly spaced at micrometric level.

The local reflectivity showed an optical contrast at 6.1 and 9.2 μm, probably due to the periodic modification of the refractive index caused by the colouration process along regularly spaced arrays. The optical images are completely not correlated to the path followed by the tip during the scan and displayed in the shear-force image (surface topography). A lateral resolution below 50 nm was observed in the topography while a resolution of the order of 200-300 nm was obtained from the optical images.

2   Experimental set-up

Good quality LiF films were grown by thermal evaporation on untreated silicon substrate [13] kept at 250 °C during the deposition. The films total thickness, measured by a stylus profilometer after the growth, was about 800 nm. The EUV and soft X-ray radiation has been obtained by focusing a powerful excimer laser beam on a solid target placed in a vacuum chamber. The plasma-point source produced by the laser and its characteristics have been described in detail elsewhere [14]. The plasma plume is generated by a non-commercial large-volume excimer laser (Hercules), developed at the ENEA Laboratories in Frascati, and operating at the emission wavelength of 308 nm. Using a 10 ns pulse, the laser intensity on the target is 3×10^{13} W/cm^2, giving rise to a plasma temperature of 200 eV. The solid target is a tape, moved after each shot by a step motor, and tilted by 45 degrees with respect to the laser-beam axis in order to utilize the strongest X-rays emission normal to the target. The X-ray emission covers the spectral interval of 0.8 – 60 nm (1.5 keV – 20 eV), that is the full EUV region and part of the soft X-rays region [15]. The LiF sample is placed inside the vacuum chamber at a distance of ~12 cm from the plasma source. This irradiation method allows the opportunity to expose a large area, several squared centimetres, in a single shot of the excimer laser.

In order to obtain regularly spaced coloured patterns on the LiF samples, a copper mesh, with wire diameter of 10 μm and period 12.7 μm, was placed side by side in contact with the LiF film, so that it masked the surface from the incoming radiation. The so masked LiF sample was irradiated by the above described radiation source with a copper target and 2000 shots. The total EUV and soft X-rays fluence was roughly 100 mJ/cm^2, 90% of which in the EUV.

The experiments were performed with a multi-technique SNOM module [16]. This instrument can deliver shear force (topographic) images as well as reflectivity SNOM images. The FEL facility at Vanderbilt (Nashville, USA) is active in the 1-10 μm spectral range, which cover the relevant absorption bands of many materials. IR photons from the unfocussed FEL beam are directed by a mirror on to the sample surface at approximately a 75° angle to the surface normal. Reflected photons are detected through a narrow-point optical fibre tip mounted on a SNOM module, which also measures shear-force (topographic) images [16]. The shear-force signal is used to keep the tip-surface distance constant while taking SNOM images, so that the applied force is constant for all data points. Topographical and optical images were taken simultaneously, with the FEL illuminating the specimen over a broad area (~1 mm spot diameter) and the SNOM probe collecting the reflected light. The direct comparison of topographic and spectroscopic SNOM images is essential to prove that no artifacts exist in the SNOM images.

The fabrication of extremely high quality infrared fibre tips [17] was a crucial step for the success of the experiment: infrared SNOM probes were obtained from single-mode, one meter long, arsenic selenide fibers with one end interfaced to the detector. The remaining end of the fiber was chemically etched using a protective layer etching system and coated with gold using a thermal evaporator. To create an aperture, the tips were angled 25° to 30° above the evaporation point source. The tips were rotated to achieve a uniform coating with an approximate coating thickness of 100-125 nm.

In the presented SNOM images bright areas correspond to higher values in the topography and reflectivity while dark ones to lower ones. No filtering has been performed on the images but a background and plane align.

3 Results and discussion

Figure 1 is an image of the X-ray irradiated LiF film grown on a silicon substrate obtained with a fluorescence optical microscope under illumination with a photon wavelength of 458 nm.

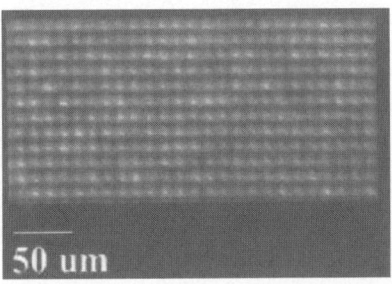

Fig. 1 Fluorescence image of a LiF film, 790 nm thick, thermally evaporated on a silicon substrate and irradiated by soft X-rays, under illumination with the 458 nm line of an argon laser. Clearly visible (black areas) the shadow of the mesh, which masked the film during the X-ray exposure.

The white spots represent the fluorescence parts of the irradiated surface of the film, while the uncoloured black parts represent the dark shadow of the meshes. The detected light is emitted from the F_2 and F_3^+ complex CCs, consisting in two electrons bound to two and three anion vacancies, respectively which have almost overlapping absorption bands at 450 nm [18] and, therefore, can be optically excited simultaneously with the same pumping wavelength.

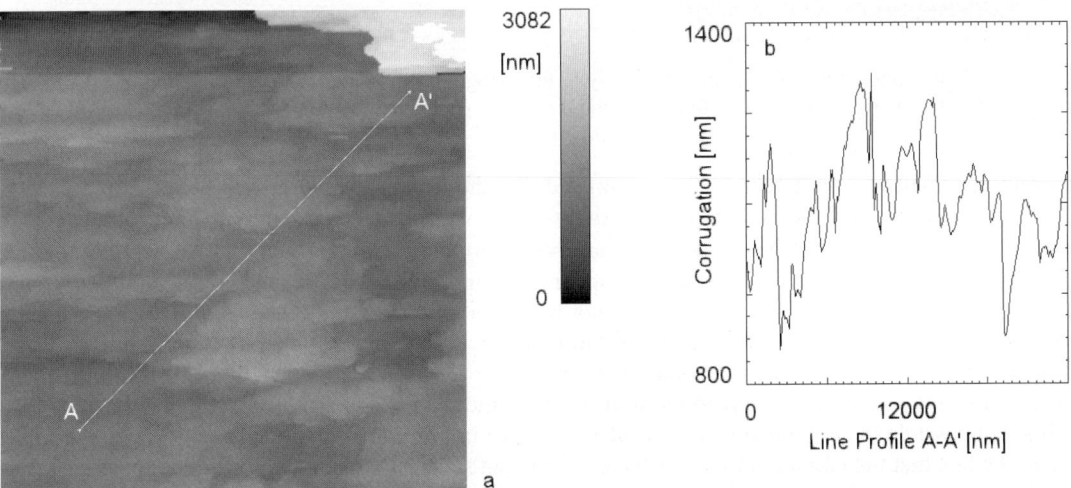

Fig. 2 (a) Shear-force image (24 μm x 24 μm) of a LiF film silicon substrate. The grey scale varies from black for the deepest zones to white for the highest; (b) corrugation along the AA' line.

Figure 2 (a) is a 24 μm x 24 μm shear force image taken on the LiF surface together with the height variation along the AA' line (Fig. 2b). The appearance of the topology as measured by shear force is generally quite smooth with variation of few hundreds of nanometers, and small features of the individual LiF grains are quite well resolved. A big structure (corrugation around 3000 nm) is present in the upper right corner. We choose this region since the above big structure is a good reference point for comparing the shear-force image with the optical image.

Figure 3 (a) is the reflectivity image of the same zone as in Fig. 2 (a), taken with a wavelength for the incident photons of 6.1 μm (similar result is obtained with λ=9.2 μm) , together with the height variation along the AA' line (Fig. 3b). The big structure does not give any clear contrast while two valleys, 10 μm wide (full width) and 5.2 mV amplitude, are crossing the image. It is important to stress the fact that the two features are totally uncorrelated with the topographical structures in Fig. 2. It is worth noting that the reflectivity valleys are present in zones where there is no contribution in the topographical image, and, due to the deep penetration of the infrared light, must be located well below the surface. This is in agreement with theoretical simulation about the fact that the creation of colour centres is located in a spatial region, about 50 nm deep, below the surface [19]. In this case, one can strongly consider that the near-field optical profile is purely optical and that the relief contrast of the sample surface has not taken part in the variation of the SNOM signal.

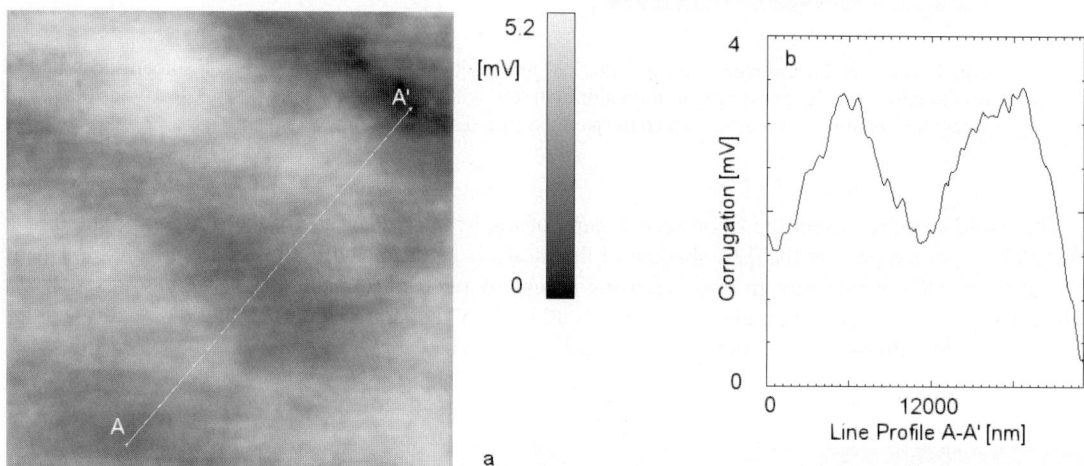

Fig. 3 (a) Reflectivity image, taken simultaneously to the topography of Fig. 2 (a), with a photon of 6.1 μm. The grey scale is set as in Fig. 2; (b) corrugation along the AA' line.

From the smallest visible features, we estimated that the lateral resolution was better than 50 nm in topographic images, and 200-300 nm in reflectivity. These values are consistent with a much larger set of data not explicitly presented here (taken also at λ=9.2 μm) and are better than the classical limit of λ/2 (with λ=6.1μm). We note, however, that the topography- and reflectivity-image values could be an overestimate, since even smaller structures might not be discernible from the noise level. By increasing the signal-to-noise ratio, the ultimate lateral resolution achievable with our apparatus might be improved. Furthermore, while superb lateral resolution is achievable for a deep structure in a SNOM experiment, this resolution is limited by the spreading of the beam in the sample.

It is not possible to correlate directly the SNOM image in Fig. 3 with the fluorescence image in Fig. 1 due to the fact that the two measurements have been taken separately using two different apparatus. Also the "big structure" visible in the Shear-force image in Fig. 2 cannot be used to correlate the SNOM image with the luminescent color centers features shown in Fig. 1, because such defect structure will not give any contribution in fluorescence. Therefore, we must rely on the fact that the spacing between the two reflectivity minima observed in the SNOM image of Fig. 3 is comparable to the 12.7 micron spacing between the color centers of Fig. 1. Moreover, no such minima and maxima as observed in Fig. 3 have been observed at the same wavelength on an un-irradiated LiF film with no color centers.

Pure LiF is quite transparent to the infrared light in the investigated spectral range. The optical contrast observed in the infrared SNOM images on such peculiar LiF samples could be attributed to a modification of the material optical properties associated with the high concentration of active defects, which can modify the refractive index [20] of the irradiated volume. This effect can be enhanced from the spatial periodicity of the peculiar CCs based patterns. In a sense, therefore, an on-absorption image like that of Fig. 3 is a case of enhanced SNOM regime, yielding high chemical contrast.

4 Conclusions

In conclusion we have observed strong variations of the local reflectivity in the infrared region between 6.1 and 9.2 μm on LiF films grown on silicon substrate and irradiated with soft x-ray to create fluorescent patterns based on active colour centres. These SNOM observations could be ascribed to a modification of the refractive index of the coloured zone with respect to that of the uncoloured LiF. The spatial periodicity of the peculiar patterns transferred to the sample surface trough a masking procedures could further enhance these effect.

The presented results confirm the fact that the future development of Near-Field techniques will be especially related to their ability of supplying useful tools for the analysis of local characteristics of new materials in the field of opto-electronics and, generally speaking, of optical integrated circuits.

Acknowledgements We would like to thank the entire staff of the W.M. Keck Foundation FEL Center at Vanderbilt University for their able assistance. We also thank E. Giovenale and the FEL group of ENEA Frascati for providing the mesh. The precious technical assistance of A. Pace is also acknowledged. This work is supported by the Italian National Research Council (CNR), Ecole Polytechnique Fédérale de Lausanne, the Fonds National Suisse de la Recherche Scientifique.

References

[1] R.M. Montereali, M. Piccinini, and F. Burattini, App. Phys. Lett. **78**, 4082 (2001).
[2] R. M. Montereali, Handbook of Thin Film Materials (Academic Press, New York, 2002), Vol. 3, Cap. 7.
[3] K. Schwartz, C. Trautmann, T. Steckenreiter, O. Greiss, and M. Kramer, Phys. Rev. B **58**, 17 (1998).
[4] C. Vigreux, P. Loiseau, L. Binet, and D. Gourier, Phys. Rev. B **61**, 13, (2000).
[5] G. Baldacchini, F. Bonfigli, F. Flora, R.M. Montereali, D. Murra, E. Nichelatti, A. Faenov, and T. Pikuz, Appl. Phys. Lett. **80**, 4810 (2002).
[6] P.Adam, S. Benrezzak, J.L. Bijeon, P. Royer, S. Guy, B. Jacquier, P. Moretti, R.M. Montereali, M. Piccinini, F. Menchini, F. Somma, C. Seassal, and H. Rigneault, Opt. Express **9**, 353 (2001).
[7] E. Betzig and J.K. Trautman, Science **257**, 189 (1992).
[8] E. Betzig, P.L. Finn, and J.S. Wiener, Appl. Phys. Lett. **60**, 2484 (1994).
[9] H. Heinzelmann and D.W. PohL, Appl. Phys. A **59**, 89 (1994).
[10] NATO ASI Series "Near Field Optics", eds. D.W. Pohl and D. Courjon (Kluwer Acad. Press, 1992), Vol. 262.
[11] V. Mussi, A. Cricenti, R.M. Montereali, E. Nichelatti, S. Scaglione, and F. Somma, Phys. Chem. Chem. Phys. **4**, 2742 (2002).
[12] G.L. Carr, P. Dumas, C.J. Hirschmugl, AND G.P. Williams, Il Nuovo Cimento **20**, 375 (1998) and references therein.
[13] G. Baldacchini, E. Burattini, L. Fornarini, A. Mancini, S. Martelli, and R.M. Montereali, Thin Solid Films **330**, 67 (1998).
[14] S. Bollanti, P. Albertano, M. Belli, P. Di Lazzaro, A. Ya. Faenov, F. Flora, G. Giordano, A. Grilli, F. Ianzini, S.V. Kukhlevsky, T. Letardi, A. Nottola, L. Palladino, T. Pikuz, A. Reale, L. Reale, A. Scafati, M.A. Tabocchini, I.C.E. Turcu, K. Vigli-Papadaki, and G. Schina, Il Nuovo Cimento **20D**, 1685 (1998).

[15] David Attwood, "Soft X-rays and Extreme Ultraviolet Radiation: Principles and Applications" (Cambridge University Press, 1999).

[16] A. Cricenti, R. Generosi, C. Barchesi, M. Luce, AND M. Rinaldi, Rev. Sci. Instrum. **69,** 3240 (1998).

[17] D. Talley, L.B. Shaw, J. S. Sanghera, I. D. Aggarwal, A. Cricenti, R. Generosi, M. Luce, G. Margaritondo, J. M. Gilligan, and N. H. Tolk , Mater. Lett. **42**, 339, (2000).

[18] J. Nahum and D.A. Wiegand, Phys. Rev. **154**, 817 (1967).

[19] B. L. Henke, E. M. Gullikson, and J.C. Davis, X-rays interactions: photoabsorption, scattering, transmission and reflection at E=50-30000 eV, Z=1-92, Atomic Data and Nucl. Data Tables **54**, 181 (1993).

[20] M.Montecchi, E.Nichelatti, A.Mancini, and R.M.Montereali, J. Appl. Phys. **86**, 3745 (1999).

phys. stat. sol. (c) **0**, No. 8, 3081–3085 (2003) / **DOI** 10.1002/pssc.200303853

Electric-field-induced second-harmonic microscopy

K. Wu[1], **J. D. Canterbury**[*1], **P. T. Wilson**[1], and **M. C. Downer**[**1]

[1] Texas Materials Institute, University of Texas at Austin, Austin, TX 78712-1081, USA

Received 30 May 2003, revised 4 August 2003, accepted 11 August 2003
Published online 10 November 2003

PACS 07.60.Pb, 42.65.Ky, 42.70.Nq

Electric-field-induced second-harmonic (EFISH) generation is used as the basis for constructing a nonlinear optical microscope to image electric field distributions near metal- Si(001) junctions with sub-micron lateral resolution. Two configurations – stationary projection imaging and confocal scanning – are demonstrated and comparatively evaluated.

1 Introduction

Metal-semiconductor contacts in silicon integrated electronic circuits and membranes of living cells represent two widely divergent contexts in which strong dc electric fields pervade centrosymmetric media over sub-micron spatial scales. Optical second-harmonic generation (SHG) enables sensitive, noninvasive probing of such fields with spatial resolution limited in the far-field by the wavelength λ_{SH} of SH light. SHG is dipole-forbidden in a centrosymmetric medium (*e.g.* silicon, water), but is locally enhanced in regions pervaded by an electric field $\vec{E}_0$, which, as a polar vector, lifts inversion symmetry. This electric field-induced second-harmonic (EFISH) process, first demonstrated in the 1960s [1], has been widely employed as a probe of electric fields in semiconductor devices [2, 3, 4, 5, 6, 7] and in aqueous environments [8, 9, 10]. However, despite extensive development of second-harmonic microscopy in other contexts [11, 12, 13], the potential of EFISH for diffraction-limited microscopy of electric fields remains largely untapped [14]. In this paper, we demonstrate EFISH microscopy of electric fields at metal-silicon junctions. Projection imaging and scanning microscopy are compared. These EFISH imaging techniques can be extended by employing homodyne detection [15], electro- [16] and photo-modulation [17] to enhance sensitivity and suppress background contributions, and immersion and near-field techniques to enhance resolution.

2 Experimental set-up

Samples were prepared on nominally undoped Si substrates (700 μm thick, resistivity $8 \times 10^3 \Omega cm$). Patterned metal-oxide-semiconductor (MOS) samples were prepared by: (1) growing a 90 $\overset{\circ}{A}$ thermal oxide; (2) patterning an array of several-μm wide gate electrodes using standard photo-lithographic techniques; (3) evaporating a semitransparent 64 $\overset{\circ}{A}$ Cr film onto the oxide, and (4) lifting off the pattern with Aceton. A uniform 3000 $\overset{\circ}{A}$ Al electrode was evaporated onto the backside of the substrate. Bias was applied between Cr and Al electrodes. In these samples, fields were oriented primarily perpendicular to the surface, and were located beneath, and probed through, the semitransparent Cr electrodes. Cr generates much weaker

---

* Current address: Department of Physics, U. Washington, Box 351560, Seattle, WA 98195-1560 USA.
** Corresponding author: e-mail: downer@physics.utexas.edu, Phone: +01 512 471 6054, Fax: +01 512 471 9637

SH signals than Si, and thus does not obscure EFISH generation from the underlying space-charge region (SCR) significantly [3, 4]. Coplanar interdigital arrays of Schottky-contacted electrodes were prepared by direct photo-lithographic treatment of the bare Si substrate, followed by evaporation of a 1000 $\overset{\circ}{A}$ thick optically opaque Al film onto the front side, and pattern lift-off. In these samples, fields were oriented primarily in the plane of the surface, in a μm-thick SCR fringe outlining the electrodes. Estimated field strengths detected in both samples ranged from 10^3 to 10^4 V/cm, but were not precisely calibrated.

Figure 1a illustrates the set-up for stationary projection EFISH microscopy. 25 fs, 5-10 nJ fundamental pulses at $\lambda = 0.8\mu m$, 76 MHz repetition rate from a mode-locked Ti:S laser were focused at f/12.5, 45° incidence, p-polarization to a spot radius $\sim 10\mu m$ on the sample surface. SH light ($\lambda_{SH} = 0.4\mu m$) was generated from the sample collinearly with reflected fundamental light, spectrally filtered, then imaged with about 140× magnification onto a charge-coupled device (CCD) with a microscopic objective. Focusing was optimized by imaging the electrode pattern using an incident test beam at λ_{SH}. Depth-of-field ($\sim 50\mu m$) was sufficient that the entire illuminated region was in focus. The stationary beam method acquires images in parallel with no moving parts, and achieves essentially diffraction-limited lateral resolution. However, spot radius (and thus field of view) was usually limited to $\sim 10\mu m$ by low count rates (0.1 to 1 count/pixel/s over the illuminated area of metallized silicon), which necessitated data acquisition times of 10 to 100 s to obtain high quality EFISH images. Spot radii several times larger could be used with higher count rates and/or longer integration times. In either case, projection microscopy is incompatible with modulation techniques, but could be enhanced by using a second SH reference beam for homodyne detection or interferometry. Coherent artifacts often appear in projection SH images when sharp boundaries are present in the imaged area.

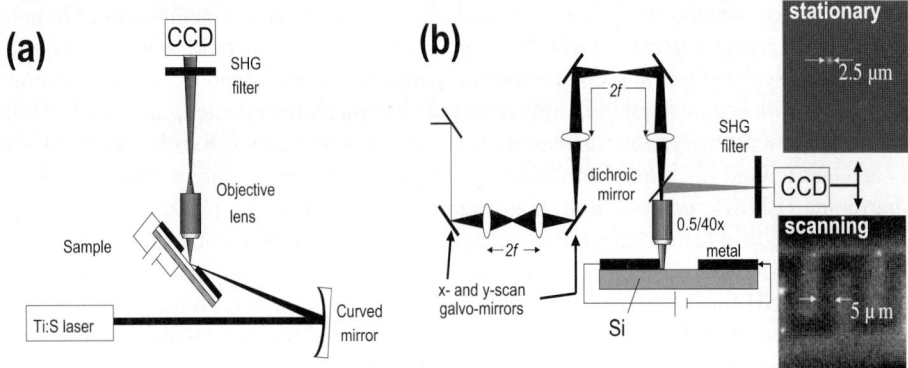

Fig. 1 Experimental set-up for (a) stationary projection and (b) confocal scanning EFISH microscopy. Insets: images of SHG with scan mirrors stationary (upper right) and scanning (lower right) over interdigital pattern of Au electrodes (light regions) with 10 V bias between upper an lower array.

Fig. 1b illustrates an alternative confocal scanning configuration, commonly used in linear optical microscopes [18]. The incident beam impinged on a scan mirror that swiveled about an axis parallel to the mirror surface. A unity magnification telescope imaged the scanned spot onto a second scan mirror that swiveled around a perpendicular axis. These mirrors were driven by galvanometer drivers (GSI-Lumonics AX-200 and CX-660). A second telescope imaged this scanned point to a stationary spot in the back pupil plane of a 40× objective lens, which focused at normal incidence onto the sample to a near-diffraction-limited spot radius $w_0 \sim 1\mu m$ (Fig. 1b, upper right inset). The focused spot raster scanned with the galvo-mirrors over a $30 \times 30\mu m$ area on the stationary sample. The same microscope objective then collected the generated SH light in an approximately confocal geometry and imaged it onto the CCD camera, where it traced out a SHG image of the scanned sample area. Fig. 1b, lower right inset shows a typical scanned image of the total SHG signal from interdigital Al electrodes (lighter regions) on Si (darker regions). Adequate signals were obtained with total scan time $\sim 100s$ comparable to the stationary projection method. Confocal scanning provided much larger count rates (~ 1000 SH cts/s) at each spot location. Thus modulation techniques could be used. Disadvantages of the scanning method are: (1) spot-size-limited resolution, which is typically somewhat less than obtainable with projection imaging with high quality microscope optics; (2) smaller depth of field, making this method difficult to implement at oblique incidence; (3) significant generation of carriers that partially screen [19] the dc fields that are the subject of the measurements.

3 Results and Discussion

3.1 Stationary projection images

Figure 2 shows examples of projection EFISH images acquired with 150 s averaging time near the tip of a pointed Cr gate electrode biased at (a) +5.5 V and (b) -5.5 V. The enhanced SHG intensity from the SCR under the positively-biased electrode is clearly evident in Fig. 2a. Similarly decreased SHG intensity relative to the adjacent oxidized Si is evident under the negatively biased electrode in Fig. 2b. Near the sharp edges of the electrodes, interference fringes with period $\sim \lambda_{SH}$ are clearly visible. These result from diffraction of part of the generated SH light out of the collection aperture of the imaging lens by the sharp edge of the electrode, and are a common feature of microscopy with coherent light [20].

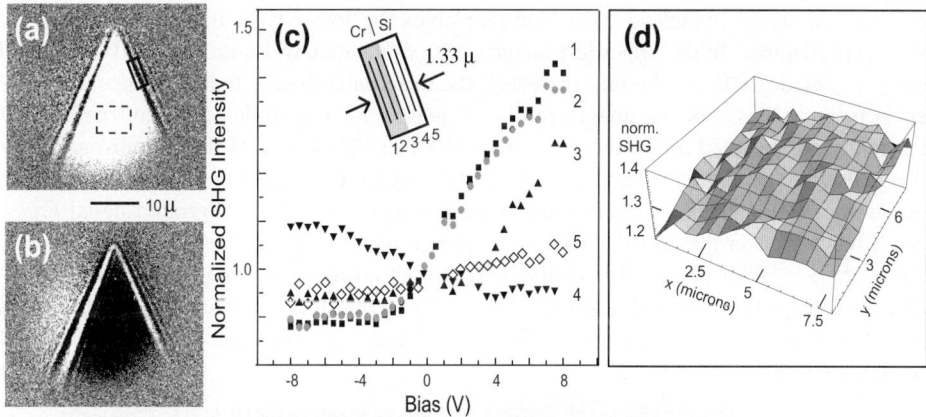

Fig. 2 Stationary projection EFISH microscopic images of a pointed Cr electrode biased at (a) +5.5 V and (b) -5.5 V; (c) Bias-dependence of normalized SHG intensity $I_{SH}(V)/I_{SH}(0V)$ along five lines within a $1.3 \times 3\mu m$ area (solid box in panel (a) and inset of (c)) straddling the edge of the Cr electrode; (d) spatial variation of $I_{SH}(V)/I_{SH}(0V)$ at +5.5 V within the $2.5 \times 3\mu m$ area indicated by dashed box in panel (a).

Figure 2c shows the bias-dependent, normalized SHG intensity $I_{SH}(x, V)/I_{SH}(x, V = 0)$ at 5 different perpendicular distances $-0.66 < x < +0.66\mu m$ from the edge ($x = 0$) of the electrode at the location indicated by the solid box in panels (a) and (c). For each curve, signal from a line of ~ 10 pixels parallel to the edge of the electrode was averaged, in order to suppress artifacts caused by pixel-to-pixel variations in the detector. Normalizing $I_{SH}(x, V)$ to $I_{SH}(x, V = 0)$ removes the dependence of the signal on the Gaussian intensity profile of the focused laser spot, leaving variations with bias V and position x on the sample. Curves 1 and 2 were acquired from beneath the electrode, and show nearly identical dependence on bias: enhanced SHG at $V > 0$, reduced at $V < 0$. In this area, the total SH field $E_{2\omega}^{TOT}(\vec{r}) = E_{2\omega}^{EFISH}(\vec{r}) + E_{2\omega}^{FI}(\vec{r})$ is a coherent sum of the EFISH field and the field-independent (FI) background SH field. The enhanced (reduced) signal at positive (negative) bias results from constructive (destructive) interference between $E_{2\omega}^{EFISH}$ and $E_{2\omega}^{FI}$ [16]. The saturation of the signal $|E_{2\omega}^{TOT}|^2$, and thus of $E_{2\omega}^{EFISH}$, at large negative bias is consistent with the low (unintentional) doping, and thus lack of accumulation regime, in our substrate.

Curve 3 was acquired at the edge of the electrode, and clearly shows a weaker bias-dependence than curves 1 and 2. Curve 4 was acquired at the first dark fringe, and shows a still weaker bias dependence of reversed sign. Curve 5 was acquired at the first bright fringe, where bias dependence has nearly disappeared. The scale length 0.5 μm over which bias-dependence disappears is consistent with the length W of the depletion region for nominally undoped Si. However, interpretation is complicated by the presence of interference fringes. A more detailed analysis of this edge behavior will be presented elsewhere.

Figure 2d shows the spatial variations in $I_{SH}(\vec{r}, V = +8V)/I_{SH}(\vec{r}, V = 0)$ in the area under the Cr electrode indicated by the dashed box in panel (a). The signal is quite uniform, as expected, but $\pm10\%$ variations are observed. These variations may result from μm-scale variations in defect density at the Si/SiO$_2$ interface or in dopant density, which would alter the distribution $E_0(z)$ of the dc field in the SCR, and thus $E_{2\omega}^{EFISH}$. Such fine-scale variations are a unique output of EFISH microscopy that cannot be observed in macroscopic electrical tests.

3.2   Confocal scanning images

Figure 3 shows differential confocal scanning EFISH images of an interdigital, coplanar pattern of Al electrodes (see inset), with biases from -20 to $+20$ V applied between the upper and lower arrays. For each image in Fig. 3, the scanned SHG signal from the *un*-biased structure has been subtracted from the total scanned signal (which in each case resembles the lower right inset of Fig. 1b), leaving only bias-dependent features. In the upper left image ($+20$ V) the electrodes are outlined by the EFISH signal from the $\sim \mu m$-wide SCR. As the bias increases, the differential image fades, disappearing near 0 V, then recovers at positive bias. For the images in Fig. 3, neither the electrode edges nor the linearly-polarized incident field were oriented along the crystalline [100] or [010] axes of the Si substrate. Consequently EFISH signals were generated with comparable efficiency from dc fields of all orientations via the various non-zero tensor components of $\chi^{(3)}$. Because of normal incidence, the background signal $E_{2\omega}^{FI}$ from the Si substrate is much weaker than for the oblique-incidence data in Fig. 2. Consequently an enhanced signal $|E_{2\omega}^{TOT}|^2 \sim E_{2\omega}^{EFISH}|^2$ is observed from the SCR at both negative and positive biases. Spatial resolution of the electrode edges ($\sim 0.6\mu m$) is poorer than in projection images, because it is limited by the focal spot size rather than the microscope optics. On the other hand, interference fringe artifacts are absent from the scanning images.

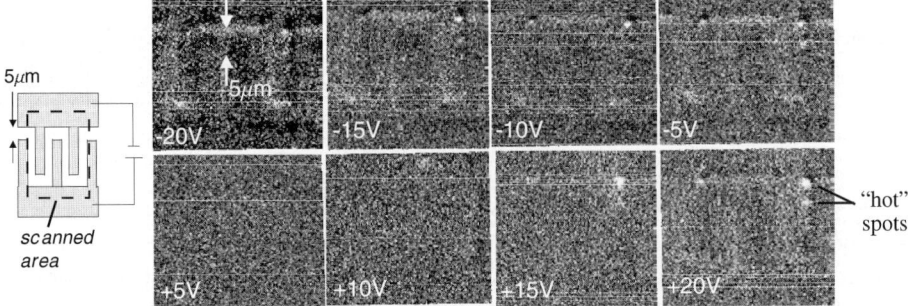

Fig. 3   Scanning confocal EFISH microscopic images of an interdigital electrode pattern on Si biased at -20 V (upper left), and at increasing 5 V increments, ending at +20 V (lower right).

The contrast between the EFISH and background field-independent SH signals was consistently lower in scanning images (Fig. 3) than in projection images (Fig. 2), even though dc field strength was similar. We believe this is caused by greater generation of electron-hole pairs by the more intense tightly-focused beam and EFISH screening [19]. While this effect limits EFISH signal levels, it does suggest the potential of applying powerful photomodulation techniques, which have been shown to yield accurate characterization of flat band potential and other properties of the Si/SiO_2 interface [17].

Several bias-dependent"hot spots" are clearly evident in the images of Fig. 3. Two of the most prominent are marked by arrows in the Fig. 3. These spots result from local whiskering, or other damage, in the metallization. Such high-field defects represent potential failure points in integrated circuits, and are not easily detected by macroscopic electrical tests. Their straightfoward detection in these images may therefore be an important practical application of EFISH microscopy.

4 Conclusion

We have demonstrated near-diffraction-limited imaging of electric fields near metal-Si junctions by two types of EFISH microscopy using an unamplified fs laser. Stationary projection imaging yields diffraction-limited lateral spatial resolution, involves no moving parts, and detects μm-scale lateral field variations related to Si/SiO_2 interface characteristics such as defect and doping density. However, the images offer a limited field of view, are complicated by coherent artifacts and require acquisition times too long for modulation techniques. Confocal scanning EFISH microscopy uses a tightly focused beam and thus yields higher signal and the potential for modulation microscopy. However it requires a more complicated scanning mirror configuration, yields somewhat lower spatial resolution, and tends to screen the dc fields by carrier generation. Both methods can be improved by incorporating techniques developed in non-microscopic contexts. Introducing a strong,independent SH local oscillator pulse (homodyne detection) that coincides spatially and temporally with the EFISH beam generates a signal that is linear in E_0 and can be calibrated straightforwardly to yield absolute field strengths[15]. Temporally separating the EFISH and local oscillator pulses and using spectroscopic detection enables rapid measurement of the phase of the EFISH pulse by frequency-domain interferometric second-harmonic (FDISH) spectroscopy[21]. Careful control of incident and EFISH polarizations with respect to crystalline axes will enable microscopic mapping of selected vector components of $\vec{E}_0$[15].

Acknowledgements This work was supported by the Robert Welch Foundation (Grant F-1038), the U. S. National Science Foundation (Grants DMR-0207295 and PHY-0114336) and the Office of Naval Research (Grant N00014-03-1-0639).

References

[1] C. H. Lee, R. K. Chang, N. Bloembergen, Phys. Rev. Lett. **18**, 167 (1967).
[2] J. Qi, M. S. Yeganeh, I. Koltover, A. G. Yodh, W. M. Theis, Phys. Rev. Lett. **71**, 633 (1993).
[3] P. Godefroy, W. de Jong, C. W. van Hasselt, M. A. C. Devillers, Th. Rasing, Appl. Phys. Lett. **68**, 1982 (1996).
[4] J. I. Dadap et al., Phys. Rev. B **53**, R7607 (1996).
[5] T. A. Germer, K. W. Kolasinski, J. C. Stephenson, L. J. Richter, Phys. Rev. B **55**, 10694 (1997).
[6] C. Ohloff, G. Lupke, C. Meyer, H. Kurz, Phys. Rev. B **55**, 4596 (1997).
[7] A. Nahata, T. F. Heinz, Opt. Lett. **23**, 67 (1998).
[8] G. L. Richmond, J. M. Robinson, V. L. Shannon, Prog. Surf. Sci. **28**, 1 (1988).
[9] R. M. Corn, D. A. Higgins, Chem. Rev. **94**, 107 (1994).
[10] K. B. Eisenthal, Chem. Rev. **96**, 1343 (1996).
[11] K. A. Schultz, I. I. Suni, and E. G. Seebauer, J. Opt. Soc. Am. B **10**, 546 (1993).
[12] V. Kirilyuk, A. Kirilyuk, Th. Rasing, Appl. Phys. Lett. **70**, 2306 (1997).
[13] K. Pedersen et al., J. Appl. Phys. **88**, 3872 (2000).
[14] C. K. Sun, S. W. Chu, S. P. Tai, S. Keller, U. K. Mishra, S. P. DenBaars, Appl. Phys. Lett. **77**, 2331 (2000).
[15] J. I. Dadap, J. Shan, A. S. Weling, J. A. Misewcih and T. F. Heinz, Appl. Phys. B **68**, 333 (1999).
[16] O. A. Aktsipetrov et al., Phys. Rev. B **60**, 8924 (1999).
[17] E. D. Mishina, S. Nakabayashi, O. A. Aktsipetrov and M. C. Downer, Jpn. J. Appl. Phys., in press (2003).
[18] T. Corle and G. Kino, *Confocal Scanning Microscopes and Related Imaging Systems* (San Diego, Academic Press, 1996).
[19] J. I. Dadap, P. T. Wilson, M. H. Anderson, M. C. Downer, and M. ter Beek, Opt. Lett. **22**, 901 (1997).
[20] Max Born and Emil Wolf, *Principles of Optics*, 7th ed. (Cambridge U. Press, 1999), chap. 8.
[21] P. T. Wilson, Y. Jiang, O. A. Aktsipetrov, E. D. Mishina, M. C. Downer, Opt. Lett. **24**, 496 (1999).

Author Index

DOI: **The fastest way to find an article online** is the *Digital Object Identifier* (DOI).
Starting in Volume 198, issue 2 (January 2003), DOIs have been printed in the header of the first page of every article in physica status solidi (b). On the WWW, one can find an article for example with a DOI of 10.1002/pssa.200306608 at **http://dx.doi.org/**10.1002/pssa.200306608.

Please use the DOI of the article to link from your home page to the articles in Wiley Interscience.

The DOI is a result of a cross-publisher initiative to create a system for the persistent identification of documents on digital networks. More information is available from **www.doi.org**.

Information for conference organizers and guest editors

The third journal section *physica status solidi (c) – conferences and critical reviews* is devoted to the publication of proceedings, ranging from large international meetings to specialized workshops, as well as collections of topical reviews on various areas of current solid state physics research. The new series has been launched in December 2002 and will appear with several issues in its starting volume **0** (2002/03). It is available both as an online journal and in hardcover print volumes, to be delivered to conference contributors and participants (upon arrangement with the organizers). Single copies of pss (c) may be ordered as a book using its ISBN number. Regular subscriptions to pss (c) will be offered from 2004. In 2003, a free trial access is available online through Wiley InterScience (www.interscience.wiley.com). Subscribers to pss (a) or (b) will receive free pss (c) sample copies.

Essential details concerning layout and organization of the new journal series are:

- pss (c) is published as a full hardcover-bound series, carrying a standard green-coloured cover design, individually adapted according to the organizers' request which includes conference designation, logo, names of Guest Editors etc.

- Proceedings issues contain all conference contributions which have been peer-reviewed and accepted by the Guest Editors. Upon special agreement between the pss journal editors and the Guest Editors, part of the conference papers may also be published simultaneously in an issue of pss (a) or (b). For all papers, strict criteria for journal publications, i.e. positive peer-review by independent referees, are obligatory. All papers are unambiguously citable as phys. stat. sol. (a), (b), or (c) journal articles and will be covered by standard reference databases.

- All articles are published online in PDF format at Wiley InterScience. Access for registered users (e. g. conference participants with special password) may be installed. The online version contains colour figures at no additional cost, regardless of their colour or black/white representation in print.

- The Editorial Office provides document templates and style files for Word and LaTeX, respectively, to be used by all authors, allowing an easy manuscript preparation and length estimate of their paper with respect to the page limits given by the organizers.

- The issue is completed by a table of contents in topical order, an author index, a preface, listings of conference committee members, organizers and sponsors, and any additional material, if desired.

- The usual service of the Editorial Office is available and includes support in the refereeing process, acceptance messages, PDF proofs (for typesetted papers), free PDF reprints (hardcopy reprints may be ordered) as well as individual communication with authors and organizers. The use of a Web-based software system for online submission and refereeing of papers is offered to Guest Editors.

- The editors of pss (c) aim at a timely, professional, and high-quality print and online publication of proceedings, typically within only four to six months after a conference.

- Various service packages for production are available, including either full typesetting of papers using electronic manuscript data or publication-ready delivery of manuscript files (prepared using the template/style files) by the organizers.

For further details as well as an individual offer for the publication of the proceedings of your forthcoming conference or of a special issue containing topical reviews, please contact the Editorial Office at pss@wiley-vch.de (for other contact information see the title page).